IET ENERGY ENGINEERING 114

Lightning-Induced Effects in Electrical and Telecommunication Systems

Other volumes in this series:

Lightning-Induced Effects in Electrical and Telecommunication Systems

Yoshihiro Baba and Vladimir A. Rakov

The Institution of Engineering and Technology

Published by The Institution of Engineering and Technology, London, United Kingdom

The Institution of Engineering and Technology is registered as a Charity in England & Wales (no. 211014) and Scotland (no. SC038698).

© The Institution of Engineering and Technology 2021

First published 2020

The Institution of Engineering and Technology
Michael Faraday House
Six Hills Way, Stevenage
Herts, SG1 2AY, United Kingdom

www.theiet.org

British Library Cataloguing in Publication Data
A catalogue record for this product is available from the British Library

ISBN 978-1-78561-353-1 (Hardback)
ISBN 978-1-78561-354-8 (PDF)

Typeset in India by MPS Limited

Contents

About the authors

Yoshihiro Baba is a professor at Doshisha University, Japan. He has (co)written two other books, seven book chapters, and more than 100 papers published in reviewed international journals. He received the Technical Achievement Award from the IEEE EMC Society in 2014. He has previously been an editor of the IEEE Transactions on Power Delivery and a guest associate editor of the IEEE Transactions on EMC. He is a Fellow of the IEEE and IET.

Vladimir A. Rakov is a professor at the University of Florida, USA, and co-director of the International Center for Lightning Research and Testing. He has (co)written four other books and over 300 papers published in reviewed journals. He has been an associate editor of the IEEE Transactions on EMC since 2003 and the Journal of Geophysical Research since 2017. He is a Fellow of the IEEE, the AGU, the American Meteorological Society, and the IET.

Preface

This book aims at providing an introduction to full-wave electromagnetic computation methods with a focus on the finite-difference time-domain (FDTD) method for simulation of lightning-induced surges in power and telecommunication systems. At the same time, it contains a considerable amount of background information on lightning, lightning models, and lightning electromagnetics, including electromagnetic coupling models based on the distributed-circuit theory approach. The FDTD method uses the central difference approximation to Maxwell's curl equations, Faraday's law and Ampere's law, in the time domain. Gauss' law is also satisfied. The method solves the update equations for electric and magnetic fields at each time step and at each discretized space point in the working volume using the leapfrog method. The FDTD method is based on a simple procedure and, therefore, its programming is relatively easy. It is capable of treating complex geometries and inhomogeneities, as well as incorporating nonlinear effects and components. Further, it can handle wideband quantities in one run with a time-to-frequency transforming tool.

The first peer-reviewed paper, in which the FDTD method was used for simulation of lightning-induced surges, was published in 2006. About 30 journal papers and a large number of conference papers, in which the FDTD method is employed in simulations of lightning-induced surges, have been published during the last 15 years. Interest in using the FDTD method continues to grow because of the availability of both commercial and noncommercial software and increased computer capabilities.

The reader is expected to be familiar with fundamental electromagnetics and engineering mathematics. The book is suitable for senior undergraduate and graduate students specializing in electrical engineering, as well as for electrical engineers and researchers, who are interested in studying lightning-induced effects.

The book is composed of five chapters. In Chapter 1, we give an overview of lightning discharges, which can cause electromagnetically-induced effects in power and telecommunication systems. Then, we describe modeling of the lightning return stroke and explain "engineering" models, in which the spatial and temporal distribution of current or line charge density along the lightning channel is specified based on such observed lightning return-stroke characteristics as current at the channel base and return-stroke speed. Also, we discuss the equivalency between the lumped-source and distributed-source lightning representations, and the extension of engineering models to include a tall grounded strike object. Further, we describe two primary approaches to test lightning return-stroke models that are referred to as "typical-event" and "individual-event" approaches. In Chapter 2, we derive the exact electric and magnetic field equations for the vertical lightning channel and flat, perfectly conducting ground, using the so-called dipole (Lorentz condition) approach. Then, we discuss the difference between the radiation field component on the one hand and electrostatic and induction electric field components on the other hand and the non-uniqueness of electric field components. Also, we provide the mathematical function that is most often used for representing the channel-base current waveform. Further, we discuss calculation of lightning electric and magnetic fields propagating over lossy ground. In Chapter 3, we introduce three different sets of telegrapher's equations with source terms and corresponding equivalent circuits that can be used for studying voltage and current surges induced on an overhead conductor by transient electromagnetic fields such as those produced by lightning. The source terms (forcing functions) incorporated into the classical telegrapher's equations are derived, using the electromagnetic theory, following works of Taylor *et al.* (1965), Agrawal *et al.* (1980), and Rachidi (1993), who arrived at different coupling model formulations. It is shown that, since all three formulations are based on Maxwell's equations, they yield identical results. In Chapter 4, we give update equations for electric and magnetic fields used in FDTD computations in the 3D Cartesian, 2D cylindrical, and 2D spherical coordinate systems. Then, we describe a subgridding technique, which allows one to employ locally finer grids, and representations of lumped sources and lumped circuit elements such as resistor, inductor, and capacitor. We discuss representations of a thin-wire conductor and the lightning sources. Also, we explain representations of nonlinear elements such as surge arrester and nonlinear phenomena such as corona on a horizontal conductor. Further, we review absorbing boundary conditions, which are needed for the analysis of electromagnetic fields in an unbounded space. Finally, in Chapter 5, we classify about 30 journal papers published in the last 15 years, which use the FDTD method in simulations of lightning-induced surges, in terms of spatial dimension (2D or 3D), lightning representation, and application. We also give an overview of these works. Further, we describe six representative works in detail, which cover the following topics: (i) voltages induced on a single overhead conductor by lightning strikes to a nearby tall grounded object, (ii) lightning-induced voltages on an overhead two-conductor line, (iii) lightning-induced voltages on a single overhead conductor in the presence of corona, (iv) lightning-induced voltages on overhead multi-conductor lines with surge arresters and pole transformers, (v) lightning-induced voltages on overhead multi-conductor lines in the presence of nearby buildings, and (vi) lightning-induced currents in buried cables.

The authors would like to thank the many colleagues and former students and postdocs for useful discussions of topics presented in this book, including, in alphabetical order, Prof. Akihiro Ametani, Prof. Amedeo Andreotti, Mr. Takashi Asada, Dr. Celio F. Barbosa, Dr. Carl Baum, Prof. Yazhou Chen, Prof. Vernon Cooray, Prof. Alfred A. Dulzon, Mr. Shunsuke Imato, Prof. Masaru Ishii, Dr. Naoki Itamoto, Prof. Matti Lehtonen, Prof. Leonid Grcev, Prof. Grzegorz Maslowski, Prof. Dan D. Micu, Prof. Amitabh Nag, Prof. Naoto Nagaoka, Mr. Masaki

Nakagawa, Prof. Carlo Alberto Nucci, Prof. Alexandre Piantini, Dr. Shigemitsu Okabe, Prof. Ramesh K. Pokharel, Prof. Farhad Rachidi, Prof. Mohammad E.M. Rizk, Prof. Marcos Rubinstein, Mr. Hiroki Saito, Mr. Hiroshi Sumitani, Dr. Jun Takami, Mr. Toshiki Takeshima, Mr. Nobuyuki Tanabe, Mr. Hiroki Tanaka, Mr. Tokuya Tanaka, Mr. Yohei Taniguchi, Dr. Akiyoshi Tatematsu, Dr. Frederick M. Tesche, Prof. Rajeev Thottappillil, Prof. Thang H. Tran, Mr. Hiroyuki Tsubata, Mr. Ryosuke Tsuge, Dr. Naoyuki Tsukamoto, Prof. Martin A. Uman, Mr. Junya Yamamoto, and Prof. Peerawut Yutthagowith.

Chapter 1

Lightning return stroke and its modeling

In this chapter, we give an overview of lightning discharges, which can cause electromagnetically induced effects in power and telecommunication systems. Then, we describe modeling of the lightning return stroke and explain "engineering" models, in which the spatial and temporal distribution of the current or the line charge density along the lightning channel is specified based on such observed lightning return-stroke characteristics as current at the channel base, the speed of the upward-propagating front, and the channel luminosity profile. Also, we discuss the equivalency between the lumped-source and distributed-source representations, and the extension of engineering models to include a tall grounded strike object. Further, we describe two primary approaches to test lightning return-stroke models that are referred to as "typical-event" and "individual-event" approaches.

Key Words: Lightning; lightning return stroke; physical models; electromagnetic models; distributed-circuit models; engineering models; lumped source; distributed source; tall strike object

1.1 Introduction

The primary source of lightning discharge is a thundercloud, which can accumulate electric charges of the order of tens of coulombs or more. Lightning discharges are classified into two categories: cloud-to-ground discharges and cloud discharges. Cloud-to-ground discharges are of primary interest in studying and designing lightning protection of manmade structures located on the ground surface such as electrical and telecommunication systems. Cloud-to-ground discharges are classified, on the basis of the polarity of the charge effectively transferred to ground and the direction of the initial leader that creates a conducting path between the thundercloud and ground prior to high-current return strokes, into four types: downward negative lightning, upward negative lightning, downward positive lightning, and upward positive lightning. It is believed that about 90% of all cloud-to-ground lightning are downward negative lightning discharges and about 10% are downward positive lightning discharges. It is thought that upward lightning discharges occur only from tall objects that are higher than about 100 m or from objects of moderate height located on mountaintops.

Lightning return-stroke models are needed in studying lightning effects on various objects and systems, and in characterizing the lightning electromagnetic environment.

Four classes of lightning return-stroke models are usually defined (Rakov and Uman 1998). These classes are primarily distinguished by the type of governing equations: (1) gas dynamic or "physical" models, (2) electromagnetic models, (3) distributed-circuit models, and (4) "engineering" models.

Outputs of the electromagnetic, distributed-circuit, and engineering models can be used directly for the computation of electromagnetic fields, while the gas-dynamic models can be used for finding R (series resistance per unit length) as a function of time, which is one of the parameters of the electromagnetic and distributed-circuit models. Since the distributed-circuit and engineering models generally do not consider lightning channel branches, they best describe subsequent strokes or first strokes before the first major branch has been reached by the upward-propagating return-stroke front, a time that is usually longer than the time required for the formation of the initial current peak at ground level.

In this chapter, we give an overview of lightning discharges, whose electromagnetic fields can cause induced effects in power and telecommunication systems. Then, we describe modeling of the lightning return stroke and explain engineering models, in which the spatial and temporal distribution of the current or the line charge density along the lightning channel is specified based on such observed lightning return-stroke characteristics as current at the channel base, the speed of the upward-propagating front, and the channel luminosity profile. Also, we discuss the equivalency between the lumped-source and distributed-source representations, and the extension of engineering models to include a tall grounded strike object. Further, we describe two primary approaches to test model validity, the so-called "typical-event" and "individual-event" approaches.

1.2   Lightning

1.2.1   Categories of lightning discharges

All lightning discharges could be divided into two categories: those that bridge the gap between the cloud charge region and the ground, and those that do not. The former discharges are referred to as cloud-to-ground discharges. The latter discharges are referred to as cloud discharges and they account for approximately three-quarters of all lightning discharges. The cloud discharges include (a) intracloud discharges that occur within the confines of a single thundercloud, (b) intercloud discharges that occur between thunderclouds or thundercloud cells, and (c) air discharges that occur between a thundercloud and the surrounding air. It is thought that the majority of cloud discharges are of the intracloud type.

Since most manmade structures, such as electric power systems, telecommunication systems, and buildings that are exposed to lightning and require lightning protection, are located on the ground surface, the properties of cloud-to-ground lightning discharges are of primary interest in studying and designing lightning protection. Note that cloud discharges are of interest in evaluating the interaction of lightning with airborne vehicles and designing their protection against lightning.

1.2.2   Classification of cloud-to-ground lightning discharges

Cloud-to-ground lightning discharges are classified, on the basis of the polarity of the charge effectively transferred to ground and the direction of the initial leader, into four types. These are (a) downward negative lightning, (b) upward negative lightning, (c) downward positive lightning, and (d) upward positive lightning, each illustrated in Figure 1.1(a) to (d). The leader is a process that precedes the return stroke. It creates a conducting path between the cloud charge source and ground, and deposits charge along this path. It is believed that downward negative lightning discharges, type (a), account for about 90% of all cloud-to-ground lightning, and that about 10% of cloud-to-ground lightning are downward positive lightning discharges, type (c). It is thought that upward lightning discharges, types (b) and (d), occur only from tall objects that are higher than 100 m or so or from objects of moderate height located on mountaintops.

1.2.3   Downward negative lightning discharges to ground

In this subsection, a general explanation of downward negative lightning flashes, which account for about 90% of all cloud-to-ground lightning, is presented. Figure 1.2(a) and (b) schematically shows still and time-resolved optical images of a downward negative lightning flash containing three strokes, respectively. Figure 1.2 (c) shows the corresponding current at the channel base. In Figure 1.2(b) and (c), time advances from left to right. Each of the three strokes is composed of a downward-moving process termed "leader" and an upward-moving process termed "return stroke." The leader creates a conducting path between the cloud negative charge region and ground and deposits negative charge along this path. The return stroke traverses the leader path upward from ground to the cloud charge region and neutralizes the negative leader charge. Thus, both leader and return-stroke processes contribute to transporting negative charge from the cloud to ground. The leader initiating the first return stroke develops in virgin air and appears to be an optically intermittent process. Therefore, it is termed "stepped leader." The stepped-leader branches in Figure 1.2 are directed downward, which indicates that the stepped leader (and the flash) is initiated in the cloud and develops downward. The leaders initiating the two subsequent return strokes in Figure 1.2 move continuously, as a downward-moving dart, along the preconditioned path of the preceding return stroke or strokes. Hence, these leaders are termed "dart leaders." Note that each downward negative lightning flash typically contains three to five strokes.

In the following, a sequence of the processes involved in a typical downward negative lightning flash is presented in more detail with reference to Figure 1.3. Figure 1.3 can be viewed as a sequence of snapshots with the corresponding times indicated. The source of lightning (cloud charge distribution) is shown at $t = 0$. The generally accepted features of the thundercloud charge structure include a net positive charge near the top, a net negative charge below it, and an additional positive charge at the bottom. The stepped leader is preceded by an in-cloud process called the initial breakdown or preliminary breakdown (see $t = 1.00$ ms in Figure 1.3). There is no consensus on the mechanism of the initial breakdown. It can be viewed as a discharge

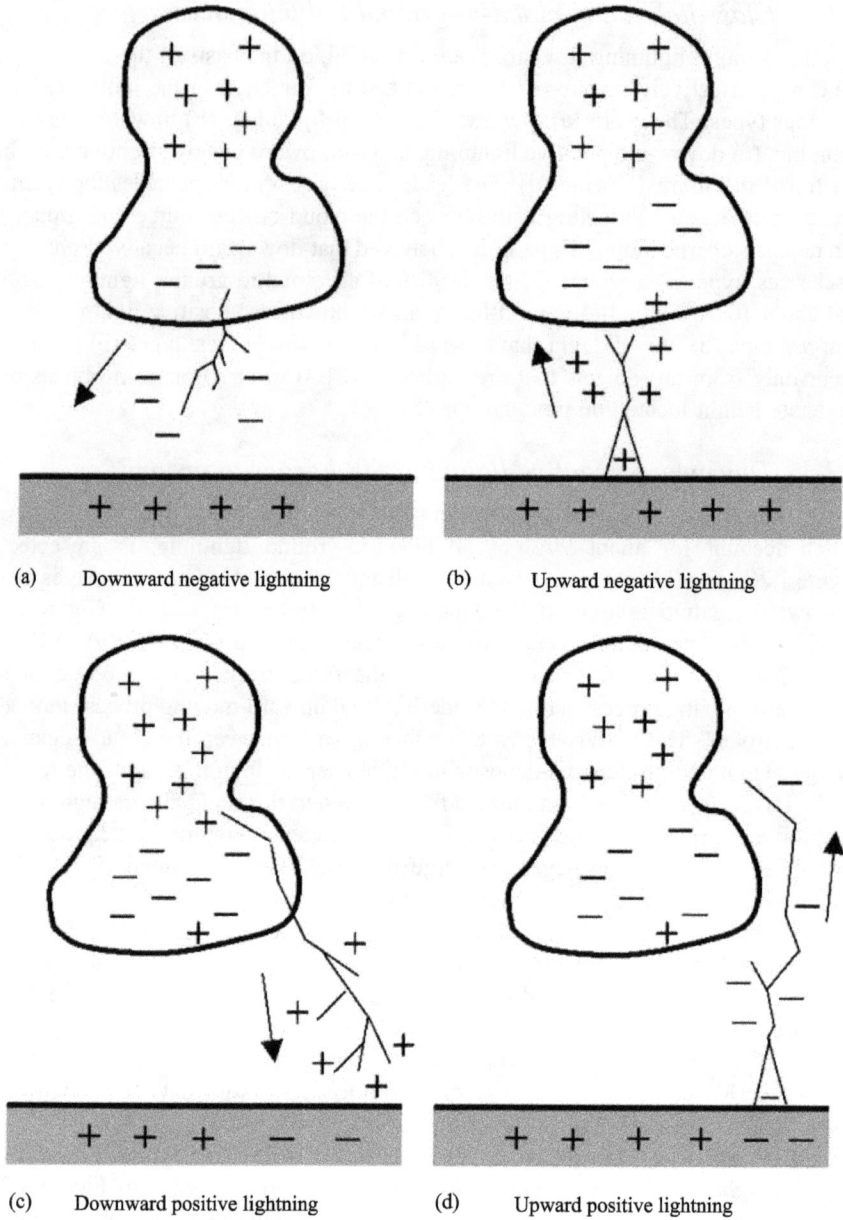

Figure 1.1   *Four types of lightning effectively lowering cloud charge to ground. Only the initial leader is shown for each type. In each lightning-type name given below the sketch, the direction of progression of the initial leader and the polarity of the cloud charge effectively lowered to ground (not necessarily the same as the leader polarity) are indicated. Adapted from Rakov and Uman (2003, Figure 1.1)*

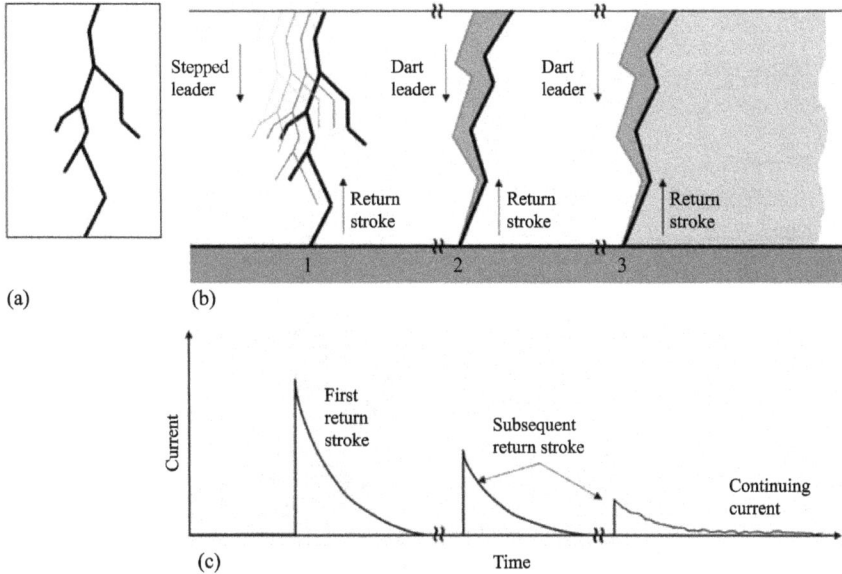

*Figure 1.2   Diagram showing the luminosity of a downward negative lightning
flash to ground containing three strokes and the corresponding
current at the channel base: (a) still camera image, (b) streak-camera
image, and (c) channel-base current. Adapted from Rakov and Uman
(2003, Figure 4.2)*

process between the negative and lower positive charge regions, but it can also involve
a sequence of channels extending in random directions from the cloud charge source.
One of these events (in the case of multiple channels) evolves into the stepped leader
which is a negatively charged channel that bridges the cloud charge source and the
ground (see $t = 1.10$ to 19.00 ms in Figure 1.3). The stepped leader extends toward
ground at an average speed of 2×10^5 m/s in a series of discrete steps, with each step
being typically 1 μs in duration and some meters to tens of meters in length, with the
interval between steps being 20 to 50 μs (e.g., Rakov and Uman 2003). The peak value
of the current pulse associated with an individual step is inferred to be 1 kA or greater.
Several coulombs of negative charge is distributed along the stepped-leader channel.
The stepped-leader duration is typically some tens of milliseconds, and the average
leader current is some hundreds of amperes. The stepped-leader channel is likely to
consist of a thin highly conducting plasma core that carries the longitudinal channel
current, surrounded by a low-conductivity corona sheath with a diameter of several
meters that contains the bulk of the leader charge.

As the stepped leader approaches ground, the electric field at the ground surface
or grounded objects increases until it exceeds the critical value for the initiation of
upward connecting leader. The initiation of upward connecting leader from ground
in response to the descending stepped leader marks the beginning of the attachment

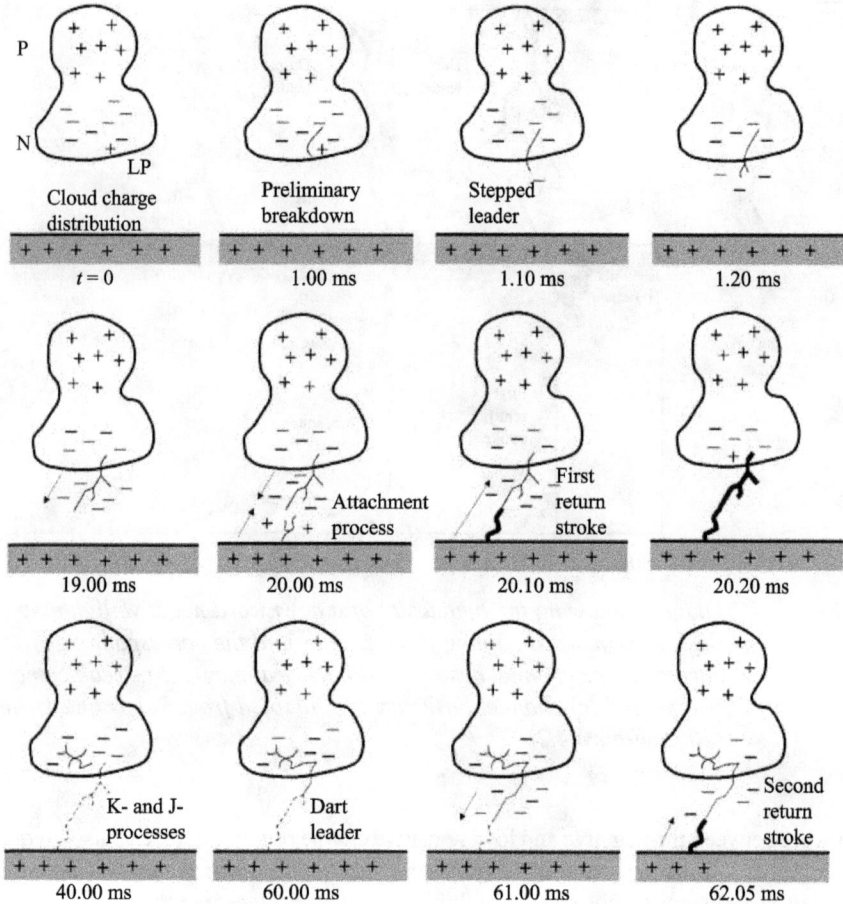

Figure 1.3   Various processes comprising a negative cloud-to-ground lightning flash. In the figure, P, N, and LP stand for positive, negative, and lower positive charge regions in the cloud, respectively. Adapted from Rakov and Uman (2003, Figure 4.3)

process (see $t = 20.00$ ms in Figure 1.3). The process by which the extending plasma channels of the upward and downward leaders make contact, via forming common streamer zone, is called the breakthrough phase or final jump. The breakthrough phase can be viewed as a switch-closing operation that serves to launch two waves from the junction point. The length of an upward connecting leader involved in a first stroke is some tens of meters or less if that leader is launched from the ground, and it can be several hundred meters long if it is initiated from a tall object.

The return stroke (see $t = 20.10$ to 20.20 ms in Figure 1.3) serves to neutralize the leader charge, although it may not neutralize all the leader charge or may deposit some excess positive charge onto the leader channel and into the cloud charge source region. The speed of the return stroke, averaged over the visible channel, is typically between one-third and one-half of the speed of light (e.g., Rakov 2007). The speed decreases with increasing height, dropping abruptly after passing each major branch. The first return-stroke current measured at ground rises to an initial peak of about 30 kA in some microseconds and decays to half-peak value in some tens of microseconds while exhibiting a number of subsidiary peaks, probably associated with the branches (e.g., Rakov and Uman 2003). This impulsive component of current may be followed by a slowly varying current of some hundreds of amperes lasting for some milliseconds. The return stroke effectively lowers to ground the several coulombs of charge originally deposited on the stepped-leader channel, including that residing on all the branches. The high-current return-stroke wave rapidly heats the channel to a peak temperature near or above 30,000 K and creates a channel pressure of 10 atm or more (e.g., Rakov and Uman 2003), which results in channel expansion, intense optical radiation, and an outward propagating shock wave that eventually becomes the thunder.

When the first return stroke ceases, the flash may end. In this case, the lightning is called a single-stroke flash. However, more often the residual first-stroke channel is traversed by a downward leader that appears to move continuously, a dart leader (see $t = 60.00$ ms in Figure 1.3). During the time interval between the end of the first return stroke and the initiation of dart leader, J- and K-processes occur in the cloud (see $t = 40.00$ ms in Figure 1.3). The J-process is often viewed as a relatively slow positive leader extending from the flash origin into the negative charge region, the K-process then being a relatively fast "recoil leader" that begins at the tip of the positive leader and propagates toward the flash origin. Both the J-processes and the K-processes in cloud-to-ground discharges serve to transport additional negative charge into and along the existing channel, although not all the way to ground. In this respect, K-processes may be viewed as attempted dart leaders. The processes that occur after the only stroke in single-stroke flashes and after the last stroke in multiple-stroke flashes are sometimes termed final (F) processes. These are similar, if not identical, to J-processes.

The dart leader progresses downward at a typical speed of 10^7 m/s, typically ignores the first stroke branches, and deposits along the channel a total charge of the order of 1 C (e.g., Rakov and Uman 2003). The dart-leader current peak is about 1 kA. Some leaders exhibit stepping near ground while propagating along the path traversed by the preceding return stroke. These leaders are termed "dart-stepped leaders." Additionally, some dart or dart-stepped leaders deflect from the previous return-stroke path, become stepped leaders, and form a new termination on the ground.

When a dart leader or dart-stepped leader approaches the ground, an attachment process similar to that described for the first stroke takes place, although it typically occurs over a shorter distance and consequently takes less time, with the upward connecting-leader length being of the order of some meters. Once the bottom of the dart leader or dart-stepped leader channel is connected to the ground, the second or any subsequent return-stroke wave is launched upward (see $t = 62.05$ ms in Figure 1.3) and serves to neutralize the leader charge. The subsequent return-stroke current at ground typically rises to a peak value of 10 to 15 kA in less than a microsecond and decays to half-peak value in a few tens of microseconds. The upward propagation speed of such a subsequent return stroke is similar to or slightly higher than that of the first return stroke (e.g., Rakov 2007). Note that due to the absence of branches, the speed variation along the channel for subsequent return strokes does not exhibit abrupt drops.

The impulsive component of the current in a subsequent return stroke is often followed by a continuing current that has a magnitude of tens to hundreds of amperes and a duration up to hundreds of milliseconds. Continuing currents with a duration in excess of 40 ms are traditionally termed "long continuing currents." Between 30 and 50% of all negative cloud-to-ground flashes contain long continuing currents. The source for continuing current is the cloud charge, as opposed to the charge distributed along the leader channel, the latter charge contributing to at least the initial few hundred microseconds of the return-stroke current observed at ground. Continuing current typically exhibits a number of superimposed surges that rise to a peak in some tens to hundreds of microseconds, with the peak being generally in the hundreds of amperes range but occasionally in the kiloamperes range. These current surges are associated with enhancements in the relatively faint luminosity of the continuing-current channel and are called "M-components."

The time interval between successive return strokes in a flash is usually several tens of milliseconds, although it can be as large as many hundreds of milliseconds if a long continuing current is involved and as small as one millisecond or less. The total duration of a flash is typically some hundreds of milliseconds, and the total charge lowered to ground is some tens of coulombs. The overwhelming majority of negative cloud-to-ground flashes contain more than one stroke. Although the first stroke is usually a factor of 2 to 3 larger in terms of current magnitude than a subsequent stroke, about one-third of multiple stroke flashes have at least one subsequent stroke that is larger than the first stroke in the flash (e.g., Rakov *et al.* 1994). Note that terms "lightning," "lightning discharge," and "lightning flash" are used interchangeably to refer to the overall lightning discharge process.

In the rest of this subsection, parameters of downward negative lightning return-stroke currents are presented in more detail, since they are of great importance in studying lightning surge protection of various electrical power and communication systems. Figure 1.4 shows, on two time scales, A and B, the average impulsive current waveforms for downward negative first and subsequent strokes, based on Berger *et al.*'s direct measurements on instrumented towers. The rising portion of the first-stroke waveform has a characteristic concave shape. Figure 1.5 shows the cumulative statistical distributions (the solid-line curves) of return-stroke peak currents for (1) negative first strokes, (2) negative subsequent strokes, and (3) positive strokes (each of which was the only stroke in a flash), the latter not being further discussed in this subsection. These experimental curves are approximated by log-normal distributions which are represented in Figure 1.5 by broken slanted lines. The ordinate gives the percentage of peak currents exceeding the corresponding value on the horizontal axis (abscissa value). The lightning peak current distributions for negative first and negative subsequent strokes shown in Figure 1.5 are characterized by 95, 50, and 5% values (based on the log-normal approximations) given in Table 1.1, which contains a number of other parameters derived from the current oscillograms (Berger *et al.* 1975).

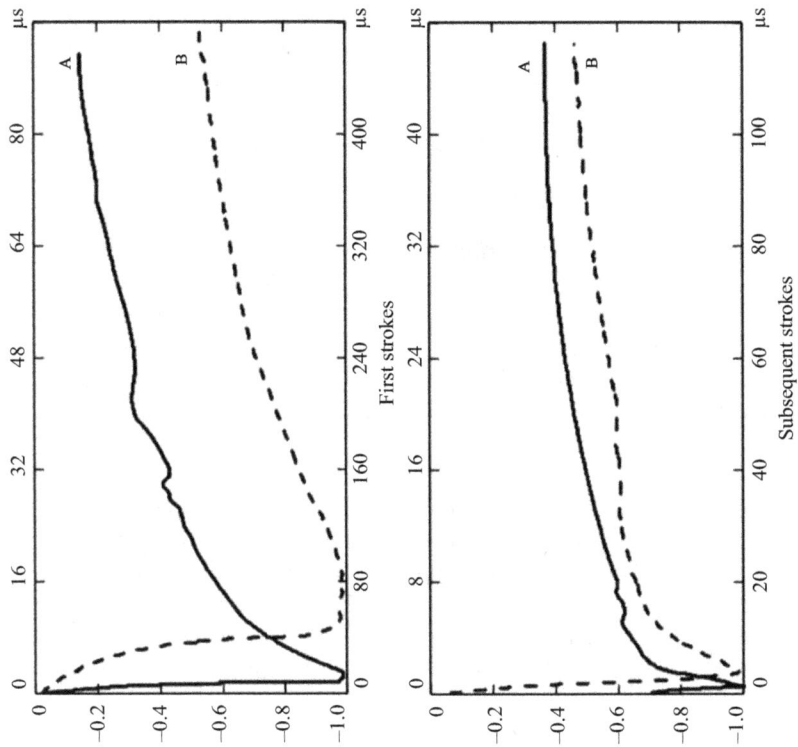

Figure 1.4   Average negative first- and subsequent-return-stroke currents, each shown on two time scales, A and B, as reported by Berger et al. (1975). The lower time scales correspond to the solid-line curves (A), while the upper time scales correspond to the broken-line curves (B). The vertical scale is in relative units, with the peak values being equal to negative unity. Reprinted, with permission, from Berger et al. (1975, Figures 12 and 13)

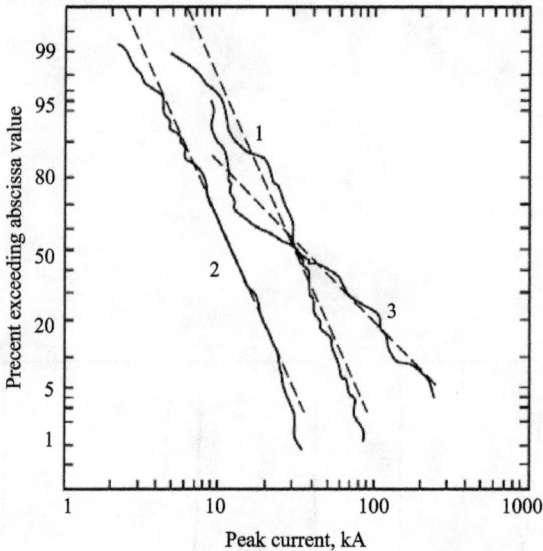

Figure 1.5   Cumulative statistical distributions of return-stroke peak current from direct measurements at tower top (solid-line curves) and their log-normal approximations (broken lines) for (1) negative first strokes, (2) negative subsequent strokes, and (3) positive first (and only) strokes, as reported by Berger et al. (1975). Reprinted, with permission, from Berger et al. (1975, Figure 1)

Note from Figure 1.5 and Table 1.1 that the median return-stroke current peak for first strokes is two to three times higher than that for subsequent strokes. Also, negative first strokes transfer about a factor of five larger impulse charge than negative subsequent strokes do. On the other hand, subsequent return strokes are characterized by three to four times higher current maximum rate of rise. Note that the smallest measurable time in Berger *et al.*'s current oscillograms was 0.5 μs versus the 95% value of 0.22 μs for the front duration for subsequent strokes in Table 1.1, which is a prediction of the log-normal approximation. The maximum dI/dt in Table 1.1 is likely to be an underestimate: 50% value for subsequent strokes is 40 kA/μs vs. 100 kA/μs obtained using modern instrumentation for triggered-lightning strokes (see Section 1.2.6). Only a few percent of negative first strokes are expected to exceed 100 kA. The action integral in Table 1.1 represents the energy that would be dissipated in a 1-Ω resistor if the lightning current were to flow through it. Note that Rakov *et al.* (2013) have presented an updated review of lightning parameters for engineering applications.

    Typical values of return-stroke wavefront speed (based on optical measurements) are in the range of one-third to one-half of the speed of light (Rakov 2007). The equivalent impedance of the lightning return-stroke channel is expected to be in the range from 0.6 to 2.5 kΩ (Gorin and Shkilev 1984), as estimated from

*Table 1.1   Parameters of downward negative lightning derived from channel-base
current measurements, as reported by Berger et al. (1975)*

Parameters	Units	Sample size	Percentage exceeding tabulated value		
			95%	**50%**	**5%**
Peak current (minimum 2 kA)	kA				
First stroke		101	14	30	80
Subsequent stroke		135	4.6	12	30
Charge (total charge)	C				
First stroke		93	1.1	5.2	24
Subsequent strokes		122	0.2	1.4	11
Complete flash		94	1.3	7.5	40
Impulse charge (excluding continuing current)	C				
First strokes		90	1.1	4.5	20
Subsequent strokes		117	0.22	0.95	4
Front duration (2 kA to peak)	μs				
First strokes		89	1.8	5.5	18
Subsequent strokes		118	0.22	1.1	4.5
Maximum dI/dt	kA μs^{-1}				
First strokes		92	5.5	12	32
Subsequent strokes		122	12	40	120
Stroke duration (2 kA to half-peak value on the tail)	μs				
First strokes		90	30	75	200
Subsequent strokes		115	6.5	32	140
Action integral	A^{2}s				
First strokes		91	6.0×10^3	5.5×10^4	5.5×10^5
Subsequent strokes		88	5.5×10^2	6.0×10^3	5.2×10^4
Time interval between strokes	ms	133	7	33	150
Flash duration	ms				
All flashes		94	0.15	13	1100
Excluding single-stroke flashes		39	31	180	900

Reprinted, with permission, from Berger *et al.* (1975, Table 1).

measurements of lightning current at different points along the 530-m-high Ostankino Tower in Moscow. The radius of the lightning return-stroke channel is expected to be about 3 cm, and the resistance per unit length of a lightning channel is estimated to be about 0.035 Ω/m behind the return-stroke front and about 3.5 Ω/m ahead of the return-stroke front (e.g., Rakov 1998).

1.2.4   Positive lightning discharges

Positive lightning discharges, defined as those transferring positive charge from cloud to ground, account for only about 10% of all lightning discharges taking place between cloud and ground, but they have lately attracted considerable attention of scientists and engineers. This is, in part, because positive lightning discharges more often than their negative counterparts have higher currents and larger charge transfers to ground and, as a result, can cause more severe damage to various objects and systems. It is thought that positive lightning discharges tend to occur more often in the following five situations (Rakov and Uman 2003): (a) the dissipating stage of an individual thunderstorm, (b) winter thunderstorms, (c) shallow clouds such as the trailing stratiform regions of mesoscale convective systems, (d) severe storms, and (e) thunderclouds formed over forest fires or contaminated by smoke.

According to the parameters reported from direct current measurements by Berger *et al.* (1975) for positive and negative lightning discharges, the 5% peak current for positive discharges is significantly greater than that for negative first return strokes (250 vs. 80 kA), while the median peak current for positive discharges is only slightly higher than that for negative first return strokes (35 vs. 30 kA) (see Figure 1.5). Also, the median charge transfer for positive discharges is about an order of magnitude greater than that for negative discharges (complete flashes). All current waveforms observed by Berger *et al.* (1975) for positive lightning can be divided into two types. The first type includes microsecond-scale waveforms similar to those for negative lightning (see Figure 1.6(a)) and the second type includes millisecond-scale waveforms with risetimes up to hundreds of microseconds (see Figure 1.6(b)) (Rakov 2003). While microsecond-scale waveforms are probably formed in a manner similar to that in downward negative lightning, millisecond-scale waveforms are likely to be a result of the M-component mode of charge transfer to ground (Rakov *et al.* 2001). Indeed, if a downward current wave originates at a height of 1 to 2 km as a result of connection of the upward connecting leader to a charged in-cloud channel, the charge transfer to ground associated with this wave is likely to be a process of M-component type, which is characterized by a relatively slow current front at the channel base.

It is thought that positive discharges have the following characteristics (Rakov and Uman 2003): (a) positive flashes are usually composed of a single stroke, whereas about 80% of negative flashes contain two or more strokes; (b) positive return strokes tend to be followed by considerable continuing currents; (c) positive return strokes often appear to be preceded by significant in-cloud discharge activity; (d) positive lightning discharges often involve long horizontal channels, up to tens of kilometers in extent; (e) positive leaders can move either continuously or intermittently (as seen in time-resolved optical records), while negative leaders are always stepped when they progress in virgin air.

1.2.5   Upward lightning discharges

Upward lightning, as opposed to downward lightning, would not occur if the grounded strike object were not present. Hence, it can be considered to be initiated by the object. Objects with heights, ranging from approximately 100 to 500 m, experience both downward and upward lightning flashes. The fraction of upward flashes increases with increasing the height of the object. Structures having heights less than 100 m or so are usually assumed to be struck only by downward lightning and structures with heights greater than 500 m or so are usually assumed to

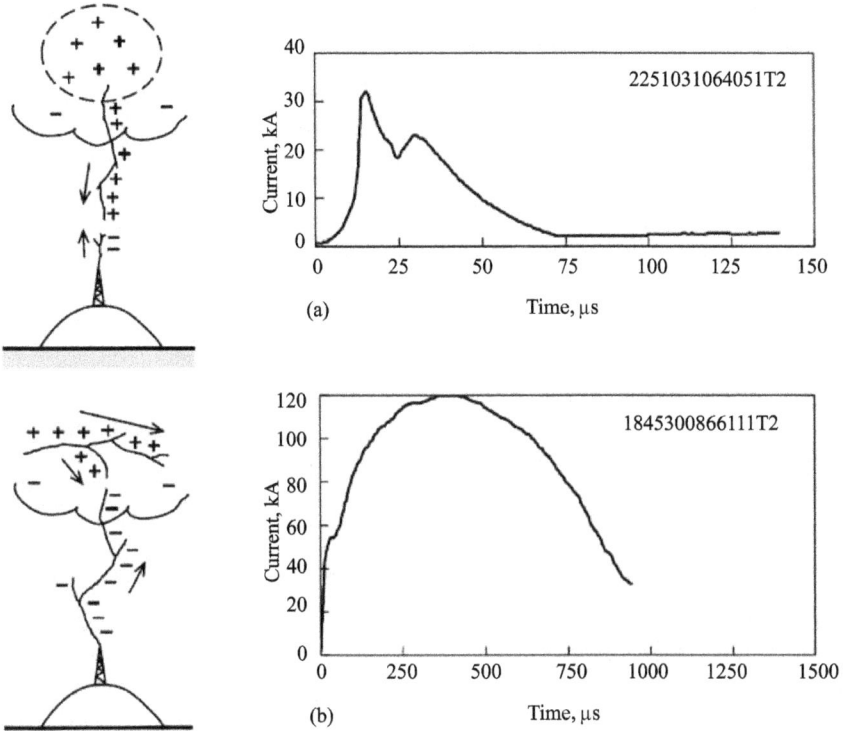

Figure 1.6   *Examples of two types of positive lightning current waveforms observed by Berger et al. (1975): (a) a microsecond-scale waveform (right-hand panel) and a sketch (left-hand panel) illustrating the type of lightning that might have led to its production; (b) a millisecond-scale waveform (right-hand panel) and a sketch (left-hand panel) illustrating the type of lightning that might have led to its production. Adapted from Rakov (2003, Figure 2)*

experience only upward flashes. If a structure is located on the top of a mountain, then an effective height that is greater than the structure's physical height is often assigned to the structure in order to account for the additional field enhancement due to the presence of the mountain on which the structure is located. For example, each of the two towers used by Berger in his lightning studies on Monte San Salvatore in Switzerland had a physical height of about 70 m, while their effective height was estimated to be 350 m by Eriksson (1978). Eriksson's estimate is based on the observed percentage of upward flashes initiated from the towers. Note that upward flashes more often transport negative than positive charge to ground.

Figure 1.7 shows schematic diagrams that illustrate still and time-resolved photographic records along with the corresponding current record at the channel base of upward lightning. Upward negative discharges are initiated by upward positive leaders

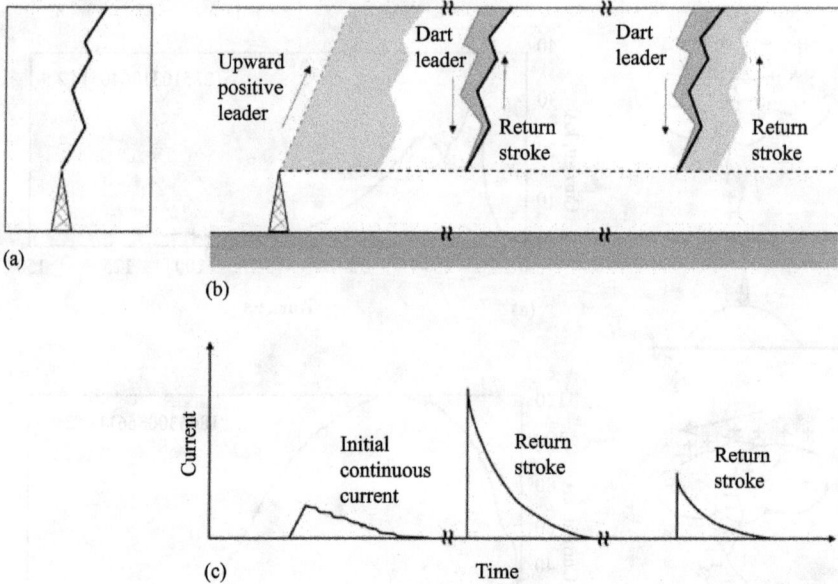

Figure 1.7   Schematic diagram showing the luminosity of an upward negative flash and the corresponding current at the channel base; (a) still-camera image, (b) streak-camera image, and (c) current record. The flash is composed of an upward positive leader followed by an initial continuous current, and two downward-dart-leader/upward-return-stroke sequences. The upward positive leader and initial continuous current constitute the initial stage of upward negative flash. Adapted from Rakov and Uman (2003, Figure 6.1)

from the tops of grounded objects. The upward positive leader bridges the gap between the object and the negative charge region in the cloud, and serves to establish an initial continuous current, typically lasting for some hundreds of milliseconds. The upward positive leader and initial continuous current constitute the initial stage of an upward flash. The initial stage can be followed, after a no-current interval, by one or more downward-leader/upward-return-stroke sequences, as illustrated in Figure 1.7. Downward-leader/upward-return-stroke sequences in upward lightning are similar to subsequent leader/return-stroke sequences in downward lightning. Upward lightning flashes can be also induced (triggered) by other lightning discharges occurring near the tall object.

1.2.6   Rocket-triggered lightning discharges

Lightning discharge from a natural thundercloud to ground can be stimulated to occur by enhancing the electric field below the cloud. The most effective technique for triggering lightning involves the launching of a small rocket trailing a thin-grounded wire toward a charged cloud overhead. This triggering method is usually

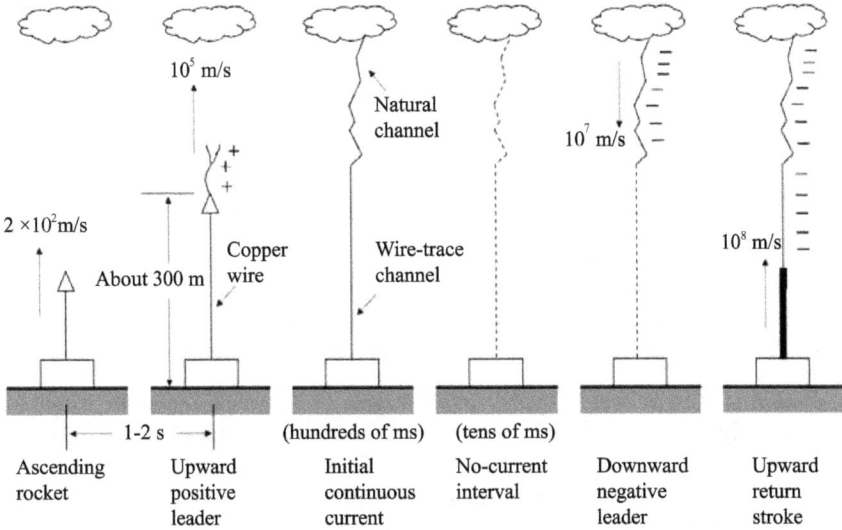

Figure 1.8   Sequence of events in classical triggered lightning. The upward positive leader and initial continuous current constitute the initial stage. Adapted from Rakov et al. (1998, Figure 1)

called classical triggering and is illustrated in Figure 1.8. To decide when to launch a rocket, the cloud charge is indirectly sensed by measuring the electric field at ground; for example, field values of 4 to 10 kV/m are generally good indicators of favorable conditions for triggering negative lightning in Florida (e.g., Rakov *et al.* 1998).

When the rocket, ascending at 150 to 200 m/s, is about 200- to 300-m high, the enhanced field near the wire top launches an upward positive leader (for the case of dominant negative charge overhead). This leader establishes an initial continuous current with a duration of some hundreds of milliseconds that effectively transports negative charge from the cloud charge region to the triggering facility. The initial continuous current can be viewed as a continuation of the upward-positive leader when the latter has reached the main negative charge region in the cloud. At that time, the upper extremity of the upward positive leader is likely to become heavily branched. The upward positive leader and initial continuous current constitute the initial stage of a classical triggered-lightning discharge. After cessation of the initial continuous current, one or more dart-leader/return-stroke sequences may traverse the same path to the triggering facility. The dart leaders and the following return strokes in triggered lightning are similar to dart-leader/return-stroke sequences in natural lightning, although the initial processes in natural downward lightning and in triggered lightning are distinctly different.

Triggered-lightning experiments have provided considerable insights into natural lightning processes that would not have been possible from studies of natural lightning due to its random occurrence in space and time. Also, they have contributed significantly to testing the validity of various lightning models and to providing ground-truth data for testing the performance characteristics of lightning detection networks. Further, triggered lightning has been used to study the interaction of lightning with various objects and systems.

1.3   Lightning models

1.3.1   Overview of modeling of the lightning return stroke

Lightning return-stroke models are needed in studying lightning effects on various objects and systems, and in characterizing the lightning electromagnetic environment. Four classes of lightning return stroke models can be defined (Rakov and Uman 1998). These classes are primarily distinguished by the type of governing equations:

1. The first class of models is the gas-dynamic or "physical" models, which are primarily concerned with the radial evolution of a short segment of the lightning channel and its associated shock wave. These models typically involve the solution of three gas-dynamic equations representing the conservation of mass, of momentum, and of energy, coupled to two equations of state. Principal model outputs include temperature, pressure, and mass density as a function of the radial coordinate and time.

2. The second class of models is the electromagnetic models. These models involve a numerical solution of Maxwell's equations to find the current distribution along the channel from which the remote electric and magnetic fields can be computed. Electromagnetic return-stroke models are reviewed by Baba and Rakov (2012) and in Section 4.6 of this book.

3. The third class of models is the distributed-circuit models, which can be viewed as an approximation to the electromagnetic models and represent the lightning return stroke as a transient process on a transmission line character- ized by resistance (R), inductance (L), and capacitance (C), all per unit length. The governing equations are the telegrapher's equations. The distributed- circuit models, which are also called R-L-C transmission line models, are used to determine the channel current versus time and height and can, therefore, be used to compute remote electric and magnetic fields.

4. The fourth class of models is the "engineering" models, in which a spatial and temporal distribution of the channel current or the channel line charge density is specified based on such observed lightning return-stroke characteristics as current at the channel base, the speed of the upward-propagating front, and the channel luminosity profile. In these models, the physics of the lightning return stroke is deliberately downplayed, and the emphasis is placed on achieving agreement between the model-predicted electromagnetic fields and those observed at different distances from the lightning channel. A characteristic feature of the engineering models is the small number of adjustable parameters, usually only one or two besides the specified channel-base current. Engineering return-stroke models have been reviewed, among others, by Nucci *et al.* (1990), Thottappillil and Uman (1993), Rakov and Uman (1998), and Gomes and Cooray (2000).

Outputs of the electromagnetic, distributed-circuit, and engineering models can be used directly for the computation of electromagnetic fields, while the gas-dynamic models can be used for finding R as a function of time, which is one of the parameters of the electromagnetic and distributed-circuit models. Since the distributed-circuit and engineering models generally do not consider lightning channel branches, they best describe subsequent strokes or first strokes before the first major branch has been reached by the upward-propagating return-stroke front, a time that is usually longer than the time required for the formation of the initial current peak at ground level.

Engineering models are discussed in more detail in the following sections of this chapter.

1.3.2   Engineering models

An engineering return-stroke model is a simple equation relating the longitudinal channel current $I(z', t)$ at any height z' and any time t to the current $I(0, t)$ at the channel origin, $z' = 0$. An equivalent expression in terms of the line charge density $\rho_L(z', t)$ on the channel can be obtained using the continuity equation (Thottappillil *et al.* 1997). Thottappillil *et al.* (1997) defined two components of the charge density at a given channel section, one component being associated with the return-stroke charge transferred through the channel section and the other with the charge deposited at that channel section. As a result, their charge density formulation reveals new aspects of the physical mechanisms behind the models that are not apparent in the longitudinal current formulation: for example, the existence of radial current associated with the neutralization of leader charge stored in the corona sheath.

We first consider the mathematical and graphical representations of some simple models and then categorize and discuss the most used engineering models based on their implications regarding the principal mechanism of return-stroke process. Rakov (1997, 2016) expressed several engineering models by the following generalized current equation:

$$I(z',t) = u(t - z'/v_f)P(z')I(0, t - z'/v) \tag{1.1}$$

where u is the Heaviside function equal to unity for $t \geq z'/v_f$ and zero otherwise, $P(z')$ is the height-dependent current attenuation factor, v_f is the upward-propagating return-stroke front speed, and v is the current-wave propagation speed (depending on the model, v may or may not be equal to v_f). Table 1.2 summarizes $P(z')$ and v for five representative engineering models: the transmission line model, TL (Uman and Mclain 1969) (not to be confused with the *R-L-C* transmission line models); the modified transmission line model with linear current decay with height, MTLL (Rakov and Dulzon 1987); the modified transmission line model with exponential current decay with height, MTLE (Nucci *et al.* 1988); the Bruce-Golde model, BG (Bruce and Golde 1941); and the traveling current source model, TCS (Heidler 1985). In Table 1.2, H is the total channel height, λ is the current decay height constant (assumed by Nucci *et al.*

Table 1.2   P(z') and v in (1.1) for five engineering models

Model	$P(z')$	v
TL (Uman and McLain 1969)	1	v_f
MTLL (Rakov and Dulzon 1987)	$1 - z'/H$	v_f
MTLE (Nucci *et al.* 1988)	$\exp(-z'/\lambda)$	v_f
BG (Bruce and Golde 1941)	1	∞
TCS (Heidler 1985)	1	$-c$

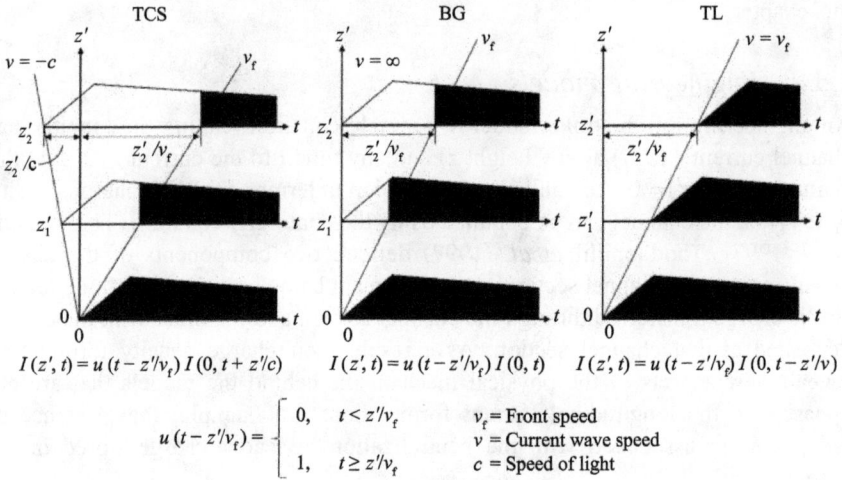

$$I(z', t) = u(t - z'/v_f)\,I(0, t + z'/c) \qquad I(z', t) = u(t - z'/v_f)\,I(0, t) \qquad I(z', t) = u(t - z'/v_f)\,I(0, t - z'/v)$$

$$u(t - z'/v_f) = \begin{cases} 0, & t < z'/v_f \\ 1, & t \geq z'/v_f \end{cases} \qquad \begin{aligned} v_f &= \text{Front speed} \\ v &= \text{Current wave speed} \\ c &= \text{Speed of light} \end{aligned}$$

Figure 1.9   Current versus time waveforms at ground (z' = 0) and at two heights
z'$_1$ and z'$_2$ above ground for the TCS, BG, and TL return-stroke
models. Slanted lines labeled v$_f$ represent upward speed of the return-
stroke front and lines labeled v represent speed of the return-stroke
current wave. The dark portion of the waveform indicates current that
actually flows through a given channel section. Note that the current
waveform at z' = 0 and v$_f$ are the same for all three models.
Adapted from Rakov and Uman (1998, Figure 8)

(1988) to be 2000 m), and c is the speed of light. If not specified otherwise, v_f is assumed to be constant. The three simplest models, TCS, BG, and TL, are illustrated in Figure 1.9, and the TCS and TL models additionally in Figure 1.10. We consider first Figure 1.9. For all three models, we assume the same current waveform at the channel base ($z' = 0$) and the same front speed represented in z'-t coordinates by the slanted line labeled v_f. The current-wave speed is represented by the line labeled v, which coincides with the vertical axis for the BG model and with v_f line for the TL model. Shown for each model are current versus time waveforms at the channel base ($z' = 0$) and at heights z'_1 and z'_2. Because of the

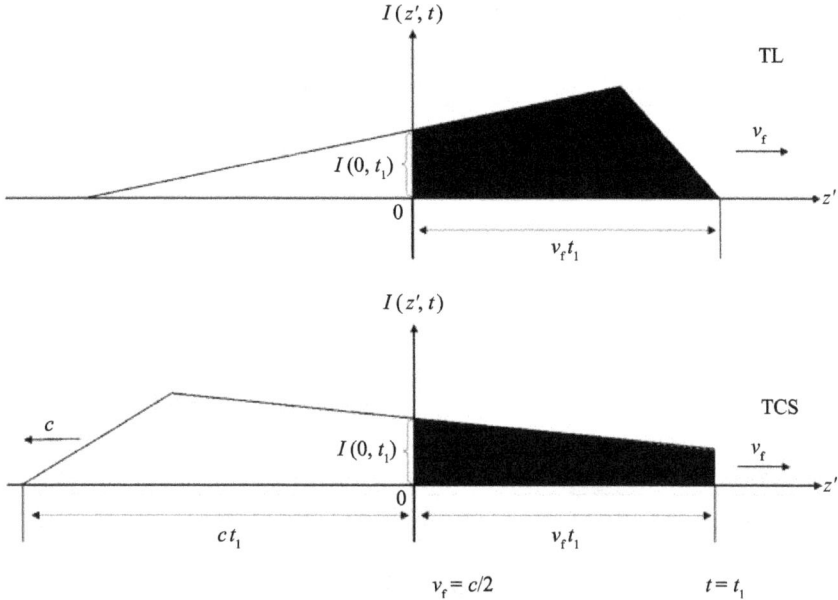

Figure 1.10   Current versus height z′ above ground at an arbitrary fixed instant of time t = t₁ for the TL and TCS models. Note that the current at z′ = 0 and v_f are the same for both models. Adapted from Rakov and Uman (1998, Figure 9)

Table 1.3   Transmission-line-type models for $t \geq z'/v_f$

Model	$I(z', t)$ and $\rho_L(z', t)$
TL (Uman and McLain 1969)	$I(z',t) = I(0, t - z'/v)$
	$\rho_L(z',t) = \frac{I(0,t-z'/v)}{v}$
MTLL (Rakov and Dulzon 1987)	$I(z',t) = \left(1 - \frac{z'}{H}\right)I(0, t - z'/v)$
	$\rho_L(z',t) = \left(1 - \frac{z'}{H}\right)\left(\frac{I(0,t-z'/v)}{v}\right) + \frac{Q(z',t)}{H}$
MTLE (Nucci *et al.* 1988)	$I(z',t) = e^{-z'/\lambda}I(0, t - z'/v)$
	$\rho_L(z',t) = e^{-z'/\lambda}\frac{I(0,t-z'/v)}{v} + \frac{e^{-z'/\lambda}}{\lambda}Q(z',t)$

$Q(z',t) = \int_{z'/v}^{t} I(0, \tau - z'/v)\mathrm{d}\tau, \quad v = v_f = \text{const}, \quad H = \text{const}, \quad \lambda = \text{const}.$

finite front propagation speed v_f, current at a height, for example, z'_2, begins with a delay z'_2/v_f with respect to the current at the channel base. The dark portion of the waveform indicates the current that actually flows in the channel, the blank portion being shown for illustrative purpose only.

    The most used engineering models can be grouped in two categories: the transmission-line-type models and the traveling-current-source-type models,

Table 1.4   Traveling-current-source-type models for $t \geq z'/v_f$

Model	$I\,(z',\,t)$ and $\rho_L\,(z',\,t)$
BG (Bruce and Golde 1941)	$I(z',t) = I(0,t)$ $\rho_L(z',t) = \dfrac{I\left(0, z'/v_f\right)}{v_f}$
TCS (Heidler 1985)	$I(z',t) = I(0, t + z'/c)$ $\rho_L(z',t) = -\dfrac{I(0, t + z'/c)}{c} + \dfrac{I(0, z'/v*)}{v*}$
DU (Diendorfer and Uman 1990)	$I(z',t) = I(0, t + z'/c) - e^{-\left(t - z'/v_f\right)/\tau_D} I(0, t - z'/v*)$ $\rho_L(z',t) = -\dfrac{I(0, t + z'/c)}{c}$ $\quad - e^{-\left(t - z'/v_f\right)/\tau_D}\left[\dfrac{I(0, z'/v*)}{v_f} + \dfrac{\tau_D}{v*}\dfrac{dI(0, z'/v*)}{dt}\right]$ $\quad + \dfrac{I(0, z'/v*)}{v*} + \dfrac{\tau_D}{v*}\dfrac{dI(0, z'/v*)}{dt}$

$v* = v_f/(1 + v_f/c)$, $v_f = $ const, $\tau_D = $ const.

summarized in Tables 1.3 and 1.4, respectively. Each model in Tables 1.3 and 1.4 is represented by both current and line charge density equations. Table 1.3 includes the TL model and its two modifications: the MTLL and MTLE models. The transmission-line-type models can be viewed as incorporating a current source at the channel base which injects a specified current wave into the channel, the wave propagating upward (1) without either distortion or attenuation (TL), or (2) without distortion but with specified attenuation (MTLL and MTLE), as seen from the corresponding current equations given in Table 1.3. The TL model is often portrayed as being equivalent to an ideal (lossless), uniform transmission line, which is not accurate. First, for such a transmission line in air, the wave propagation speed is equal to the speed of light, while in the TL model it is set to a lower value, in order to make it consistent with observations. Second, a vertical conductor of nonzero radius above ground is actually a nonuniform transmission line whose characteristic impedance increases with increasing height. The resultant distributed impedance discontinuity causes distributed reflections back to the source that is located at ground level. For this reason, even in the absence of ohmic losses, the current amplitude appears to decrease with increasing height (Baba and Rakov 2005a), the effect neglected in the TL model.

Table 1.4 includes the BG model, the TCS model, and the Diendorfer-Uman (DU) model (Diendorfer and Uman 1990). In the traveling-current-source-type models, the return-stroke current may be viewed as generated at the upward-moving return-stroke front and propagating downward, or equivalently as resulting from the cumulative effect of shunt current sources that are distributed along the lightning channel and progressively activated by the upward-moving return-stroke front. In the TCS model, the current at a given channel section turns on instantaneously as the front passes this section, while in the DU model, the current turns on gradually (exponentially with a time constant τ_D if $I(0, t+z'/c)$ were a step function). The channel current in the TCS model may be viewed as a single downward-propagating wave, as illustrated in

Figure 1.10. The DU model involves two terms (see Table 1.4), one being the same as the downward-propagating current in the TCS model that exhibits an inherent discontinuity at the upward-moving front (see Figures 1.9 and 1.10) and the other being an opposite polarity current which rises instantaneously to a value equal in magnitude to the current at the front and then decays exponentially with a time constant τ_D. The second current component in the DU model may be viewed as a "front modifier." It propagates upward with the front and eliminates any current discontinuity at that front. The time constant τ_D can be viewed as the time during which the charge per unit length deposited at a given channel section by the preceding leader reduces to $1/e$ (about 37%) of its original value after this channel section is passed by the upward-moving front. Thottappillil *et al.* (1997) assumed that $\tau_D = 0.1$ μs. Diendorfer and Uman (1990) considered two components of charge density, each released with its own time constant in order to match model predicted fields with measured fields. If $\tau_D = 0$, the DU model reduces to the TCS model. In both the TCS and DU models, the downward-propagating current wave speed is set to be equal to the speed of light. The TCS model reduces to the BG model if the downward current propagation speed is set equal to infinity instead of the speed of light. Although the BG model could be also viewed mathematically as a special case of the TL model with v replaced by infinity, we choose to include the BG model in the traveling-current-source-type model category.

The principal distinction between the two types of engineering models formulated in terms of current is the direction of propagation of the current wave: upward for the transmission-line-type models ($v = v_f$) and downward for the traveling-current-source-type models ($v = -c$), as seen for the TL and TCS models, respectively, in Figures 1.9 and 1.10. As noted earlier, the BG model includes a current wave propagating at an infinitely large speed and, as a result, the wave propagation direction is indeterminate. As in all other models, the BG model includes a front moving at a finite speed v_f. Note that, even though the direction of propagation of current wave in a model can be either up or down, the direction of current is the same; that is, charge of the same sign is transported to ground in both types of engineering models.

The TL model predicts that, as long as (1) the height above ground of the upward-moving return-stroke front is much smaller than the distance r between the observation point on the ground and the channel base, so that all contributing channel points are essentially equidistant from the observer, (2) the return-stroke front propagates at a constant speed, (3) the return-stroke front has not reached the top of the channel, and (4) the ground conductivity is high enough that the associated propagation effects are negligible, the vertical component E_z^{rad} of the electric radiation field and the azimuthal component of the magnetic radiation field are each proportional to the channel-base current I (e.g., Uman *et al.* 1975). The equation for the electric radiation field E_z^{rad} is as follows:

$$E_z^{rad}(r,t) = -\frac{v}{2\pi\varepsilon_0 c^2 r} I(0, t - r/c) \qquad (1.2)$$

where ε_0 is the permittivity of free space, v is the upward propagation speed of the current wave, which is the same as the front speed v_f in the TL model, and c is the speed of light. For the most common return stroke lowering negative charge to ground, the sense of the positive charge flow is upward so that current I, assumed to be upward-directed in deriving (1.2), by convention is positive, and E_z^{rad} by (1.2) is negative; that is the electric field vector points in the negative z direction. Taking the derivative of this equation with respect to time, one obtains

$$\frac{\partial E_z^{rad}(r,t)}{\partial t} = -\frac{v}{2\pi\varepsilon_0 c^2 r}\frac{\partial I(0, t - r/c)}{\partial t} \tag{1.3}$$

Equations (1.2) and (1.3) are commonly used, particularly the first one and its magnetic radiation field counterpart, found from $|B_\varphi^{rad}| = |E_z^{rad}|/c$, where B_φ^{rad} is the radiation component of the azimuthal magnetic flux density, for the estimation of the peak values of return-stroke current and its time derivative, subject to the assumptions listed prior to (1.2).

1.3.3   Equivalency between the lumped-source and distributed-source representations

Maslowski and Rakov (2007) showed that any engineering return-stroke model can be expressed, using an appropriate continuity equation, in terms of either a lumped current source placed at the bottom of the channel or multiple shunt current sources distributed along the channel, with the resultant longitudinal current and the total charge density distribution along the channel being the same in both formulations. This property can be viewed as the duality of engineering models.

In general, any engineering model includes (explicitly or implicitly) both the longitudinal and transverse (radial) currents, with the radial current in the TL model being zero. This is illustrated in Figure 1.11. In the distributed-source-type (DS) models, the radial current, supplied by distributed current sources, enters the channel, with the current sink being at the bottom of the channel. In the lumped-source-type (LS) models, the radial current leaves the channel to compensate the leader charge stored in the corona sheath; the current source is at the bottom of the channel and the partial radial currents can be viewed as current sinks distributed along the channel. Although the directions of the actual radial current in the LS and DS models are opposite (out of the channel and into the channel, respectively), charge of the same sign is effectively transported into the channel core in both types of models. It is also important to note that for either type of model the radial current is distributed along the channel.

Conversion of, for example, an LS model to its equivalent DS model amounts to replacement of the actual corona current of the model by an equivalent one that results in a reversal of direction of the longitudinal current (the source becomes a sink and sinks become sources), while the longitudinal current and the total charge density distribution along the channel remain the same. For the TL model (no longitudinal-current attenuation with height and, hence, no radial current), conversion from the LS to its equivalent DS formulation leads to the introduction of a purely fictitious bipolar radial current which is required to make the lumped source a sink. Conversion of a DS model to its equivalent LS model is done in a

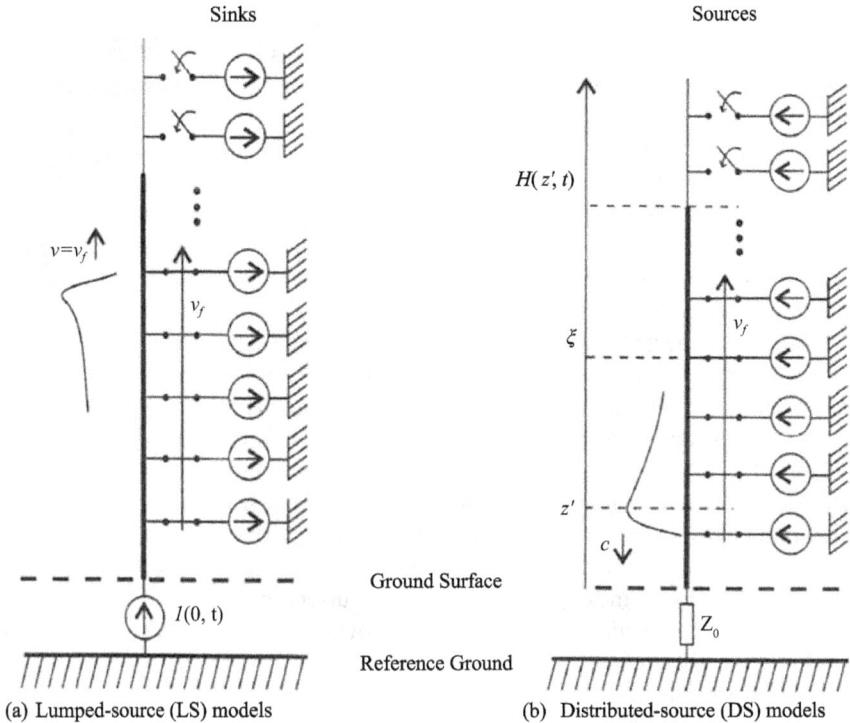

Figure 1.11   Schematic representation of engineering return-stroke models that employ (a) a lumped current source at the lightning channel base (LS models) and (b) distributed current sources along the channel (DS models). Here, v_f is the upward return-stroke front speed (equal to the current wave speed v in LS models), c is the speed of light, and Z_0 is the characteristic impedance of the lightning channel (matched conditions at ground are implied in DS models). LS models with longitudinal-current decay with height imply current sinks distributed along the channel, as shown in (a). Reprinted, with permission, from Maslowski and Rakov (2007, Figure 1)

similar manner. Conversion of LS models to their DS models is particularly useful in extending return-stroke models to include a tall strike object (see Section 1.3.4), as done, for example, by Rachidi *et al.* (2002).

1.3.4   Extension of models to include a tall strike object

Some engineering models have been extended to include a grounded strike object modeled as an ideal, uniform transmission line that supports the propagation of waves at the speed of light and without attenuation or distortion (e.g., Baba and Rakov 2005b). Such an extension results in a second current wavefront, which propagates from the top of the object toward ground at the speed of light, produces reflection on its arrival there, is allowed to bounce between the top and bottom ends of the object, and, in general, produces transmitted waves at either end. The transient behavior of tall objects under direct lightning strike conditions can be illustrated as follows. For the simple example of a nonideal current source (ideal current source connected in parallel with source impedance) attached to the top of the object generating a step-function current wave, the magnitude of the wave injected into the object depends on the characteristic impedance of the object. Specifically, the ideal source current initially divides between the source impedance and the characteristic impedance of the object. However, after a sufficiently long period of time, the current magnitude at any point on the object will be equal to the magnitude of current that would be injected directly into the grounding impedance of the object from the same current source in the absence of the object. In other words, at late times, the ideal source current will in effect divide between the source impedance and the grounding impedance, as if the strike object were not there. Note that the above-mentioned example applies only to a step-function current wave, with the current distribution along the object being more complex for the case of an impulsive current waveform characteristic of the lightning return stroke. If the lightning current wave round-trip time on the strike object is appreciably longer than the risetime of current measured at the top of the object, the peak of the current reflected from the ground is separated from the incident-current peak in the overall current waveform, at least in the upper part of the object.

Model-predicted lightning electromagnetic environment in the presence of tall (electrically long) strike object was studied by many researchers. According to Baba and Rakov (2007), for a typical subsequent stroke, the vertical electric field due to a lightning strike to a 100-m high object is expected to be reduced relative to that due to the same strike to flat ground at distances ranging from 30 to 200 m from the object and enhanced at distances greater than 200 m. The azimuthal magnetic field for the tall object case is larger than that for the flat ground case at any distance. Beyond about 3 km, the field peak is essentially determined by its radiation component and the so-called far-field enhancement factor becomes insensitive to distance change and is expected to be about 2.3.

Note that when the shortest significant wavelength in the lightning current is much longer than the height of the strike object, there is no need to consider the distributed-circuit behavior of such an object. For example, if the minimum significant wavelength is 300 m (corresponding to a frequency of 1 MHz), objects whose heights are about 30 m or less may be considered as lumped, in most cases a short-circuit between the lightning channel base and grounding impedance of the object.

1.4 Testing model validity

The overall strategy in testing the validity of engineering models is illustrated in Figure 1.12. For a given set of input parameters, including $I(0, t)$ and v_f, the model is used to find the distribution of current $I(z', t)$ along the channel, which is then used for computing electric and magnetic fields at different distances from the

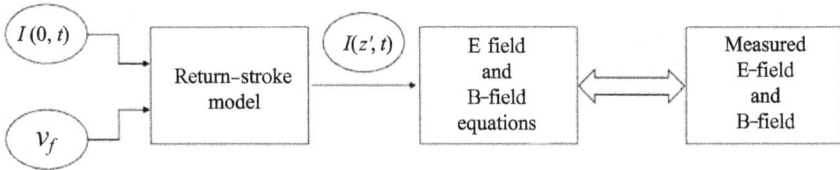

Figure 1.12   Illustration of the overall strategy in testing the validity of engineering models. Adapted from Rakov (2016, Figure 6.4)

lightning channel. The model-predicted fields are compared with corresponding measured fields. Current and field derivatives can be also used for model testing. Ideally, a good model should reproduce measurements at all distances for which the employed electric and magnetic field equations are valid. However, in practice it may be sufficient for a model to be capable of reasonably reproducing measured fields only within a certain range of distances or for certain times (for example, only for the first few microseconds, when the current and radiation field peaks usually occur).

Two primary approaches to model testing have been used: the so-called "typical-event" and "individual-event" approaches. The typical-event approach involves the use of a typical channel-base current waveform $I(0, t)$ and a typical front propagation speed v_f as inputs to the model and a comparison of the model-predicted fields with typical observed fields. In the individual-event approach, $I(0, t)$ and v_f, both measured for the same individual event, are used to compute fields that are compared with the measured fields for the same event. When v_f is not available, a range of reasonable values (measured values are typically in the $c/3$ to $2c/3$ range) can be used to see if a match with measured fields can be achieved for any of those speed values. The individual-event approach is capable of providing a more definitive answer regarding model validity, but it is feasible only in the case of triggered-lightning return strokes or natural lightning strikes to tall towers where channel-base current can be measured.

In the field calculations, the channel is generally assumed to be straight and vertical with its origin at ground level ($z' = 0$): conditions which are expected to be valid for subsequent strokes, but not necessarily for first strokes. The channel length is usually not specified unless it is an inherent feature of the model, as is the case for the MTLL model. As a result, the model-predicted fields and associated model validation may not be meaningful after 25 to 75 μs, the expected time it takes for the return-stroke front to traverse the distance from ground to the cloud charge source region.

1.4.1   "Typical-event" approach

This approach has been adopted, among others, by Nucci *et al.* (1990), Rakov and Dulzon (1991), and Thottappillil *et al.* (1997).

Nucci *et al.* (1990) identified four characteristic features in the fields at 1 to 200 km measured by Lin *et al.* (1979) and used those features as a benchmark for their validation of the TL, MTLE, BG, and TCS models (also of the MULS model (Master, Uman, Lin, and Standler 1981), not considered here). The characteristic features include (1) a sharp initial peak that varies approximately as the inverse distance beyond a kilometer or so in both electric and magnetic fields, (2) a slow ramp following the initial peak and lasting in excess of 100 µs for electric fields measured within a few tens of kilometers, (3) a hump following the initial peak in magnetic fields within a few tens of kilometers, the maximum of which occurs between 10 and 40 µs, and (4) a zero crossing within tens of microseconds of the initial peak in both electric and magnetic fields at 50 to 200 km. For the current shown in Figure 1.13 and other model characteristics assumed by Nucci *et al.* (1990), feature (1) is reproduced by all the models examined, feature (2) by all the models except for the TL model, feature (3) by the BG, TL, and TCS models, but not by the MTLE model, and feature (4) only by the MTLE model, but not by the BG, TL, and TCS models, as illustrated in Figures 1.14 and 1.15. Diendorfer and Uman (1990) showed that the DU model reproduces features (1), (2), and (3), and Thottappillil *et al.* (1991) demonstrated that a relatively insignificant change in the channel-base current waveform (well within the range of typical waveforms) allows the reproduction of feature (4), the zero crossing, by the TCS and DU models. Rakov and Dulzon (1991) showed that the MTLL model reproduces features (1), (2), and (4). Nucci *et al.* (1990) conclude from their study that all the models evaluated by them using measured fields at distances ranging from 1 to 200 km predict reasonable fields for the first 5 to 10 µs, and all models, except for the TL model, do so for the first 100 µs.

There is another typical-event method of testing the validity of return-stroke models, which is based on using net electrostatic field changes produced by leader, ΔE_L, and return-stroke, ΔE_{RS}, processes. The ratio of these field changes, $\Delta E_L/\Delta E_{RS}$, depends on the distribution of charge along the fully formed leader channel. For a uniformly charged channel, this ratio at far distances is equal to 1, if one assumes that return stroke completely neutralizes the leader charge and deposits no additional charge anywhere in the system (this assumption is discussed later in this section). If the leader charge density distribution is skewed toward the ground, the ratio will be greater than 1, and smaller than 1 if it is skewed toward the cloud. Now, the leader charge density distribution is related to the return-stroke current decay along the channel (Rakov and Dulzon 1991). Specifically, the uniform charge density distribution corresponds to a linear current decay with height (the MTLL model), a linear decrease of charge density with increasing height to a parabolic current decay with height, and an exponential decrease of charge density with increasing height to an exponential current decay with height (the MTLE model). In other words, a return-stroke model predicting a variation of current magnitude with height also implicitly specifies the distribution of leader charge density along the channel. It follows that computing $\Delta E_L/\Delta E_{RS}$ at a far distance (some tens of kilometers) for different return-stroke models and comparing model predictions with measurements can be used for testing model validity. Thottappillil *et al.* (1997) assumed that the charge source height is 7.5 km and found that at 100 km, the field change ratio is equal to 0.99 (0.81 at 20 km) for a uniformly charged leader (the MTLL model) and 3.1 (2.6 at 20 km) for a leader with charge

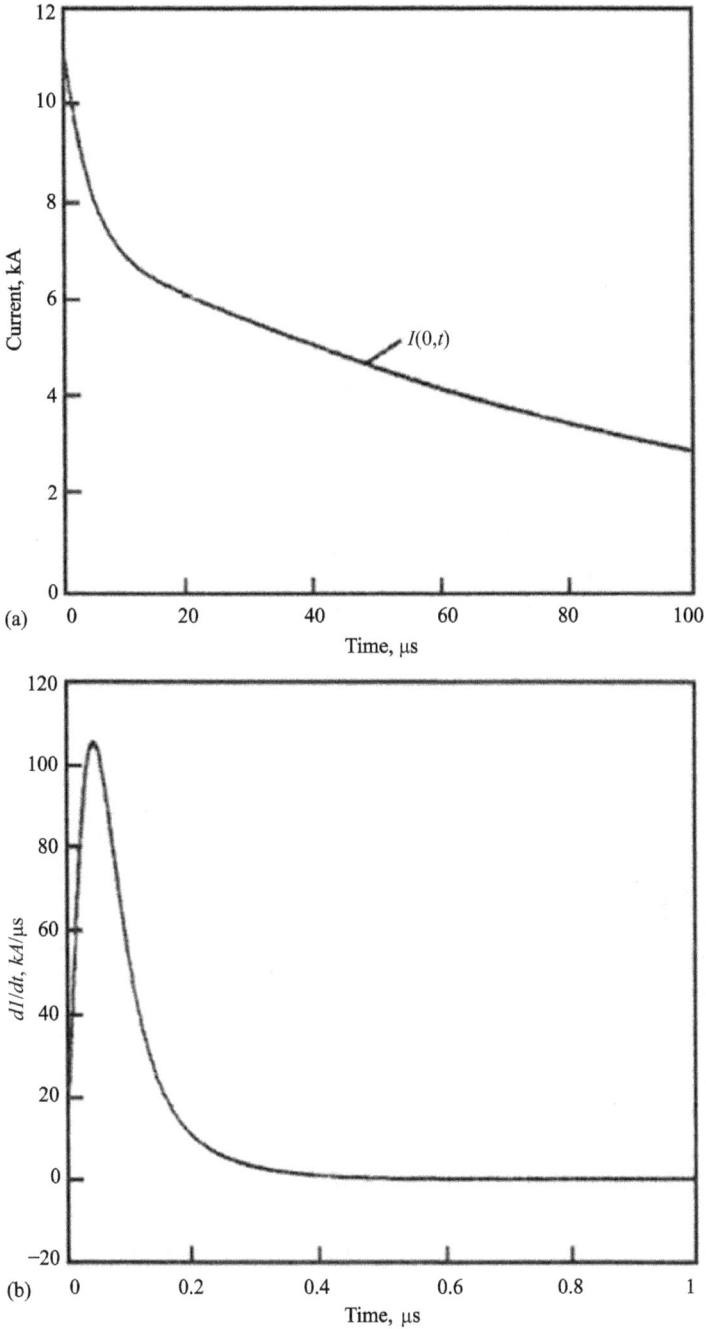

Figure 1.13 (a) Current at ground level and (b) corresponding current derivative used by Nucci et al. (1990), Rakov and Dulzon (1991), and Thottappillil et al. (1997) for testing the validity of return-stroke models by means of the typical-event approach. The peak current is about 11 kA, and peak current rate of rise is about 105 kA/μs. Reprinted, with permission, from Nucci et al. (1990, Figure 4)

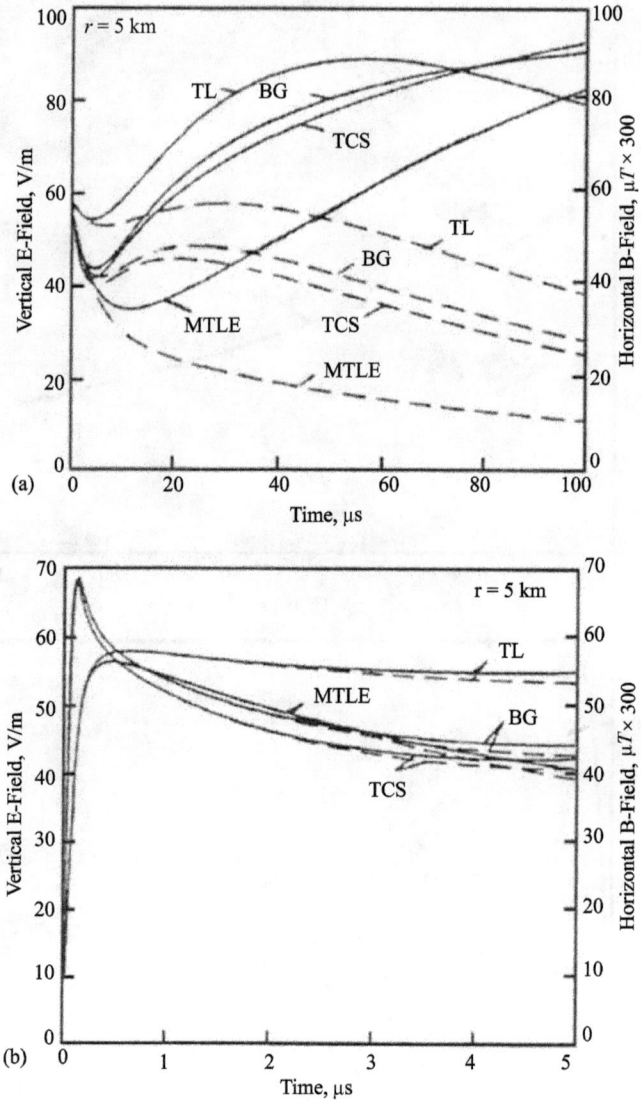

Figure 1.14   Calculated vertical electric (left scaling, solid lines) and horizontal
(azimuthal) magnetic (right scaling, dashed lines) fields for different
return-stroke models at a distance r = 5 km displayed on (a) 100-μs
and (b) 5-μs time scales. Reprinted, with permission, from Nucci
et al. (1990, Figures 11 and 12)

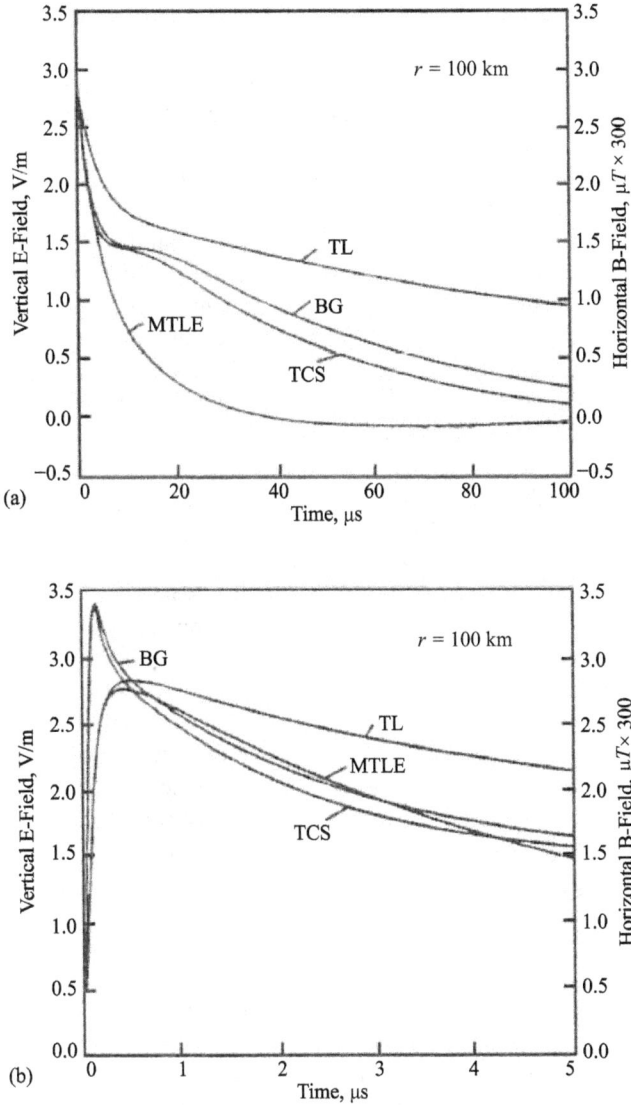

Figure 1.15   Same as Figure 1.14, but for r = 100 km. Reprinted, with permission, from Nucci et al. (1990, Figures 8 and 9)

density exponentially decaying with height (the MTLE model). Ratios of ΔE_L and ΔE_{RS} close to 1 were also computed by Thottappillil *et al.* (1997) for the BG, TCS, and DU models. These results are to be compared to the observations of Beasley *et al.* (1982), who, for 97 first strokes at distances of approximately 20 to 50 km, reported a mean value for the $\Delta E_L/\Delta E_{RS}$ ratio of 0.8. It follows that the BG, MTLL, TCS, and DU models are supported by the available $\Delta E_L/\Delta E_{RS}$ measurements at far ranges from the lightning channel, whereas the MTLE model (whose leader charge density distribution is strongly skewed toward ground) is not.

The $\Delta E_L/\Delta E_{RS}$ test can be also applied to the bidirectional leader model of Mazur and Ruhnke (1993), who simulated the upper section of the leader system by a single-channel positive leader extending vertically upward at the same speed as the negative section of the leader system. They found that the charge density on such a vertically symmetrical bidirectional leader, regarded as a vertical conductor polarized in a uniform electric field, varied linearly with height, with zero charge density at the origin. The model of Mazur and Ruhnke predicts $\Delta E_L/\Delta E_{RS}$ at 20 to 50 km to be between approximately 0.2 and 0.3, which is significantly lower than the average value of 0.8 observed for the first strokes at these distances by Beasley *et al.* (1982). Thus, the vertically symmetrical bidirectional leader model of Mazur and Ruhnke (1993) is not supported by measurements. A more realistic bidirectional leader model was developed by Tran and Rakov (2016).

In calculations of the $\Delta E_L/\Delta E_{RS}$ ratio discussed earlier, the leader charge is assumed to be exactly equal to the return-stroke charge. In general, the total positive charge that enters the leader channel at the strike point (or the negative charge that goes into the ground) during the return stroke can be divided into three components. The first part is the positive charge that is necessary to neutralize the negative charge stored in the leader channel. The second part is the positive charge induced in (sucked from the ground into) the return stroke channel due to the background electric field produced by remaining negative cloud charges. The third part is the additional positive charge spent to neutralize negative cloud charge that was not involved in the leader process. Note that the return stroke charge is often defined as the first part only (as done in the calculations of the $\Delta E_L/\Delta E_{RS}$ ratio discussed above), while all three parts can materially contribute to the measured electric field change. Thus, caution is to be exercised in using the $\Delta E_L/\Delta E_{RS}$ ratio method of testing model validity. Note also that this method is not applicable to the TL model, in which the implicit leader charge is zero.

Thottappillil *et al.* (1997) noted that measured electric fields at tens to hundreds of meters from triggered lightning exhibited a characteristic flattening within 15 µs or so. Electric fields predicted at 50 m by the BG, TL, MTLL, TCS, MTLE, and DU models are shown in Figure 1.16. As follows from Figure 1.16, the BG, MTLL, TCS, and DU models, but not the TL and MTLE models, more or less reproduce the characteristic field flattening at later times. Thus, the TL and MTLE models should be viewed as inadequate for computing close electric fields at later times (beyond the initial 10 µs or so).

1.4.2   *"Individual-event" approach*

This approach has been adopted by Thottappillil and Uman (1993) who compared the TL, TCS, MTLE, and DU models (also the MDU model (Thottappillil *et al.* 1993), not considered here) and Schoene *et al.* (2003) who compared only the TL and TCS models. Thottappillil and Uman (1993) used 18 sets of three simultaneously measured features of triggered-lightning return strokes: channel-base

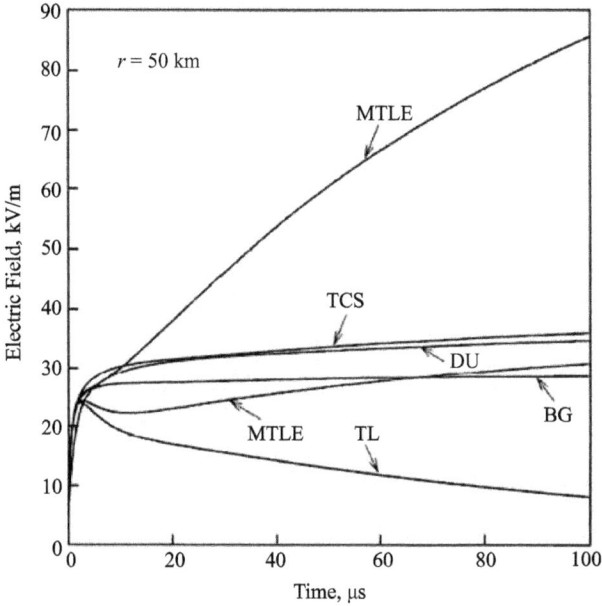

Figure 1.16   Calculated vertical electric field for six return-stroke models at a distance r = 50 m. Adapted from Thottappillil et al. (1997, Figure 4)

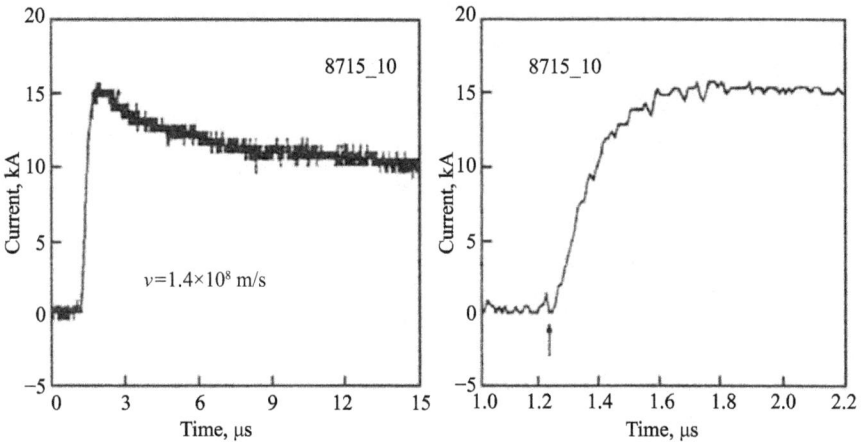

Figure 1.17   An example of current waveform at the base of the channel (left-hand panel) and a close-up of the current wavefront on an expanded timescale (right-hand panel) used by Thottappillil and Uman (1993) for testing the validity of return-stroke models by means of the individual-event approach. Also given (in the left-hand panel) is the measured return-stroke speed v. Reprinted, with permission, from Thottappillil and Uman (1993, Figure 2)

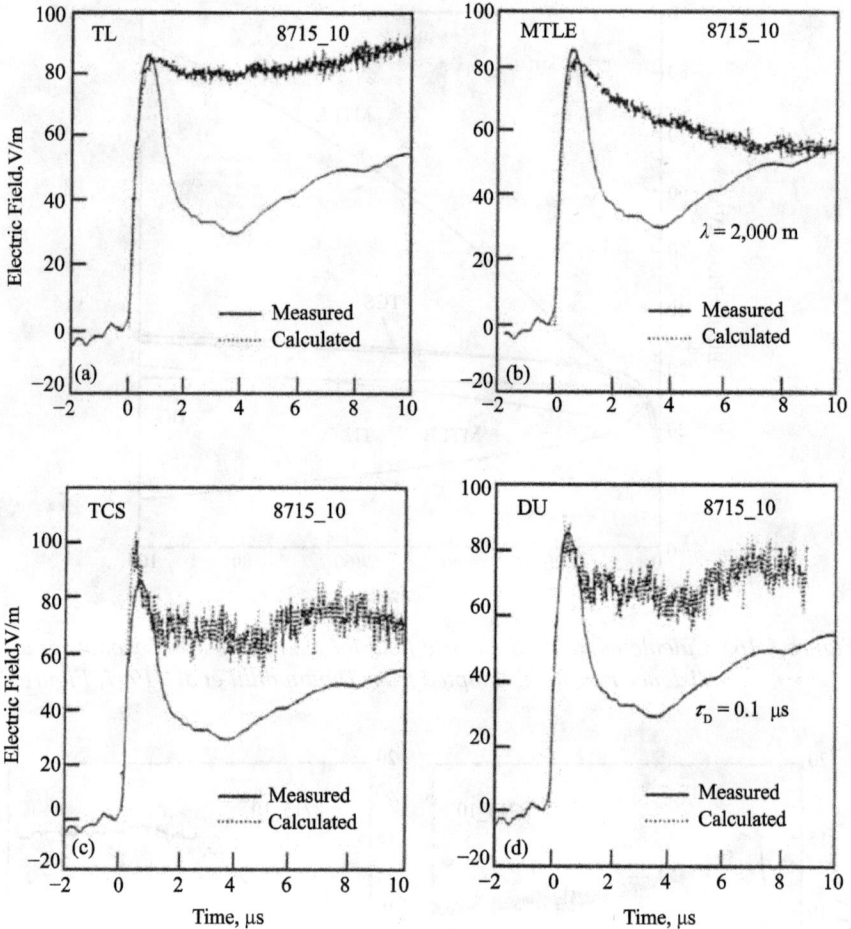

Figure 1.18   *The vertical electric fields calculated using the TL, MTLE, TCS, and DU models ("noisy" dotted lines), shown together with the measured field (solid lines) at about 5 km for the return stroke whose measured current at the channel base and the measured return-stroke speed are given in Figure 1.17. Reprinted, with permission, from Thottappillil and Uman et al. (1993, Figure 4)*

current, return-stroke propagation speed, and electric field at about 5 km from the channel base, all obtained by Willett *et al.* (1989). An example of the comparison for one return stroke, whose measured channel-base current and measured speed are found in Figure 1.17, is given in Figure 1.18. It was found for the overall data set that the TL, MTLE, and DU models each predicted the measured initial electric field peaks with an error whose mean absolute value was about 20%, while the TCS model had a mean absolute error of about 40%. From the standpoint of the overall field waveforms at 5 km, all the tested models should be considered less than adequate.

Schoene *et al.* (2003) tested the TL and TCS models by comparing the first microsecond of model-predicted electric and magnetic field waveforms and field derivative waveforms at 15 and 30 m with the corresponding measured waveforms for triggered-lightning return strokes. The electric and magnetic fields were calculated using the measured current or current derivative at the channel base, an assumed return stroke front speed (three values, $v_f = 1 \times 10^8$ m/s, $v_f = 2 \times 10^8$ m/s, and $v_f = 2.99 \times 10^8$ m/s, were considered), and the temporal and spatial distribution of the channel current specified by the return-stroke model. This was a somewhat lesser testing method than the individual approach discussed earlier, since the speeds were not measured and had to be assumed. The TL model was found to work reasonably well in predicting the measured electric and magnetic fields (and field derivatives) at 15 and 30 m if return-stroke speeds during the first microsecond were chosen to be between 1×10^8 and 2×10^8 m/s. The TCS model did not adequately predict either the measured electric fields or the measured electric and magnetic field derivatives at 15 and 30 m during the first microsecond or so. The TCS model deficiency is related to the fact that it implicitly assumes that the channel is terminated in its characteristic impedance (see Figure 1.11); that is, the current reflection coefficient at ground level is zero. In most cases, this assumption is invalid since the impedance of the lightning channel is typically much larger than the impedance of the grounding, resulting in a current reflection coefficient close to one (conditions at ground are usually close to short-circuit rather than to matched conditions).

On the basis of the entirety of the testing results and mathematical simplicity, the engineering models were ranked by Rakov and Uman (1998) in the following descending order: MTLL, DU, MTLE, TCS, BG, and TL. However, the TL model is recommended for the estimation of the initial field peak from the current peak or conversely the current peak from the field peak, since it is the mathematically simplest model with a predicted peak field/peak current relation that is not less accurate than those of the more mathematically complex models.

1.5 Summary

Lightning return-stroke models can be assigned to one, sometimes two, of the following four classes: (1) gas-dynamic models, (2) electromagnetic models, (3) distributed-circuit models, and (4) engineering models. The most used engineering models can be grouped in two categories: the transmission-line-type models (lumped current source at the bottom of the channel) and the traveling-current-source-type models (multiple equivalent current sources distributed along the channel). Any lumped-source model can be converted to its equivalent distributed-source model and vice versa, with the resultant longitudinal current and the total charge density distribution along the channel being the same. This property can be viewed as the duality of engineering models. Conversion of lumped-source models to distributed-source models is particularly useful in extending return-stroke models to include a tall strike object. Testing model validity is a necessary component of modeling. The engineering models are most conveniently tested using measured electric and magnetic fields from natural and triggered lightning. On the basis of the entirety of the testing results and mathematical simplicity, the engineering models are ranked in the following descending order: MTLL, DU, MTLE, TCS, BG, and TL. When only the relation between the initial peak values of the channel-base current and the remote electric or magnetic fields is concerned, the TL model is preferred.

References

Baba, Y., and Rakov, V. A. (2005a), On the mechanism of attenuation of current waves propagating along a vertical perfectly conducting wire above ground: application to lightning, *IEEE Transactions on Electromagnetic Compatibility*, vol. 47, no. 3, pp. 521–532.

Baba, Y., and Rakov, V. A. (2005b), On the use of lumped sources in lightning return stroke models, *Journal of Geophysical Research*, vol. 110, no. D03101, doi:10.1029/2004JD005202.

Baba, Y., and Rakov, V. A. (2007), Influences of the presence of a tall grounded strike object and an upward connecting leader on lightning currents and electromagnetic fields, *IEEE Transactions on Electromagnetic Compatibility*, vol. 49, no. 4, pp. 886–892.

Baba, Y., and Rakov, V. A. (2012), Electromagnetic models of lightning return strokes, in *Lightning Electromagnetics* (edited by Cooray), IET, UK, Ch. 8, pp. 263–313.

Beasley, W. H., Uman, M. A., and Rustan, P. L. (1982), Electric fields preceding cloud to ground lightning flashes, *Journal of Geophysical Research*, vol. 87, pp. 4884–4902.

Berger, K., Anderson, R. B., and Kroninger, H. (1975), Parameters of lightning flashes, *Electra*, vol. 80, pp. 23–37.

Bruce, C. E. R., and Golde, R. H. (1941), The lightning discharge, *Journal of the Institution of Electrical Engineering*, vol. 88, no. 6, pp. 487–505.

Diendorfer, G., and Uman, M. A. (1990), An improved return stroke model with specified channel-base current, *Journal of Geophysical Research*, vol. 95, pp. 13621–13644.

Eriksson, A. J., (1978), Lightning and tall structures, *Transaction of South African IEE*, vol. 69, pp. 238–252.

Gomes, C., and Cooray, V. (2000), Concepts of lightning return stroke models, *IEEE Transactions on Electromagnetic Compatibility*, vol. 42, no. 1, pp. 82–96.

Gorin, B. N., and Shkilev, A. V. (1984), Measurements of lightning currents at the Ostankino tower (in Russian), *Electrichestrvo*, vol. 8, pp. 64–65.

Heidler, F. (1985), Traveling current source model for LEMP calculation, In Proceedings of 6th International Zurich Symposium on Electromagnetic Compatibility, Zurich, Switzerland, 157–162.

Lin, Y. T., Uman, M. A., Tiller, J. A., *et al.* (1979), Characterization of lightning return stroke electric and magnetic fields from simultaneous two-station measurements, *Journal of Geophysical Research*, vol. 84, pp. 6307–6314.

Maslowski, G., and Rakov, V. A. (2007), Equivalency of lightning return stroke models employing lumped and distributed current sources, *IEEE Transactions on Electromagnetic Compatibility*, vol. 49, no. 1, pp. 123–132.

Master, M. J., Uman, M. A., Lin, Y. T., and Standler, R. B. (1981), Calculations of lightning return stroke electric and magnetic fields above ground, *Journal of Geophysical Research*, vol. 86, no. C12, pp. 12127–12132.

Mazur, V., and Ruhnke, L. (1993), Common physical processes in natural and artificially triggered lightning, *Journal of Geophysical Research*, vol. 98, pp. 12913–12930.

Nucci, C. A., Mazzetti, C., Rachidi, F., and Ianoz, M. (1988), On lightning return stroke models for LEMP calculations, In Proceedings of 19th International Conference on Lightning Protection, Graz, Austria, pp. 463–469.

Nucci, C. A., Diendorfer, G., Uman, M. A., Rachidi, F., Ianoz, M., and Mazzetti, C. (1990), Lightning return stroke current models with specified channel-base current: a review and comparison, *Journal of Geophysical Research*, vol. 95, no. D12, pp. 20395–20408.

Rachidi, F., Rakov, V. A., Nucci, C. A., and Bermudez, J. L. (2002), The effect of vertically-extended strike object on the distribution of current along the lightning channel, *Journal of Geophysical Research*, vol. 107, no. D23, 4699, doi:10.1029/2002JD002119.

Rakov, V. A. (1997), Lightning electromagnetic fields: Modeling and measurements, In Proceedings of 12th International Zurich Symposium on Electromagnetic Compatibility, Zurich, Switzerland, pp. 59–64.

Rakov, V. A. (1998), Some inferences on the propagation mechanisms of dart leaders and return strokes, *Journal of Geophysical Research*, vol. 103, 1879–1887.

Rakov, V. A. (2003), A review of positive and bipolar lightning discharges, *Bulletin of the American Meteorological Society*, vol. 84, no. 6, pp. 767–776.

Rakov, V. A. (2007), Lightning return stroke speed, *Journal of Lightning Research*, vol. 1, pp. 80–89.

Rakov, V. A. (2016), *Fundamentals of Lightning*, Cambridge University Press, UK, pp.1–257.

Rakov, V. A., and Dulzon, A. A. (1987), Calculated electromagnetic fields of lightning return stroke (in Russian), *Tekh. Elektrodinam.*, vol. 1, pp. 87–89.

Rakov, V. A., and Dulzon, A. A. (1991), A modified transmission line model for lightning return stroke field calculations, In Proceedings of 9th International Zurich Symposium on Electromagnetic Compatibility, pp. 229–235.

Rakov, V. A., and Uman, M. A. (1998), Review and evaluation of lightning return stroke models including some aspects of their application, *IEEE Transactions on Electromagnetic Compatibility*, vol. 40, no. 4, pp. 403–426.

Rakov, V. A., and Uman, M. A. (2003), *Lightning: Physics and Effects*, Cambridge University Press, UK, pp.1–687.

Rakov, V. A., Uman, M. A., and Thottappillil, R. (1994), Review of lightning properties from electric field and TV observations, *Journal of Geophysical Research*, vol. 99, no. D5, pp. 10745–10750.

Rakov, V. A., Uman, M. A., Rambo, K. J., *et al.* (1998), New insights into lightning processes gained from triggered-lightning experiments in Florida and Alabama, *Journal of Geophysical Research*, vol. 103, pp. 14117–14130.

Rakov, V. A., Crawford, D. E., Rambo, K. J., Schnetzer, G. H., Uman, M. A., and Thottappillil, R. (2001), M-component mode of charge transfer to ground in lightning discharges, *Journal of Geophysical Research*, vol. 106, pp. 22817–22831.

Rakov, V. A., Borghetti, A., Bouquegneau, C., *et al.* (2013), (CIGRE Working Group C4.407), Lightning parameters for engineering applications, CIGRE Technical Brochure, no. 549, pp. 1–117.

Schoene, J., Uman, M. A., Rakov, V. A., Kodali, V., Rambo, K. J., and Schnetzer, G. H. (2003), Statistical characteristics of the electric and magnetic fields and their time derivatives 15 m and 30 m from triggered lightning, *Journal of Geophysical Research*, vol. 108, no. D6, p. 4192, doi:10.1029/2002JD002698.

Thottappillil, R., and Uman, M. A. (1993), Comparison of lightning return-stroke models, *Journal of Geophysical Research*, vol. 98, pp. 22903–22914.

Thottappillil, R., McLain, D. K., Uman, M. A., and Diendorfer, G. (1991), Extension of the Diendorfer-Uman lightning return stroke model to the case of a variable upward return stroke speed and a variable downward discharge current speed, *Journal of Geophysical research*, vol. 96, no. D9, pp. 17143–17150.

Thottappillil, R., Rakov, V. A., and Uman, M. A. (1997), Distribution of charge along the lightning channel: Relation to remote electric and magnetic fields and to return-stroke models, *Journal of Geophysical Research*, vol. 102, no. D6, pp. 6987–7006.

Tran, M. D., and Rakov, V. A. (2016), Initiation and propagation of cloud-to-ground lightning observed with a high-speed video camera, *Scientific Reports*, no. 6: 39521, doi: 10.1038/srep39521.

Uman, M. A., and McLain, D. K. (1969), Magnetic field of the lightning return stroke, *Journal of Geophysical Research*, vol. 74, pp. 6899–6910.

Uman, M. A., McLain, D. K., and Krider E. P. (1975), The electromagnetic radiation from a finite antenna, *American Journal of Physics*, vol. 43, pp. 33–38.

Willett, J. C., Bailey, J. C., Idone, V. P., Eybert-Berard, A., and Barret, L. (1989), Submicrosecond intercomparison of radiation fields and currents in triggered lightning return strokes based on the transmission line model, *Journal of Geophysical Research*, vol. 94, pp. 13275–13286.

Chapter 2
Calculation of lightning electromagnetic fields

In this chapter, we derive the exact electric and magnetic field equations for the vertical lightning channel and flat, perfectly conducting ground. The so-called dipole (Lorentz condition) approach is used. Then, we discuss the difference between the radiation field component, on the one hand, and electrostatic and induction electric field components, on the other hand, and the non-uniqueness of electric field components. Also, we provide the mathematical function that is most often used for representing the channel-base current waveform. Further, we discuss calculation of lightning electric and magnetic fields propagating over lossy ground.

Key Words: Lightning; lightning electromagnetic field; electric scalar potential; magnetic vector potential; channel-base current; propagation effects

2.1 Introduction

The electric field intensity, E, and magnetic flux density, B, are usually found using the electric scalar potential, φ, and magnetic vector potential, A, as follows:

$$E = -\nabla\varphi - \frac{\partial A}{\partial t} \tag{2.1}$$

$$B = \nabla \times A \tag{2.2}$$

The potentials, φ and A, are related to the volume charge density, ρ, and current density, J, respectively, by

$$\varphi = \frac{1}{4\pi\varepsilon_0} \int_{V'} \frac{\rho(r', t - R/c)}{R} \, dV' \tag{2.3}$$

$$A = \frac{\mu_0}{4\pi} \int_{V'} \frac{J(r', t - R/c)}{R} \, dV' \tag{2.4}$$

where ε_0 and μ_0 are the electric permittivity and magnetic permeability of free space (also applicable to air), respectively, V' is the source volume, dV' is the differential source volume, r' is the position vector of the differential source volume, t is the time, c is the speed of light, and R is the inclined distance between the differential source volume and the observation point. The ratio R/c is the time required for the electromagnetic signals to propagate from the source to the observation point and $t - R/c$ is referred to as the retarded time. Note that E, B, A, and J are vectors and φ and ρ are scalars.

The scalar and vector potentials are related by the Lorentz condition:

$$\nabla \cdot A + \frac{1}{c^2}\frac{\partial \varphi}{\partial t} = 0 \tag{2.5}$$

which is equivalent to the continuity equation relating ρ and J:

$$\nabla \cdot J + \frac{\partial \rho}{\partial t} = 0 \tag{2.6}$$

It follows that the source quantities, ρ and J, cannot be specified independently; they must satisfy the continuity equation (2.6). Alternatively, one can specify J, find A from (2.4), and use the Lorentz condition (2.5) to find φ. In this latter case, there is no need to specify ρ in computing electric fields. Both approaches were used (and shown to be equivalent) by Thottappillil and Rakov (2001). Here, we will use only the Lorentz condition approach, first employed by Uman *et al.* (1975).

In this chapter, we derive the exact electric and magnetic field equations for the vertical lightning channel and flat, perfectly conducting ground. The derivation is performed in four steps corresponding to four configurations illustrated in Figure 2.1, as done by Rakov (2016): (a) differential current element at $z' = 0$ in free space; field point at $z > 0$, (b) elevated differential current element and its image; field point at $z > 0$, (c) same as (b) but the field point is at $z = 0$, and (d) vertical lightning channel above ground; field point on the ground surface. Then, we discuss dependences of electrostatic and induction electric field components of a short current element on distance and the non-uniqueness of electric field components. Also, we provide a mathematical function for representing the channel-base current waveform. Finally, we describe the effects of lossy ground on lightning electric and magnetic fields.

2.2   Derivation of equations for computing lightning electric and magnetic fields

2.2.1   *Differential current element at $z' = 0$ in free space; field point at $z > 0$*

Referring to Figure 2.1(a), the magnetic vector potential at point P due to a differential current element (Hertzian dipole) Idz' pointing in the positive z-direction and located at $z' = 0$ in free space is given by

$$dA(t) = \frac{\mu_0}{4\pi}\frac{I(0, t - R/c)}{R}dz'\ a_z \tag{2.7}$$

where a_z is the z-directed unit vector and R/c is the time required for electromagnetic signal to propagate from the source to the field point at the speed of light, c. Note that for a cylindrical source of length dz' and volume dV', Idz' is the same as

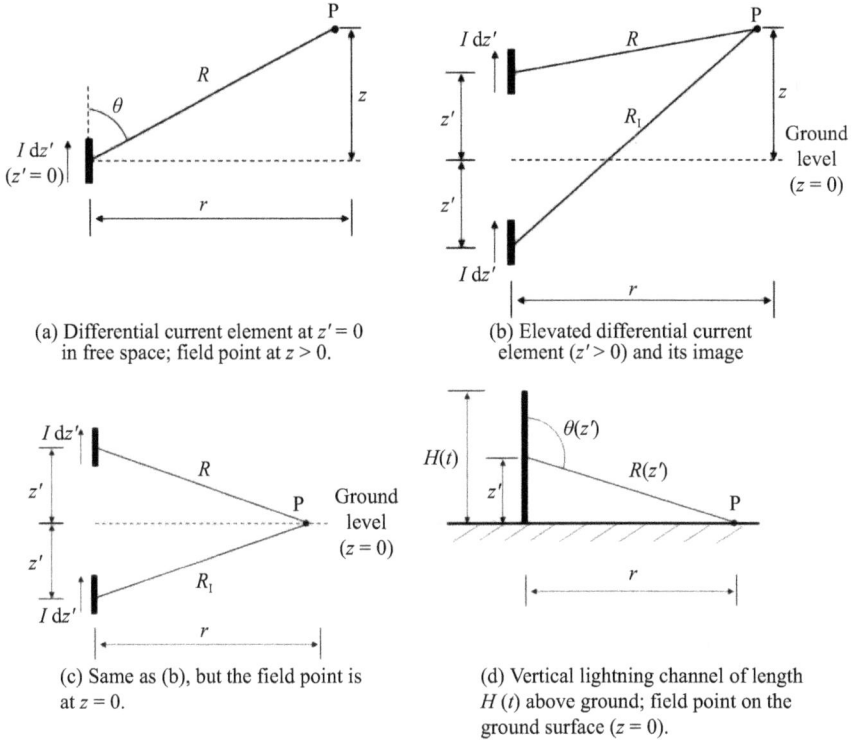

(a) Differential current element at $z' = 0$ in free space; field point at $z > 0$.

(b) Elevated differential current element ($z' > 0$) and its image

(c) Same as (b), but the field point is at $z = 0$.

(d) Vertical lightning channel of length $H(t)$ above ground; field point on the ground surface ($z = 0$).

Figure 2.1   Four configurations used in deriving exact electric and magnetic field equations. Adapted from Rakov (2016, Figure A3.1)

$J\mathrm{d}V'$, where J is the current density. The vector potential has the same direction as its causative current element. This is why $\mathrm{d}A(t)$ has only the z-component. In the following, in order to simplify notation, we use A instead of $\mathrm{d}A(t)$ and drop 0, which indicates that $z' = 0$, in the argument of the current function.

The magnetic flux density due to a z-directed current element has only the ϕ-component:

$$\mathrm{d}\boldsymbol{B} = \nabla \times \boldsymbol{A}$$
$$= \left[\frac{1}{r}\cdot\frac{\partial A_z}{\partial \phi} - \frac{\partial A_\phi}{\partial z}\right]\boldsymbol{a}_r + \left[\frac{\partial A_r}{\partial z} - \frac{\partial A_z}{\partial r}\right]\boldsymbol{a}_\phi + \left[\frac{1}{r}\cdot\frac{\partial(rA_\phi)}{\partial r} - \frac{1}{r}\cdot\frac{\partial A_r}{\partial \phi}\right]\boldsymbol{a}_z$$
$$= -\frac{\partial A_z}{\partial r}\boldsymbol{a}_\phi$$
$$= -\frac{\mu_0\mathrm{d}z'}{4\pi}\frac{\partial}{\partial r}\left[\frac{I(t - R/c)}{R}\right]\boldsymbol{a}_\phi$$

$$(2.8)$$

where $\boldsymbol{a}_r$ is the radial unit vector and $\boldsymbol{a}_\phi$ is the azimuthal unit vector. Note that the second line of (2.8) is the curl operation in the cylindrical coordinate system (see Figure 2.2).

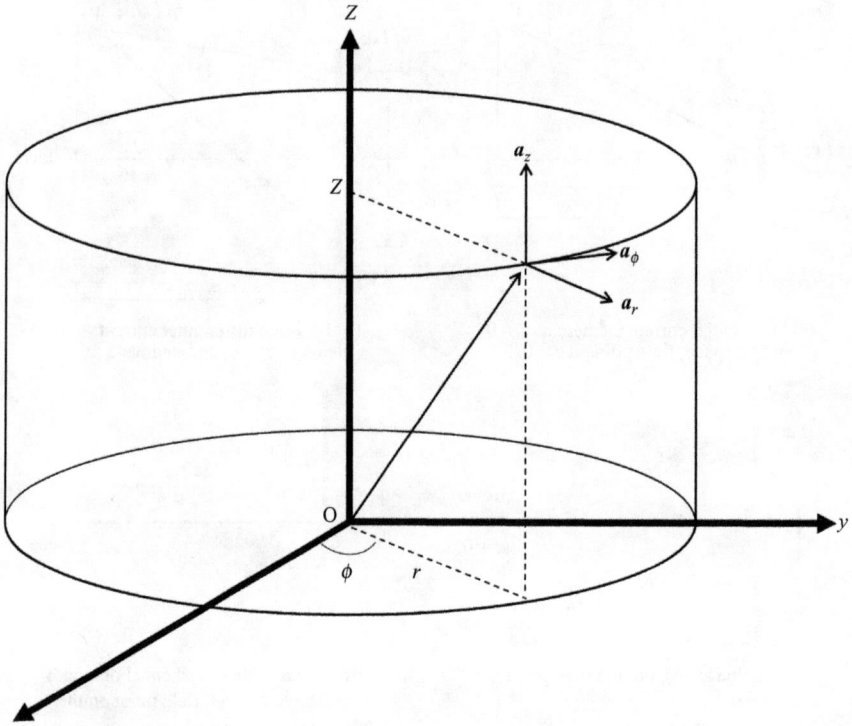

Figure 2.2   Cylindrical coordinate system

Since $R = (r^2 + z^2)^{1/2}$, we use the product rule to take the derivative with respect to r. Then, noting that the derivative of $1/R$ with respect to r is $-r/R^3$, we get

$$d\boldsymbol{B} = -\frac{\mu_0 dz'}{4\pi}\left[\frac{1}{R}\frac{\partial I(t-R/c)}{\partial r} + I(t-R/c)\left(-\frac{r}{R^3}\right)\right]\boldsymbol{a}_\phi \tag{2.9}$$

The spatial derivative $\partial I/\partial r$ can be converted to the time derivative $\partial I/\partial t$ by comparing the following two equations, in which the chain rule was used to take the partial derivatives with respect to r and t and I' stands for the derivative of I with respect to $(t-R/c)$:

$$\frac{\partial I(t-R/c)}{\partial r} = \frac{\partial I(t-R/c)}{\partial(t-R/c)}\frac{\partial(t-R/c)}{\partial r} = -\frac{r}{cR}I'(t-R/c) \tag{2.10}$$

$$\frac{\partial I(t-R/c)}{\partial t} = \frac{\partial I(t-R/c)}{\partial(t-R/c)}\frac{\partial(t-R/c)}{\partial t} = I'(t-R/c) \tag{2.11}$$

Thus,

$$\frac{\partial I}{\partial r} = -\frac{r}{cR}\frac{\partial I}{\partial t} \tag{2.12}$$

and

$$d\boldsymbol{B} = \frac{\mu_0 dz'}{4\pi}\left[\frac{r}{cR^2}\frac{\partial I(t-R/c)}{\partial t} + \frac{r}{R^3}I(t-R/c)\right]\boldsymbol{a}_\phi \tag{2.13}$$

Noting that $r/R = \sin\theta$, we can also write

$$d\boldsymbol{B} = \frac{\mu_0 dz'}{4\pi}\sin\theta\left[\frac{1}{cR}\frac{\partial I(t-R/c)}{\partial t} + \frac{I(t-R/c)}{R^2}\right]\boldsymbol{a}_\phi \tag{2.14}$$

The first term, containing $\partial I/\partial t$, is the magnetic radiation component and the second term, containing I, is the magnetostatic field component. Recall that the above magnetic field equations are for a differential current element at $z' = 0$.

Next we derive, using the dipole technique (e.g., Uman *et al.* 1975), the electric field equation for the same differential current element at $z' = 0$. The total electric field is the negative of the sum of the gradient of φ and the time derivative of A, with φ and A being related by the Lorentz condition (2.5):

$$\varphi = -c^2\int_{-\infty}^{t}\nabla\cdot\boldsymbol{A}\,d\tau \tag{2.15}$$

Since the gradient operator (differentiation with respect to spatial coordinates) and integration over time are independent, we can write

$$\nabla\varphi = -c^2\int_{-\infty}^{t}\nabla(\nabla\cdot\boldsymbol{A})\,d\tau \tag{2.16}$$

In the following, we will obtain equations for $\nabla\cdot\boldsymbol{A}$, $\nabla(\nabla\cdot\boldsymbol{A})$, and $\partial\boldsymbol{A}/\partial t$. Since A has only the z-component ($A_z\,\boldsymbol{a}_z$), the divergence of A has only one term:

$$\begin{aligned}
\nabla\cdot\boldsymbol{A} &= \frac{1}{r}\frac{\partial(r\,A_r)}{\partial r} + \frac{1}{r}\frac{\partial A_\phi}{\partial\phi} + \frac{\partial A_z}{\partial z}\\[4pt]
&= \frac{\partial A_z}{\partial z}\\[4pt]
&= \frac{\mu_0 dz'}{4\pi}\frac{\partial}{\partial z}\left[\frac{I(t-R/c)}{R}\right]\\[4pt]
&= \frac{\mu_0 dz'}{4\pi}\left[\frac{1}{R}\frac{\partial I(t-R/c)}{\partial z} - \frac{z}{R^3}I(t-R/c)\right]
\end{aligned} \tag{2.17}$$

where $-z/R^3$ is the derivative of $1/R$ with respect to z.

Due to the cylindrical symmetry of the problem, $\nabla\cdot A$ is independent of azimuth and $\nabla(\nabla\cdot A)$ has only radial (r) and vertical (z) components, so that the gradient operator can be written as

$$\nabla = a_r \frac{\partial}{\partial r} + a_\phi \frac{1}{r}\frac{\partial}{\partial \phi} + a_z \frac{\partial}{\partial z} = a_r \frac{\partial}{\partial r} + a_z \frac{\partial}{\partial z} \qquad (2.18)$$

Thus,

$$\nabla\varphi = -c^2 \int_{-\infty}^{t}\left[a_r \frac{\partial(\nabla \cdot A)}{\partial r}\right]d\tau - c^2 \int_{-\infty}^{t}\left[a_z \frac{\partial(\nabla \cdot A)}{\partial z}\right]d\tau \qquad (2.19)$$
$$= \nabla\varphi_r + \nabla\varphi_z$$

In the following, we expand the integrands in (2.19), convert the spatial derivatives to time derivatives, expand second-order derivatives, and then assemble final expressions for $\nabla\varphi_r$ and $\nabla\varphi_z$ for the differential current element at $z' = 0$.

The integrands in (2.19) are expanded as follows:

$$a_r \frac{\partial(\nabla \cdot A)}{\partial r} = \frac{\mu_0 dz'}{4\pi}\frac{\partial}{\partial r}\left[\frac{1}{R}\frac{\partial I(t - R/c)}{\partial z} - \frac{z}{R^3}I(t - R/c)\right]a_r$$

$$= \frac{\mu_0 dz'}{4\pi}\left[\begin{array}{l}\dfrac{1}{R}\dfrac{\partial^2 I(t - R/c)}{\partial r \partial z} + \left(-\dfrac{r}{R^3}\right)\dfrac{\partial I(t - R/c)}{\partial z} \\[3mm] -\dfrac{z}{R^3}\dfrac{\partial I(t - R/c)}{\partial r} - z\left(-\dfrac{3r}{R^5}\right)I(t - R/c)\end{array}\right]a_r \qquad (2.20)$$

$$a_z \frac{\partial(\nabla \cdot A)}{\partial z} = \frac{\mu_0 dz'}{4\pi}\frac{\partial}{\partial z}\left[\frac{1}{R}\frac{\partial I(t - R/c)}{\partial z} - \frac{z}{R^3}I(t - R/c)\right]a_z$$

$$= \frac{\mu_0 dz'}{4\pi}\left[\begin{array}{l}\dfrac{1}{R}\dfrac{\partial^2 I(t - R/c)}{\partial z^2} + \left(-\dfrac{z}{R^3}\right)\dfrac{\partial I(t - R/c)}{\partial z} \\[3mm] -\dfrac{1}{R^3}I(t - R/c) - z\left(-\dfrac{3z}{R^5}\right)I(t - R/c) - \dfrac{z}{R^3}\dfrac{\partial I(t - R/c)}{\partial z}\end{array}\right]a_z$$
$$(2.21)$$

Following (2.10) to (2.12), the spatial derivatives in (2.20) and (2.21) are converted to time derivatives:

$$\frac{\partial}{\partial r}I(t - R/c) = -\frac{r}{cR}\frac{\partial}{\partial t}I(t - R/c)$$

$$\frac{\partial}{\partial z}I(t - R/c) = -\frac{z}{cR}\frac{\partial}{\partial t}I(t - R/c) \qquad (2.22)$$

The second-order derivatives in (2.20) and (2.21) are expanded as follows:

$$\frac{\partial^2 I(t - R/c)}{\partial z^2} = \frac{\partial}{\partial z}\left[-\frac{z}{cR}\frac{\partial I(t - R/c)}{\partial t}\right]$$

$$= \frac{\partial(-z)}{\partial z}\frac{1}{cR}\frac{\partial I(t - R/c)}{\partial t} + (-z)\frac{\partial}{\partial z}\left(\frac{1}{cR}\right)\frac{\partial I(t - R/c)}{\partial t}$$

$$+ \left(-\frac{z}{cR}\right)\frac{\partial^2 I(t - R/c)}{\partial z \partial t}$$

$$= -\frac{1}{cR}\frac{\partial I(t - R/c)}{\partial t} + \frac{z^2}{cR^3}\frac{\partial I(t - R/c)}{\partial t} + \frac{z^2}{c^2 R^2}\frac{\partial^2 I(t - R/c)}{\partial t^2}$$

$$\tag{2.23}$$

$$\frac{\partial^2 I(t - R/c)}{\partial r \partial z} = \frac{\partial}{\partial r}\left[-\frac{z}{cR}\frac{\partial I(t - R/c)}{\partial t}\right]$$

$$= -\frac{z}{c}\left[\left(-\frac{r}{R^3}\right)\frac{\partial I(t - R/c)}{\partial t} + \frac{1}{R}\frac{\partial}{\partial t}\left(-\frac{r}{cR}\frac{\partial I(t - R/c)}{\partial t}\right)\right]$$

$$= \frac{zr}{cR^3}\frac{\partial I(t - R/c)}{\partial t} + \frac{zr}{c^2 R^2}\frac{\partial^2 I(t - R/c)}{\partial t^2}$$

$$\tag{2.24}$$

The results are assembled as follows:

$$a_r\frac{\partial(\nabla \cdot A)}{\partial r} = \frac{\mu_0 dz'}{4\pi}\left\{\begin{array}{l} \dfrac{1}{R}\left[\dfrac{zr}{cR^3}\dfrac{\partial I(t - R/c)}{\partial t} + \dfrac{zr}{c^2 R^2}\dfrac{\partial^2 I(t - R/c)}{\partial t^2}\right] \\[2ex] + \left(-\dfrac{r}{R^3}\right)\left[-\dfrac{z}{cR}\dfrac{\partial I(t - R/c)}{\partial t}\right] \\[2ex] + \left(-\dfrac{z}{R^3}\right)\left[-\dfrac{r}{cR}\dfrac{\partial I(t - R/c)}{\partial t}\right] \\[2ex] + \dfrac{3zr}{R^5}I(t - R/c) \end{array}\right\}a_r$$

$$= \frac{\mu_0 dz'}{4\pi}\left[\frac{3zr}{R^5}I(t - R/c) + \frac{3rz}{cR^4}\frac{\partial I(t - R/c)}{\partial t} + \frac{rz}{c^2 R^3}\frac{\partial^2 I(t - R/c)}{\partial t^2}\right]a_r$$

$$\tag{2.25}$$

$$a_z \frac{\partial(\nabla \cdot A)}{\partial z} = \frac{\mu_0 dz'}{4\pi} \left\{ \begin{array}{l} \frac{1}{R}\left[\left(-\frac{1}{cR}+\frac{z^2}{cR^3}\right)\frac{\partial I(t-R/c)}{\partial t}+\frac{z^2}{c^2R^2}\frac{\partial^2 I(t-R/c)}{\partial t^2}\right] \\[2mm] -\frac{z}{R^3}\left[-\frac{z}{cR}\frac{\partial I(t-R/c)}{\partial t}\right] \\[2mm] -\frac{1}{R^3}I(t-R/c)+\frac{3z^2}{R^5}I(t-R/c) \\[2mm] -\frac{z}{R^3}\left[-\frac{z}{cR}\frac{\partial I(t-R/c)}{\partial t}\right] \end{array} \right\} a_z$$

$$= \frac{\mu_0 dz'}{4\pi} \left[\left(\frac{3z^2}{R^5}-\frac{1}{R^3}\right)I(t-R/c)+\left(\frac{3z^2}{cR^4}-\frac{1}{cR^2}\right)\frac{\partial I(t-R/c)}{\partial t} \right. $$
$$\left. +\frac{z^2}{c^2R^3}\frac{\partial^2 I(t-R/c)}{\partial t^2} \right] a_z$$

$$(2.26)$$

Thus, the r- and z-components of $\nabla\varphi$ become:

$$\nabla\varphi_r = -c^2 \int_{-\infty}^{t} a_r \frac{\partial(\nabla \cdot A)}{\partial r} \, d\tau$$

$$= -\frac{dz'}{4\pi\varepsilon_0} \left[\frac{3rz}{R^5}\int_0^t I(\tau-R/c)\,d\tau + \frac{3rz}{cR^4}I(t-R/c) + \frac{rz}{c^2R^3}\frac{\partial I(t-R/c)}{\partial t} \right] a_r$$

$$(2.27)$$

$$\nabla\varphi_z = -c^2 \int_{-\infty}^{t} a_z \frac{\partial(\nabla \cdot A)}{\partial z} \, d\tau$$

$$= -\frac{dz'}{4\pi\varepsilon_0} \left[\left(\frac{3z^2}{R^5}-\frac{1}{R^3}\right)\int_0^t I(\tau-R/c)\,d\tau + \left(\frac{3z^2}{cR^4}-\frac{1}{cR^2}\right)I(t-R/c) + \frac{z^2}{c^2R^3}\frac{\partial I(t-R/c)}{\partial t} \right] a_z$$

$$(2.28)$$

where the lower integration limit is changed from $-\infty$ to zero and c^2 is replaced with $(\mu_0\varepsilon_0)^{-1}$.

Finally, the time derivative of A is

$$\frac{\partial A}{\partial t} = \frac{\partial}{\partial t}\left[\frac{\mu_0 dz'}{4\pi}\frac{I(t-R/c)}{R}\right]a_z = \frac{dz'}{4\pi\varepsilon_0}\left[\frac{1}{c^2R}\frac{\partial I(t-R/c)}{\partial t}\right]a_z \qquad (2.29)$$

where μ_0 is replaced with $(\varepsilon_0 c^2)^{-1}$.

The electric field intensity expression has three terms and is given by

$$dE = -\nabla\varphi_r - \nabla\varphi_z - \frac{\partial A}{\partial t} \qquad (2.30)$$

It should be recalled that the above equations for $\nabla\varphi_r$, $\nabla\varphi_z$, and $\partial A/\partial t$ are for a differential current element at $z' = 0$. One can see from (2.30) that dE has the radial (r) and vertical (z) components and $\nabla \varphi$ contributes to both the r- and z-components of $d\mathbf{E}$, while $\partial \mathbf{A}/\partial t$ only to the z-component. The two components of $d\mathbf{E}$ can be expressed as follows:

$$dE_r = \frac{dz'}{4\pi\varepsilon_0}\left[\frac{3rz}{R^5}\int_0^t I(\tau - R/c)\ d\tau + \frac{3rz}{cR^4}I(t - R/c) + \frac{rz}{c^2 R^3}\frac{\partial I(t - R/c)}{\partial t}\right]$$
(2.31)

$$dE_z = \frac{dz'}{4\pi\varepsilon_0}\left[\left(\frac{3z^2}{R^5} - \frac{1}{R^3}\right)\int_0^t I(\tau - R/c)\ d\tau\right.$$

$$\left. + \left(\frac{3z^2}{cR^4} - \frac{1}{cR^2}\right)I(t - R/c) + \left(\frac{z^2}{c^2 R^3} - \frac{1}{c^2 R}\right)\frac{\partial I(t - R/c)}{\partial t}\right]$$

$$= \frac{dz'}{4\pi\varepsilon_0}\left[\left(\frac{3z^2 - R^2}{R^5}\right)\int_0^t I(\tau - R/c)\ d\tau\right.$$

$$\left. + \left(\frac{3z^2 - R^2}{cR^4}\right)I(t - R/c) + \left(\frac{z^2 - R^2}{c^2 R^3}\right)\frac{\partial I(t - R/c)}{\partial t}\right]$$

$$= \frac{dz'}{4\pi\varepsilon_0}\left[\left(\frac{3z^2 - z^2 - r^2}{R^5}\right)\int_0^t I(\tau - R/c)\ d\tau\right.$$

$$\left. + \left(\frac{3z^2 - z^2 - r^2}{cR^4}\right)I(t - R/c) + \left(\frac{z^2 - z^2 - r^2}{c^2 R^3}\right)\frac{\partial I(t - R/c)}{\partial t}\right]$$

$$= \frac{dz'}{4\pi\varepsilon_0}\left[\frac{2z^2 - r^2}{R^5}\int_0^t I(\tau - R/c)\ d\tau + \frac{2z^2 - r^2}{cR^4}I(t - R/c)\right.$$

$$\left. - \frac{r^2}{c^2 R^3}\frac{\partial I(t - R/c)}{\partial t}\right]$$
(2.32)

2.2.2 Elevated differential current element ($z' > 0$) and its image

Now we assume that the lower half-space ($z < 0$) is perfectly conducting ground and the differential current element is elevated to height $z' > 0$. The field point remains at the same position as in Section 2.2.1. The presence of perfectly conducting ground can be accounted for by using the image theory; that is, by placing an image current element, having the same magnitude and the same direction as the real one, at distance z' below the ground surface plane and ignoring the presence of ground. Now both the upper and lower half-spaces are air, and the vectorial sum of field contributions from

Figure 2.1(b)) is identical to the field at point P in the original configuration (a single current element above the ground plane).

In order to make the equations for dB and dE derived in Section 2.2.1 applicable to the real and image current elements located at z' and $-z'$, respectively, the following changes are to be made.

For the real current element, we simply replace z with $(z - z')$ in the electric field equations and re-insert z' (now it is not equal to zero) in the argument of the current function:

$$dE_r = \frac{dz'}{4\pi\varepsilon_0} \left[\frac{3r(z-z')}{R^5} \int_0^t I(z', \tau - R/c) \, d\tau \right.$$
$$\left. + \frac{3r(z-z')}{cR^4} I(z', t - R/c) + \frac{r(z-z')}{c^2 R^3} \frac{\partial I(z', t - R/c)}{\partial t} \right] \qquad (2.33)$$

$$dE_z = \frac{dz'}{4\pi\varepsilon_0} \left[\frac{2(z-z')^2 - r^2}{R^5} \int_0^t I(z', \tau - R/c) \, d\tau \right.$$
$$\left. + \frac{2(z-z')^2 - r^2}{cR^4} I(z', t - R/c) - \frac{r^2}{c^2 R^3} \frac{\partial I(z', t - R/c)}{\partial t} \right] \qquad (2.34)$$

where $R = [r^2 + (z - z')^2]^{1/2}$. The equation for the ϕ-component of magnetic field (B_ϕ) produced by the real current element is similar to (2.13); the only differences being the change in the argument of current function from $(t - R/c)$, where $z' = 0$ is implied, to $(z', t - R/c)$ and different expression for R ($R = [r^2 + (z - z')^2]^{1/2}$):

$$dB_\phi = \frac{\mu_0 dz'}{4\pi} \left[\frac{r}{R^3} I(z', t - R/c) + \frac{r}{cR^2} \frac{\partial I(z', t - R/c)}{\partial t} \right] \qquad (2.35)$$

Note that the two terms in the brackets in (2.35) are transposed relative to (2.13) in order to make the magnetic field equation consistent with the traditional formulation in which the $\partial I/\partial t$ term follows the I term.

Equations for the image current element can be obtained from the equations for the real one, (2.33) to (2.35), by replacing $(z - z')$ with $(z + z')$ and R with $R_I = [r^2 + (z + z')^2]^{1/2}$.

Equations (2.33) to (2.35) (and their counterparts for the image current element) are the basis for deriving field equations for the case of a field point located at an arbitrary position in space, still above perfectly conducting ground. Note that at an elevated field point the electric field will have both z- and r-components, because the inclined distances from the real and image current elements to the field point are different.

2.2.3   *Elevated differential current element above ground and its image; field point on the ground surface $(z = 0)$*

When $z = 0$ (point P on the ground surface; see Figure 2.1(c)), $R_I = R$ and, hence, contributions from the real and image sources are equal to each other. Thus, the the real and image current elements at field point P at $z > 0$ in this configuration (see effect of perfectly conducting ground plane on the magnetic field on that plane is to double the contribution from the real current element.

Similarly, the contributions from the real and image current elements to E_z are equal, causing the field doubling effect, while the contributions to E_r are equal in magnitude and opposite in sign. Hence, the radial component of electric field on perfectly conducting ground is zero, as required by the boundary condition on the tangential component of electric field on the dielectric/conductor interface; only normal component can exist on the surface of a perfect conductor.

The total electric and magnetic fields at the ground surface ($z = 0$) due to the elevated differential current element and its image are given by

$$dE_z(t) = \frac{dz'}{2\pi\varepsilon_0}\left[\frac{2z'^2 - r^2}{R^5}\int_0^t I(z', \tau - R/c)\ d\tau\right.$$

$$\left. + \frac{2z'^2 - r^2}{cR^4}I(z', t - R/c) - \frac{r^2}{c^2R^3}\frac{\partial I(z', t - R/c)}{\partial t}\right] \tag{2.36}$$

$$dB_\phi = \frac{\mu_0 dz'}{2\pi}\left[\frac{r}{R^3}I(z', t - R/c) + \frac{r}{cR^2}\frac{\partial I(z', t - R/c)}{\partial t}\right] \tag{2.37}$$

Equations (2.36) and (2.37) are applicable to an electrically short (Hertzian) vertical dipole above ground, such as the compact intracloud discharge (CID) (Nag and Rakov 2010). They can also be used (after integration over z') for an elevated vertical channel of arbitrary length. The three terms in (2.36) are named electrostatic, induction, and radiation, and in (2.37) the two terms are magnetostatic and radiation.

2.2.4 Vertical lightning channel above ground; field point on the ground surface ($z = 0$)

This configuration is usually applied to the return-stroke process in which a current wave propagates from the ground level up along the channel.

Integrating (2.36) over the radiating channel length $H(t)$, we get

$$E_z(t) = \frac{1}{2\pi\varepsilon_0}\left[\begin{array}{l}\displaystyle\int_0^{H(t)}\frac{2z'^2 - r^2}{R^5}\int_0^t I(z', \tau - R/c)\ d\tau dz' \\[3mm] \displaystyle + \int_0^{H(t)}\frac{2z'^2 - r^2}{cR^4}I(z', t - R/c)dz' \\[3mm] \displaystyle - \int_0^{H(t)}\frac{r^2}{c^2R^3}\frac{\partial I(z', t - R/c)}{\partial t}dz'\end{array}\right] \tag{2.38}$$

Alternatively, we can write

$$E_z(t) = \frac{1}{2\pi\varepsilon_0}\left[\begin{array}{l}\displaystyle\int_0^{H(t)}\frac{(2 - 3\sin^2\theta)}{R^3}\int_0^t I(z', \tau - R/c)\ d\tau dz' \\[3mm] \displaystyle + \int_0^{H(t)}\frac{(2 - 3\sin^2\theta)}{cR^2}I(z', t - R/c)dz' \\[3mm] \displaystyle - \int_0^{H(t)}\frac{\sin^2\theta}{c^2R}\frac{\partial I(z', t - R/c)}{\partial t}dz'\end{array}\right] \tag{2.39}$$

where the following relations were used: $r^2/R^2 = \sin^2(180°-\theta) = \sin^2\theta$; $z'^2/R^2 = \cos^2(180°-\theta) = (-\cos\theta)^2 = \cos^2\theta$; $(2z'^2 - r^2)/R^2 = 2\cos^2\theta - \sin^2\theta = 2 - 2\sin^2\theta - \sin^2\theta = 2 - 3\sin^2\theta$.

Integrating (2.37) over the radiating channel length $H(t)$, we get

$$B_\phi(t) = \frac{\mu_0}{2\pi}\left[\int_0^{H(t)} \frac{r}{R^3} I(z',t-R/c)dz' + \int_0^{H(t)} \frac{r}{cR^2}\frac{\partial I(z',t-R/c)}{\partial t}dz'\right]$$

(2.40)

Alternatively, using $r/R = \sin(180°-\theta) = \sin\theta$, we can write

$$B_\phi(t) = \frac{\mu_0}{2\pi}\left[\int_0^{H(t)} \frac{\sin\theta}{R^2} I(z',t-R/c)dz' + \int_0^{H(t)} \frac{\sin\theta}{cR}\frac{\partial I(z',t-R/c)}{\partial t}dz'\right]$$

(2.41)

The radiating channel length $H(t)$ in (2.38) to (2.41) for an upward-moving current wave is found from the following equation:

$$t = \frac{H(t)}{v_f} + \frac{R(H(t))}{c}$$

(2.42)

Equations (2.38) to (2.41) are exact, provided that the lightning channel is vertical, the ground is perfectly conducting, and the field point is on the ground surface.

A lightning return-stroke model is needed to specify $I(z', t)$. Equations (2.38) to (2.41) are suitable for computing fields at ground level using the electromagnetic, distributed-circuit, or "engineering" return-stroke models (Rakov and Uman 1998; also see Section 1.3 of this book). Some of the engineering models include a current discontinuity at the moving front. Such a discontinuity is an inherent feature of some traveling-current-source-type models such as the Bruce–Golde (BG) model (Bruce and Golde 1941) and the traveling current source model, TCS (Heidler 1985), even when the current at the channel base starts from zero. The transmission-line-type models may include a discontinuity at the front if the channel-base current starts from a nonzero value. The Diendorfer–Uman (DU) model (Diendorfer and Uman 1990) does not include a current discontinuity either at the upward-moving front or at the channel base. The three terms in (2.38) or (2.39) are referred to as the electrostatic, induction (or intermediate), and electric radiation field components, respectively, and the two terms in (2.40) or (2.41) are referred to as the magnetostatic (or induction) and magnetic radiation field components, respectively. For return-stroke models with current discontinuity at the moving front, (2.38) to (2.41) describe the fields only due to sources below the upward-moving front. A current discontinuity at the moving front gives rise to an additional term in each of the equations for E_z and B_ϕ:

$$E_z^{\text{disc}} = -\frac{1}{2\pi\varepsilon_0}\frac{r^2}{c^2R^3(H(t))}I\left(H(t),\frac{H(t)}{v_f}\right)\frac{dH(t)}{dt}$$

(2.43)

$$B_\phi{}^{\text{disc}} = \frac{\mu_0}{2\pi\, cR^2(H(t))} I\!\left(H(t), \frac{H(t)}{v_f}\right) \frac{\mathrm{d}H(t)}{\mathrm{d}t} \tag{2.44}$$

Note that the front discontinuity produces only a radiation field component, no electrostatic or induction field components.

We now use the transmission-line (TL) model (Uman and McLain 1969) to derive the far-field approximation to (2.38). At far distances (typically at $r \geq$ 50 km), the radiation field component is dominant, so that we can write:

$$E_z(t) \approx E_z{}^{\text{rad}}(t) = -\frac{1}{2\pi\varepsilon_0} \int_0^{H(t)} \frac{r^2}{c^2 R^3} \frac{\partial I(z',t - R/c)}{\partial t} \mathrm{d}z' \tag{2.45}$$

From (2.42), noting that for the TL model $v_f = v$ and at far ranges $R \approx r$, we have

$$H(t) = v(t - R/c) \approx v(t - r/c) \tag{2.46}$$

Thus, (2.45) becomes

$$E_z{}^{\text{rad}}(t) = -\frac{1}{2\pi\varepsilon_0 c^2 r} \int_0^{v(t-r/c)} \frac{\partial I(z',t - r/c)}{\partial t} \mathrm{d}z' \tag{2.47}$$

For a current wave traveling at constant speed v in the positive z' direction, $I(t-z'/v)$, the time derivative of current can be converted to the spatial derivative by comparing the following two equations, in which the chain rule was used to take the partial derivatives with respect to z' and t, and I' stands for the derivative of I with respect to $(t - z'/v)$:

$$\frac{\partial I(t - z'/v)}{\partial z'} = \frac{\partial I(t - z'/v)}{\partial(t - z'/v)} \frac{\partial(t - z'/v)}{\partial z'} = -\frac{1}{v} I'(t - z'/v)$$
$$\frac{\partial I(t - z'/v)}{\partial t} = \frac{\partial I(t - z'/v)}{\partial(t - z'/v)} \frac{\partial(t - z'/v)}{\partial t} = I'(t - z'/v) \tag{2.48}$$

As a result,

$$\frac{\partial I(t - z'/v)}{\partial t} = -v\frac{\partial I(t - z'/v)}{\partial z'} \tag{2.49}$$

For the TL model in terms of retarded time, $I(z', t - r/c) = I(0, t - z'/v - r/c)$ and

$$\frac{\partial I(0,t - z'/v - r/c)}{\partial t} = -v\frac{\partial I(0,t - z'/v - r/c)}{\partial z'} \tag{2.50}$$

so that (2.47) becomes

$$\begin{aligned}
E_z{}^{\text{rad}}(t) &= \frac{v}{2\pi\varepsilon_0 c^2 r} \int_{z'=0}^{z'=v(t-r/c)} \mathrm{d}I(0, t - z'/v - r/c) \\
&= \frac{v}{2\pi\varepsilon_0 c^2 r} \left[I\!\left(0, t - \frac{v(t - r/c)}{v} - r/c\right) - I(0, t - r/c) \right] \\
&= \frac{v}{2\pi\varepsilon_0 c^2 r} [I(0,0) - I(0, t - r/c)]
\end{aligned} \tag{2.51}$$

Normally $I(0, 0) = 0$, so that

$$E_z^{\text{rad}}(t) = -\frac{v}{2\pi\varepsilon_0 c^2 r} I(0, t - r/c) \tag{2.52}$$

that is, the electric radiation field component is proportional to the channel-base ($z' = 0$) current. The minus sign indicates that for a positive current (positive charge moving upward), the electric field vector is directed downward. The corresponding magnetic radiation field can be found from $\left|B_\phi^{\text{rad}}\right| = \left|E_z^{\text{rad}}\right|/c$ or $\left|H_\phi^{\text{rad}}\right| = \left|E_z^{\text{rad}}\right|/\eta_0$, where $H_\phi = B_\phi/\mu_0$ is the magnetic field intensity and $\eta_0 = 377\ \Omega$ is the intrinsic impedance of free space. Equation (2.52) and its magnetic field counterpart are further discussed in Section 2.5, along with the close field approximations derived using other than radiation field components.

Krider (1994), using the TL model, computed the peak electric fields (radiation component) in the upper half-space for different values of the return-stroke speed, v, relative to the speed of light, c. He found that the largest fields are radiated at relatively small polar angles, θ, measured from the vertical when the return-stroke speed is very close to the speed of light. Similar results were obtained by Rakov and Tuni (2003), who used the modified transmission line model with exponential current decay with height (MTLE) (Nucci *et al.* 1988). On the other hand, when $\theta = 0°$, the electric radiation field component vanishes (a dipole does not radiate along its axis) and, as a result, the electric field at small polar angles can be dominated by its induction component (Lu 2006).

Thottappillil *et al.* (1997) derived an electrostatic field equation in terms of line charge density, ρ_L, for a very close observation point, such that $r \ll H(t)$, and assuming that (1) retardation effects are negligible and (2) return-stroke line charge density does not vary appreciably with height within the channel section significantly contributing to the electric field at r,

$$E_z(z, t) \approx -\frac{\rho_L(t)}{2\pi\varepsilon_0 r} \tag{2.53}$$

Equation (2.53) indicates that the electrostatic field produced by a very close return stroke is approximately proportional to the line charge density on the bottom part of the channel.

2.3   The reversal distance for electrostatic and induction electric field components of a short current element

For an elevated short dipole that is vertical (see Figure 2.3), the peak radiation field, on the one hand, and induction (intermediate) and electrostatic field changes, on the other hand, may have opposite polarities. This follows from (2.36) in which the first (electrostatic) and second (induction) terms contain $(2z'^2 - r^2)$.

The motion of positive charge upward (or negative charge downward) produces a radiation electric field change (initial peak) directed downward at all distances, as shown in Figure 2.3 (inset). Conversely, for positive charge moving downward (or

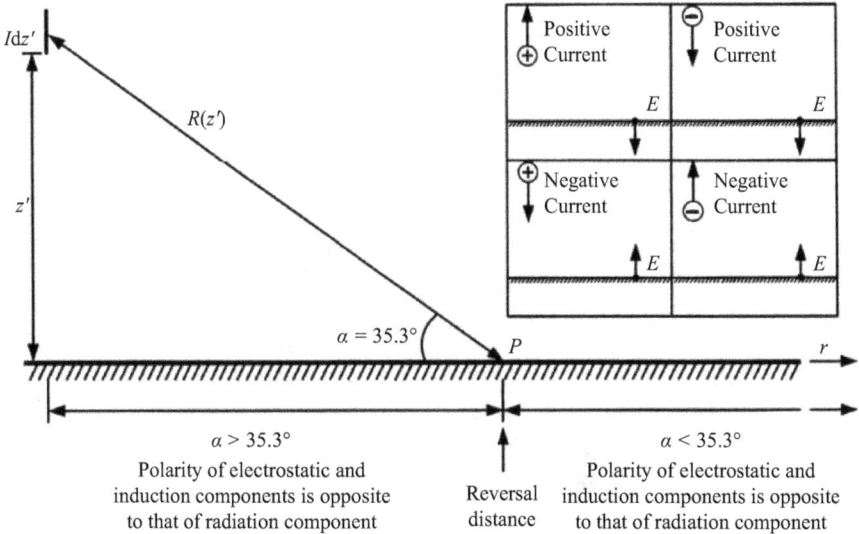

Figure 2.3 *Illustration of the reversal distance for the electrostatic and the induction field components. Inset shows the direction of the radiation component of electric field vector for different combinations of the charge polarity and the direction of its motion (also, the direction for all three components when α < 35.3°). The direction of the electric field vector refers to the initial half-cycle in the case of bipolar waveforms. Adapted from Nag and Rakov (2010, Figure A3)*

negative charge upward), the radiation electric field change is directed upward. At close distances (i.e. for angle $\alpha > 35.3°$ in Figure 2.3), the motion of positive charge upward produces electrostatic and induction field changes directed upward and the radiation electric field change is directed downward, while at far distances (i.e. for $\alpha < 35.3°$), all three electric field components are directed downward.

2.4 Non-uniqueness of electric field components

The three components of an electric field (or the change in that field), referred to as the electrostatic, induction (intermediate), and radiation components, are not unique (e.g. Rubinstein and Uman 1989). For example, these components, often identified by their $1/R^3$, $1/R^2$, and $1/R$ distance dependences, are different for the dipole (Lorentz condition) (Uman *et al.* 1975) and monopole (continuity equation) (Thomson 1999) approaches to calculating lightning electric fields. In both approaches, the electric field is found as $E = -\nabla\varphi - \partial A/\partial t$, where φ is the scalar potential and A is the vector potential. The expressions for A in the two techniques are the same, but those for φ are different: φ is found from A using Lorentz condition in the dipole technique; while in the monopole technique, it is found from the charge density, which is related to current density via the continuity equation.

Table 2.1 Comparison of electric field components at ground level based on dipole (Uman et al. 1975) and monopole (Thomson 1999) techniques for a differential current element Idz' at height z' above ground

Electric field component identified by its distance dependence	Originated from		Polarity	
	Dipole technique	Monopole technique	Dipole technique	Monopole technique
$1/R^3$ (electrostatic)	$\nabla\varphi$	$\nabla\varphi$	Negative for $\alpha < 35.3°$ Positive for $\alpha > 35.3°$	Negative
$1/R^2$ (induction)	$\nabla\varphi$	$\nabla\varphi$	Negative for $\alpha < 35.3°$ Positive for $\alpha > 35.3°$	Negative
$1/R$ (radiation)	$\nabla\varphi$ and $\partial A/\partial t$	$\partial A/\partial t$	Negative	Negative

φ is the scalar potential; $E = -\nabla\varphi - \partial A/\partial t$; $\alpha = \sin^{-1}(z'/R(z'))$. Adapted from Rakov (2016, Table 5.1).

The two approaches are equivalent: both produce identical total fields, although the individual electric field components may even have different polarities, as seen in Table 2.1. Note that in contrast to the more common dipole (Lorentz condition) technique, there is no reversal distance for the $1/R^3$ and $1/R^2$ components in the monopole technique, and the $1/R$ component originates only from the time-derivative of magnetic vector potential. The differences between the field components found using different techniques are considerable at close ranges but become negligible at far ranges.

For the dipole approach and electrically short channel segment, Thottappillil and Rakov (2001) have noted that the distance dependences of electric field components are not exactly $1/R^3$, $1/R^2$, and $1/R$, because of the additional dependence on $\sin^2\theta$, where $\sin \theta = r/R$, with r and $R = f(z')$ being the horizontal and inclined distances, respectively, between the source and field points. Only when $\sin^2\theta \approx 1$ (at relatively large distances, when $R \approx r$), the distance dependences are exactly $1/R^3$, $1/R^2$, and $1/R$.

2.5 Short channel segment vs. total radiating channel length

The electric field components discussed in Sections 2.3 and 2.4 are defined for a differential current element (an electrically short channel segment), for which the current does not vary along the radiator length. In computing lightning electric fields, the integration over the entire radiating channel (over height z') must be performed at each instant of time, with the inclined distance R being a function of z'. As a result, the horizontal distance ($r = $ const) is often used instead of $R = f(z')$ for evaluating the distance dependence of field components produced by the entire radiating channel.

Chen *et al.* (2015) have shown that, for the TL model (Uman and McLain 1969), the sum of electrostatic and induction components of electric field at close ranges and the radiation field component at far ranges each vary approximately as $1/r$, and each is proportional to the channel-base current. As the current wave propagation speed, v, approaches the speed of light, c, the approximate close-range equation (derived using only the electrostatic and induction field components) and the approximate far-range equation (derived using only the radiation field component) converge to the same exact equation for the total electric field, which is valid for any distance from the lightning channel (see Table 2.2, which also contains a similar result for magnetic field). Thus, for the transmission line model with $v = c$, the field components lose their significance. Indeed, in this case, the total electric field (and the total magnetic field) at any distance is proportional to the channel-base current and varies as $1/r$ (even at very close ranges), as expected for a spherical transverse electromagnetic (TEM) wave (Thottappillil *et al.* 2001). The approximations presented in Table 2.2 are applicable only to the initial portions of the field waveforms, since the TL model is inadequate at later times. Also, their ranges of validity depend, besides v, on the current waveshape. For example, when $v = 1.3 \times 10^8$ m/s, the far electric field approximation is valid beyond 100 km for the typical first stroke and beyond 50 km for the typical subsequent stroke. The close electric field approximation is valid within 100 m for the typical first stroke and only within 10 m for the typical subsequent stroke. The ranges of validity increase with increasing v.

Table 2.2 Approximate expressions for the electromagnetic fields based on the TL model with an arbitrary return-stroke speed v for near and far ranges, converging to exact expressions (Thottappilli et al. 2001) as v approaches c

	Approximate expressions based on the TL model with an arbitrary return-stroke speed v	Exact expressions based on the TL model with $v = c$
Near ranges	$E_z \approx -\frac{1}{2\pi\varepsilon_0 vr}I(0, t - r/c)$	$E_z = -\frac{1}{2\pi\varepsilon_0 cr}I(0, t - r/c)$ (any range)
Far ranges	$E_z \approx -\frac{v}{2\pi\varepsilon_0 c^2 r}I(0, t - r/c)$	
Near ranges	$H_\phi \approx \frac{1}{2\pi r}I(0, t - r/c)$	$H_\phi = \frac{1}{2\pi r}I(0, t - r/c)$ (any range)
Far ranges	$H_\phi \approx \frac{v}{2\pi cr}I(0, t - r/c)$	

Adapted from Chen et al. (2015, Figure 1).

Sometimes, the entire lightning channel is approximated as an electrically short dipole with $I(z', t) = I(0, t)$. In this case, the radiation field component is proportional to dI/dt, while for the more realistic transmission line model, $I(z', t) = I(0, t - z'/v)$, where v is the current wave propagation speed, the radiation field component is proportional to the product of I and v. The difference here is similar to the one between a Hertzian (electrically short) dipole and a traveling-wave antenna, and it has important implications for the estimation of peak currents from range-normalized measured peak fields, as done in modern lightning locating systems. Since the field-to-current conversion procedures in those systems are usually developed for return strokes, assuming that the current peak is proportional to the field peak, they may yield incorrect results for short cloud discharges, because the field peak for electrically short radiators is proportional to the current derivative peak, not to the current peak (Nag *et al.* 2011).

2.6   Channel-base current equation

For the engineering models, in which a vertical lightning channel and a perfectly conducting ground are often assumed, the information on the source required for computing the fields usually includes (1) the channel base current (either measured or assumed from typical measurements) and (2) the upward return-stroke front speed, typically assumed to be constant and in a range from 1×10^8 to 2×10^8 m/s. The return-stroke current waveform at the channel base is often approximated by the Heidler function (Heidler 1985):

$$I(0, t) = \frac{I_0}{k} \frac{(t/\tau_1)^n}{(t/\tau_1)^n + 1} e^{-t/\tau_2} \tag{2.54}$$

where I_0, k, n, τ_1, and τ_2 are constants. This function allows one to change conveniently the current peak, maximum current derivative, and associated electrical charge transfer nearly independently by changing I_0, τ_1, and τ_2, respectively. Equation (2.54) reproduces the observed concave rising portion of a typical current waveform, as opposed to the once more commonly used double-exponential function (Bruce and Golde 1941): $I(0, t) = I_0(e^{-\alpha t} - e^{-\beta t})$, where I_0, α, and β are constants. It is characterized by an unrealistic convex wavefront with a maximum current derivative at $t = 0$. Sometimes, the sum of two Heidler functions with different parameters or the sum of a Heidler function and a double-exponential function is used to approximate the desired current waveshape. The return-stroke current waveforms recommended by the International Electrotechnical Commission (IEC) Lightning Protection Standard IEC 62305-1 (2010) (for Lightning Protection Level 1) for first positive (10/350 μs), first negative (1/200 μs), and subsequent (0.25/100 μs) strokes are shown in Figure 2.4(a), (b), and (c), respectively. The current waveform for first positive strokes is approximated by (2.54) with $I_0 = 200$ kA, $n = 10$, $k = 0.93, \tau_1 = 19$ μs, and $\tau_2 = 485$ μs. For negative first strokes, $n = 10$, $I_0 = 100$ kA, $k = 0.986, \tau_1 = 1.82$ μs, and $\tau_2 = 285$ μs; and for subsequent strokes $I_0 = 50$ kA, $n = 10$, $k = 0.993, \tau_1 = 0.454$ μs, and $\tau_2 = 143$ μs. Note that most subsequent strokes are negative.

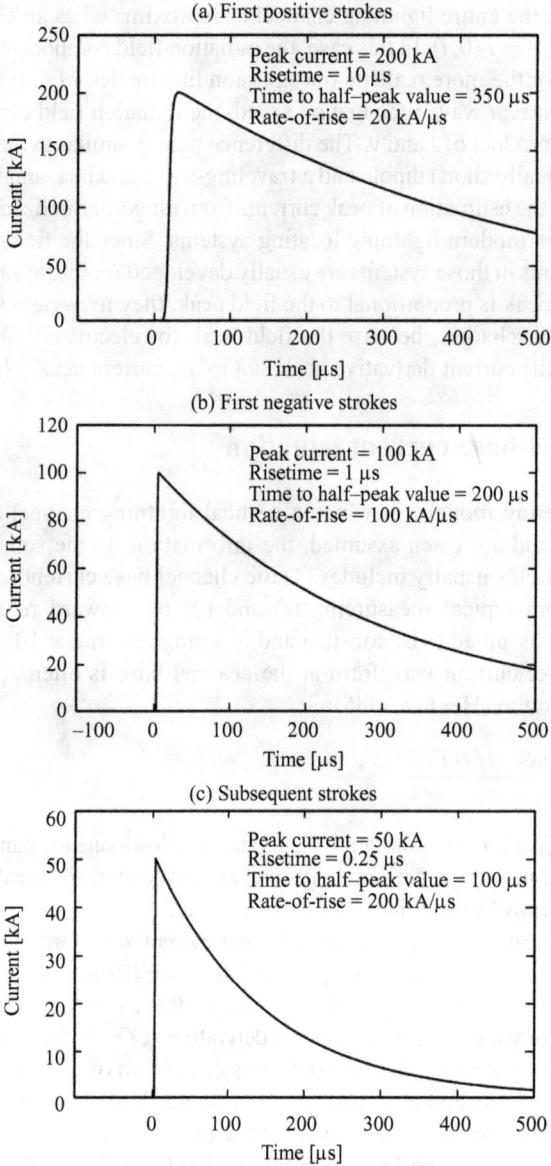

Figure 2.4   Current waveforms recommended by IEC 62305-1 (2010) for (a) first positive (10/350 μs), (b) first negative (1/200 μs), and (c) subsequent (0.25/100 μs) return strokes (for Lightning Protection Level I). The current rate-of-rise (steepness) was estimated as the peak current divided by the risetime. Most of subsequent strokes are negative. All the waveforms are computed using (2.54) and shown positive (in absolute values), regardless of stroke polarity

2.7   Propagation effects

If the observation point is located on the ground surface, and the ground is assumed to be perfectly conducting, only two field components exist: the vertical electric field and the azimuthal magnetic field. The horizontal electric field component is zero as required by the boundary condition on the surface of a perfect conductor. At an observation point above a perfectly conducting ground, a non-zero horizontal electric field component exists. A horizontal electric field exists both above ground and on (and below) its surface in the case of finite ground conductivity. The horizontal (radial) electric field at and below the ground surface is associated with a radial current flow and resultant ohmic losses in the earth. Propagation effects include preferential attenuation of the higher-frequency components in the vertical electric field and the azimuthal magnetic field waveforms. A good review of the literature on the effects of finite ground conductivity on lightning electric and magnetic fields is given by Rachidi *et al.* (1996).

Two approximate equations, namely, the wavetilt formula (Zenneck 1915) and the Cooray–Rubinstein formula (Cooray 1992; Rubinstein 1996), both in the frequency domain, are commonly used for computation of the horizontal electric field in air within 10 m or so above a finitely conducting earth. The term "wavetilt" originates from the fact that when a plane electromagnetic wave propagates over a finitely conducting ground, the total electric field vector at the surface is tilted from the vertical because of the presence of a non-zero horizontal (radial) electric field component. The tilt is in the direction of propagation if the vertical electric field component is directed upward and in the direction opposite to the propagation direction if the vertical electric field component is directed downward, with the vertical component of the Poynting vector being directed into the ground in both cases.

The wavetilt formula states that, for a plane wave, the ratio of the Fourier transform of the horizontal electric field $E_r(j\omega)$, where $j = \sqrt{-1}$, to that of the vertical electric field $E_z(j\omega)$ is equal to the ratio of the propagation constants in the air and in the ground (Zenneck 1915). Therefore,

$$E_r(j\omega) = E_z(j\omega) \frac{1}{\sqrt{\varepsilon_{rg} + \sigma_g/(j\omega\varepsilon_0)}} \qquad (2.55)$$

where σ_g and ε_{rg} are the conductivity and relative permittivity of the ground, respectively, and ω is the angular frequency. Equation (2.55) is a special case, valid for grazing incidence, of the formula giving the reflection of electromagnetic waves off a conducting surface and, hence, is a reasonable approximation only for relatively distant lightning or for the early microseconds of close lightning when the return-stroke wavefront is near ground. Typically, $E_z(j\omega)$ is computed assuming a perfectly conducting ground or is measured.

The Cooray–Rubinstein equation can be expressed as follows (Cooray 1992; Rubinstein 1996):

$$E_r(r,z,j\omega) = E_{rp}(r,z,j\omega) - H_{qp}(r,0,j\omega) \frac{c\mu_0}{\sqrt{\varepsilon_{rg} + \sigma_g/(j\omega\varepsilon_0)}} \qquad (2.56)$$

where μ_0 is the permeability of free space, $E_{rp}\,(r, z, j\omega)$ and $H_{\varphi p}\,(r, 0, j\omega)$ are the Fourier transforms of the horizontal electric field at height z above ground and the azimuthal magnetic field at ground level, respectively, both computed for the case of a perfectly (subscript "p") conducting ground. The second term is equal to zero for $\sigma_g \rightarrow \infty$ and becomes increasingly important as σ_g decreases. A generalization of the Cooray–Rubinstein formula has been offered by Wait (1997).

Cooray and Lundquist (1983) and Cooray (1987), using an analytical time-domain attenuation function proposed by Wait (1956), have calculated the effects of a finitely conducting earth in modifying the initial portion of the vertical electric field waveforms from the values expected over an infinitely conducting earth. The results are in good agreement with the measurements of Uman *et al.* (1976) and Lin *et al.* (1979). As noted earlier, Uman *et al.* (1976) observed that zero-to-peak risetimes for typical strokes increase by 1 μs or so in propagating 200 km across Florida soil, and Lin *et al.* (1979) reported that normalized peak fields were typically attenuated by 10% after propagating over 50 km of Florida soil and by 20% after propagating 200 km. It is thought that minimal distortion of the fast transition in the field wavefront and other rapidly changing portions of the measured field waveforms can be assured when the propagation path is almost entirely over salt water, a relatively good (4 S/m) conductor. Nevertheless, Ming and Cooray (1994) found from theory that for frequencies higher than about 10 MHz, the attenuation caused by the rough ocean surface can be significant. For the worst cases considered, they reported that the peak of the radiation field derivative was attenuated by about 35% in propagating 50–100 km. Cooray and Ming (1994) considered theoretically the case of propagation partly over sea and partly over land and found that propagation effects on the electric radiation field derivative are significant unless the length of the land portion of the propagation path is less than a few tens of meters. They found that propagation effects on the peak of the radiation field could be neglected if the length of the over-land propagation path is less than about 100 m.

Aoki *et al.* (2015) studied in detail the effects of finite ground conductivity on lightning electric and magnetic fields, using the 2D finite-difference time-domain method (Yee 1966). Their distance range was from 5 to 200 km and the ground conductivity range was from 10^{-4} S/m to infinity. They used the MTLL return-stroke model (Rakov and Dulzon 1987) and considered the influence of source parameters, including the return-stroke speed v (ranging from $c/2$ to c) and current risetime RT (ranging from 0.5 to 5 μs). The main results can be summarized as follows. The peaks of E_z, E_h (horizontal component of electric field), and B_ϕ are each nearly proportional to v. The peak of E_h decreases with increasing RT, while the peaks of E_z and B_ϕ are only slightly influenced by this parameter. At a distance of 5 km, the peaks of E_z and B_ϕ are essentially not affected by ground conductivity. Indeed, the difference between the 10^{-4} S/m and ∞ cases is only 1% for E_z and about 4% for B_ϕ. At 50 km, E_z and B_ϕ each reduces by 5% for 10^{-2} S/m and by about 30% for 10^{-4} S/m relative to the perfectly conducting ground case. As expected, E_h decreases with increasing ground conductivity and vanishes at perfect ground.

Additional information on calculation of lightning electromagnetic fields, including propagation effects, is found in the review paper by Rakov and Rachidi (2009, Section V.A).

2.8   Summary

The electric field intensity, E, and magnetic flux density, B, produced by lightning are usually found using the scalar, φ, and vector, A, potentials. The potentials in turn are related to the source quantities, the volume charge density, ρ, for the scalar potential φ and current density, J, for the vector potential A. The source quantities must satisfy the continuity equation. Alternatively, one can specify J, find A, and then use the Lorentz condition to find φ. In this latter case, there is no need to specify ρ in computing electric fields. The two approaches are equivalent; that is, they produce identical total fields, although the individual electric field components are different. For the transmission line model, the sum of electrostatic and induction components of electric field at close ranges and the radiation field component at far ranges each varies approximately as $1/r$, where r is the horizontal distance between the lightning channel base and the ground-level observation point, and each is proportional to the channel-base current. As the return-stroke speed v approaches the speed of light c, the approximate close-range equation (derived using only the electrostatic and induction field components) and the approximate far-range equation (derived using only the radiation field component) converge to the same exact equation for the total electric field, which is valid for any distance from the lightning channel.

If the observation point is located on the ground surface, the lightning channel is vertical, and the ground is assumed to be perfectly conducting, only two field components exist: the vertical electric field and the azimuthal magnetic field. On the finitely conducting ground, there will also be a horizontal (radial) electric field component, so that the total electric field vector will be tilted from the vertical. This radial electric field component is associated with a radial current flow and resultant ohmic losses in the ground (Poynting vector component directed into the ground).

References

Aoki, M., Baba, Y., and Rakov, V. A. (2015), FDTD simulation of LEMP propagation over lossy ground: Influence of distance, ground conductivity, and source parameters, *Journal of Geophysical Research*, vol. 120, no. 16, pp. 8043–8051, doi: 10.1002/2015JD023245.

Bruce, C. E. R., and Golde, R. H. (1941), The lightning discharge, *Journal of the Institution of Electrical Engineering*, vol. 88, no. 6, pp. 487–505.

Chen, Y., Wang, X., and Rakov, V. A. (2015), Approximate expressions for lightning electromagnetic fields at near and far ranges: Influence of return-stroke speed, *Journal of Geophysical Research*, vol. 120, no. 7, pp. 2855–2880, doi:10.1002/2014JD022867.

Cooray, V. (1987), Effects of propagation on the return stroke radiation fields, *Radio Science*, vol. 22, no.5, pp. 757–768.

Cooray, V. (1992), Horizontal fields generated by return strokes, *Radio Science*, vol. 27, no. 4, pp. 529–537.

Cooray, V., and Lundquist, S. (1983), Effects of propagation on the rise times and the initial peaks of radiation fields from return strokes, *Radio Science*, vol. 18, no. 3, pp. 409–415.

Cooray, V., and Ming, Y. (1994), Propagation effects on the lightning-generated electromagnetic fields for homogeneous and mixed sea-land paths, *Journal of Geophysical Research*, vol. 99, no. 16, pp. 10641–10652.

Diendorfer, G., and Uman, M. A. (1990), An improved return stroke model with specified channel-base current, *Journal of Geophysical Research*, vol. 95, pp. 13621–13644.

Heidler, F. (1985), Traveling current source model for LEMP calculation, In Proceedings of 6th International Zurich Symposium on Electromagnetic Compatibility, Zurich, Switzerland, 157–162.

IEC 62305-1 (2010), Protection against lightning — Part 1: General principles, Ed. 2, Geneva: International Electrotechnical Commission.

Krider, E. P. (1994), On the peak electromagnetic fields radiated by lightning return strokes toward the middle-atmosphere, *Journal of Atmospheric Electricity*, vol. 14, pp. 17–24.

Lin, Y. T., Uman, M. A., Tiller, J. A., *et al.* (1979), Characterization of lightning return stroke electric and magnetic fields from simultaneous two-station measurements, *Journal of Geophysical Research*, vol. 84, no. 10, pp. 6307–6314.

Lu, G. (2006), Transient electric field at high altitudes due to lightning: possible role of induction field in the formation of elves, *Journal of Geophysical Research*, vol. 115, no. D02103, doi:10.1029/2005JD005781.

Ming, Y., and Cooray, V. (1994), Propagation effects caused by a rough ocean surface on the electromagnetic fields generated by lightning return strokes, *Radio Science*, vol. 29, no. 1, pp. 73–85.

Nag, A., and Rakov, V. A. (2010), Compact intracloud lightning discharges: 1. Mechanism of electromagnetic radiation and modeling, *Journal of Geophysical Research*, vol. 115, no. D20102, doi:10.1029/2010JD014235, *20* pages.

Nag, A., Rakov, V. A., and Cramer, J. A. (2011), Remote measurements of currents in cloud lightning discharges, *IEEE Transactions on Electromagnetic Compatibility*, vol. 53, no. 2, pp. 407–413.

Nucci, C. A., Mazzetti, C., Rachidi, F., and Ianoz, M. (1988), On lightning return stroke models for LEMP calculations, In Proceedings of 19th International Conference on Lightning Protection, Graz, Austria, pp. 463–469.

Rachidi, F., Nucci, C. A., Ianoz, M., and Mazzetti, C. (1996), Influence of a lossy ground on lightning-induced voltages on overhead lines, *IEEE Transactions on Electromagnetic Compatibility*, vol. 38, no. 3, pp. 250–264.

Rakov, V. A. (2016), *Fundamentals of Lightning*, Cambridge University Press, UK, pp.1–257.

Rakov, V. A., and Dulzon, A. A. (1987), Calculated electromagnetic fields of lightning return stroke (in Russian), *Tekh. Elektrodinam.*, vol. 1, pp. 87–89.

Rakov, V. A., and Uman, M. A. (1998), Review and evaluation of lightning return stroke models including some aspects of their application, *IEEE Transactions on Electromagnetic Compatibility*, vol. 40, no. 4, pp. 403–426.

Rakov, V. A., and Tuni, W. G. (2003), Lightning electric field intensity at high altitudes: Inferences for production of Elves, *Journal of Geophysical Research*, vol. 108, no. D20, doi:10.1029/2003JD003618.

Rakov, V. A., and Rachidi, F. (2009), Overview of recent progress in lightning research and lightning protection, *IEEE Transactions on Electromagnetic Compatibility*, vol. 51, no. 3, pp. 428–442.

Rubinstein, M. (1996), An approximate formula for the calculation of the horizontal electric field from lightning at close, intermediate, and long range, *IEEE Transactions on Electromagnetic Compatibility*, vol. 38, no. 3, pp. 531–535.

Rubinstein, M., and Uman, M. A. (1989), Methods for calculating the electromagnetic fields from a known source distribution: application to lightning, *IEEE Transactions on Electromagnetic Compatibility*, vol. 31, no. 2, pp. 183–189.

Thomson, E. M. (1999), Exact expressions for electric and magnetic fields from a propagating lightning channel with arbitrary orientation, *Journal of Geophysical Research*, vol. 104, no. D18, pp. 22293–22300.

Thottappillil, R., and Rakov, V. A. (2001), On different approaches to calculating lightning electric fields, *Journal of Geophysical Research*, vol. 106, no. D13, pp. 14191–14205.

Thottappillil, R., Rakov, V. A., and Uman, M. A. (1997), Distribution of charge along the lightning channel: Relation to remote electric and magnetic fields and to return-stroke models, *Journal of Geophysical Research*, vol. 102, no. D6, pp. 6987–7006.

Thottappillil, R., Schoene, J., and Uman, M. A. (2001), Return stroke transmission line model for stroke speed near and equal that of light, *Geophysical Research Letters*, vol. 28, no. 18, pp. 3593–3596, doi:10.1029/2001GL013029.

Uman, M. A., and McLain, D. K. (1969), Magnetic field of the lightning return stroke, *Journal of Geophysical Research*, vol. 74, pp. 6899–6910.

Uman, M. A., McLain, D. K., and Krider E. P. (1975), The electromagnetic radiation from a finite antenna, *American Journal of Physics*, vol. 43, pp. 33–38.

Uman, M. A., Swanberg, C. E., Tiller, J. A., Lin, Y. T., and Krider, E. P. (1976), Effects of 200 km propagation in Florida lightning return stroke electric fields, *Radio Science*, vol. 11, no. 12, pp. 985–990.

Wait, J. R. (1956), Transient fields of a vertical dipole over a homogeneous curved ground, *Canadian Journal of Physics*, vol. 34, no. 1, pp. 27–35.

Wait, J. R. (1997), Concerning the horizontal electric field of lightning, *IEEE Transactions on Electromagnetic Compatibility*, vol. 39, no. 2, p. 34.

Yee, K. S. (1966), Numerical solution of initial boundary value problems involving Maxwell's equations in isotropic media, *IEEE Transactions on Antennas and Propagation*, vol. 14, no. 3, pp. 302–307.

Zenneck, J. (1915), *Wireless Telegraphy*, New York, McGraw-Hill, *443* pages.

Chapter 3

Distributed-circuit models of electromagnetic coupling to overhead conductor

We introduce three different sets of telegrapher's equations with source terms and corresponding equivalent circuits that can be used for studying voltage and current surges induced on an overhead conductor by transient electromagnetic fields such as those produced by lightning. The source terms (forcing functions) incorporated into the classical telegrapher's equations are derived, using the electromagnetic theory, following works of Taylor *et al.* (1965), Agrawal *et al.* (1980), and Rachidi (1993), who arrived at different coupling model formulations. Since all three formulations are based on Maxwell's equations, they yield identical results, as demonstrated by Nucci and Rachidi (1995). As of today, the model of Agrawal *et al.* (1980) has been most widely used for the evaluation of lightning-induced effects on power and telecommunication lines.

Key Words: Lightning-induced voltage; distributed-circuit model; electromagnetic coupling model; Faraday's law; Ampere's law; total field; incident field; scattered field; equivalent circuit

3.1 Introduction

In this chapter, we introduce three different sets of telegrapher's equations with source terms and corresponding equivalent circuits that can be used for studying voltage and current surges induced on an overhead conductor by transient electromagnetic fields such as those produced by lightning. The source terms (forcing functions) incorporated into the classical telegrapher's equations are derived, using the electromagnetic theory, following works of Taylor *et al.* (1965), Agrawal *et al.* (1980), and Rachidi (1993), who arrived at different coupling model formulations. Lightning electromagnetic fields needed for evaluating the source terms are usually computed using the expressions found in Chapter 2. The spatial and temporal distribution of current along the lightning channel needed for computing the fields is usually specified using an engineering model of lightning return stroke. Such models are discussed in Sections 1.3.2 (lightning strike to flat ground) and 1.3.4 (lightning strike to a tall object). Since all three coupling model formulations are based on Maxwell's equations, they yield identical results, as demonstrated by Nucci and Rachidi (1995) (see also Section 3.5 of this book). As of today, the model of Agrawal *et al.* (1980) has been most widely used for evaluation of lightning-induced effects on power and telecommunication lines. Distributed-circuit models of electromagnetic coupling are presented here because they help better understand the basic concept of lightning-induced effects. Application of full-wave electromagnetic computation methods such as the finite-difference time-domain (FDTD) method (Yee 1966) for solving Maxwell's equations to studying lightning-induced surges is discussed in Chapter 5.

In this section, we consider one of the simplest configurations: a single, constant-radius horizontal lossless conductor in air above flat, perfectly conducting ground. The conductor radius is much smaller than the height of the conductor above ground, to avoid the proximity effects (concentration of charge and current at the lower, closer to the ground, side of the conductor). Further, the conductor height above ground is much smaller than the shortest significant wavelength λ, which assures the transverse electromagnetic (TEM) structure of scattered fields associated with the charge and current induced on the conductor (Paul 1994, p. 37). The induced charge and current serve to satisfy the boundary condition for the tangential component of electric field on the surface of the conductor when it is illuminated by lightning electromagnetic field (for a lossless conductor, the tangential component of the total electric field must be zero). The above restrictions on the transmission line geometry are generally fulfilled for overhead power distribution and telecommunication lines. The TEM structure for scattered fields means that there is no axial electric or magnetic field due to the charge and/or current induced on the conductor. Since the lightning electromagnetic pulse (LEMP) first illuminates the nearest point of the conductor and then its progressively more distant points, the induced charge/current is in the form of surge(s) propagating away from the nearest point.

Inclusion of losses in the horizontal conductor and in the ground in the coupling model and extension of the model to the case of multiple conductors are outside of the scope of this book. Those topics are extensively covered by Rachidi and Tkachenko (2008), Nucci *et al.* (2012), Nucci and Rachidi (2012), and Cooray *et al.* (2020). Note that lossy ground and multiple-conductor configurations are considered in the framework of full-wave electromagnetic models in Chapter 5.

In the following three sections, we will derive (in the frequency domain) three sets of telegrapher's (transmission line) equations with source terms representing electromagnetic coupling. The original telegrapher's equations (without source terms) for our simplest configuration are given below:

$$\frac{dV(x)}{dx} + j\omega L'I(x) = 0 \tag{3.1}$$

$$\frac{dI(x)}{dx} + j\omega C'V(x) = 0 \tag{3.2}$$

where the horizontal conductor is assumed to be oriented along the x-axis, $V(x)$ is the voltage between the horizontal conductor and ground, $I(x)$ is the current in the horizontal conductor, L' and C' are the inductance and capacitance, each per unit length, of the two-conductor transmission line formed by the horizontal conductor and ground, and ω is the angular frequency.

It is worth noting that there exist other coupling model formulations that are equivalent to the three formulations considered here. For example, Wuyts and De Zutter (1994) derived telegrapher's equations in terms of total voltage and total current with the source terms expressed in terms of electric field only, and Cooray *et al.* (2017) presented coupling models with source terms related to the scalar and vector potentials or only to the vector potential.

3.2 Model of Taylor, Satterwhite, and Harrison (1965)

The geometry of the single-conductor-above-ground transmission line is shown in Figure 3.1. The conductor is located in the x–z plane at $y = 0$ and is parallel to the x-axis. The conductor radius is a and its height above perfectly conducting ground (x–y plane) is h. It is terminated in lumped impedances Z_s at $x = 0$, and Z_r at $x = l$, where l is the length of the line (usually $l \gg h$). This passive (no sources) transmission line (TL) is illuminated (and energized) by external electromagnetic field. Once energized, the TL will produce its own electromagnetic field, so that the total field (electric or magnetic) at any point in space will be the vectorial sum of two field components, the external one and the reaction to it of the TL. We will refer to the external fields, computed in the presence of ground, but in the absence of overhead conductor and its terminations, as incident fields, E^i (incident electric field intensity) and B^i (incident magnetic flux density). The fields representing the

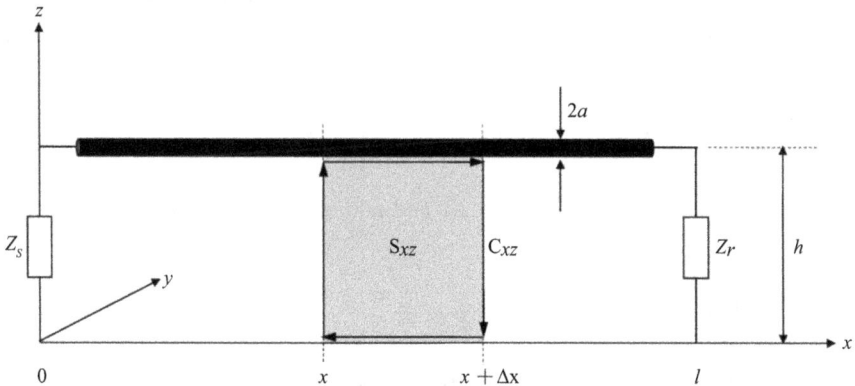

Figure 3.1 *A lossless horizontal conductor above flat perfectly conducting ground (x–y) plane excited by an external electromagnetic field. Also shown is the closed integration path (contour) C_{xz} bounding surface S_{xz} in the x–z plane used in formulating Faraday's law in integral form. Arrows indicate the direction of integration. Δx is arbitrarily small, so that variation of electric and magnetic fields between x and $x + \Delta x$ is negligible*

reaction of the TL to external excitation will be referred to as scattered fields, E^s (scattered electric field intensity) and B^s (scattered magnetic flux density). The scattered fields are generated by the charges and currents induced in the TL by the incident fields. Note that Maxwell's equations apply to both total fields and to their incident and scattered components individually.

Applying Faraday's law to contour C_{xz} enclosing surface S_{xz} in the x–z plane, as shown in Figure 3.1, we can write:

$$\oint_{C_{xz}} E \cdot dl = -j\omega \int_{S_{xz}} B \cdot dS, \qquad dS = dx\, dz\, a_y \tag{3.3}$$

where ω is the angular frequency, dl is the differential displacement vector along contour C_{xz} (the direction of integration is shown by arrows), and a_y is the y-directed unit vector that is normal to surface S_{xz}. Assuming that Δx is arbitrarily small (no integration over x is needed), (3.3) can be written as

$$\int_0^h [E_z(x,z) - E_z(x+\Delta x, z)]dz + [E_x(x,h) - E_x(x,0)]\Delta x$$
$$= -j\omega\, \Delta x \int_0^h B_y(x,z)\, dz \tag{3.4}$$

where E_x are E_z are total electric fields in the positive x- and z-directions, respectively, and B_y is the total magnetic flux density in the positive y-direction. Note that the y coordinate does not appear in (3.4); it is dropped (just to simplify notation), since $y = 0$ in all terms of (3.4). This simplification is used throughout this chapter, except for (3.12), which contains non-zero y-values. Since the horizontal conductor and the ground are perfectly conducting, the total tangential electric fields along both the conductor and the ground are zero: $E_x(x, h) = E_x(x, 0) = 0$. Thus, (3.4) becomes

$$\int_0^h [E_z(x,z) - E_z(x+\Delta x, z)]dz = -j\omega\, \Delta x \int_0^h B_y(x,z)\, dz \tag{3.5}$$

Dividing both sides of (3.5) by Δx and taking the limit as Δx approaches 0, we can write

$$-\frac{\partial}{\partial x}\int_0^h E_z(x,z)dz = -j\omega \int_0^h B_y(x,z)\, dz \tag{3.6}$$

The total voltage between the horizontal conductor at point x and ground can be found (in the quasistatic sense, since $h \ll \lambda$) as

$$V(x) = -\int_0^h E_z(x,z)dz \tag{3.7}$$

Substitution of (3.7) into (3.6) yields

$$
\begin{aligned}
\frac{dV(x)}{dx} &= -j\omega \int_0^h B_y(x,z)\, dz \\
&= -j\omega \int_0^h B_y{}^i(x,z)\, dz - j\omega \int_0^h B_y{}^s(x,z)\, dz
\end{aligned}
\tag{3.8}
$$

where the total magnetic flux density in the y-direction is decomposed into its incident $B_y{}^i$ and scattered $B_y{}^s$ components. The latter is produced by the current $I(x)$ induced in the horizontal conductor (see Figure 3.2) in response to the incident field. The integral in the last term of (3.8) is the y-directed scattered magnetic flux through the rectangular area $(\Delta x \times h)$, where $\Delta x = 1$ m; that is, flux per unit length of the conductor. This flux is proportional to its causative current $I(x)$, with the proportionality coefficient being the inductance per unit length L' of the horizontal conductor above ground ($L' = (\mu_0 / 2\pi) \ln (2h/a)$ for $a \ll h$ (Paul 1994, p. 30)). Thus, the second term on the right-hand side of (3.8) can be expressed as

$$
j\omega \int_0^h B_y{}^s(x,z)\, dz = j\omega L' I(x)
\tag{3.9}
$$

Then, after rearranging the terms, (3.8) becomes

$$
\frac{dV(x)}{dx} + j\omega L' I(x) = -j\omega \int_0^h B_y{}^i(x,z)\, dz
\tag{3.10}
$$

This is the first telegrapher's equation in Taylor *et al.*'s (1965) coupling model formulation. It differs from the first classical telegrapher's equation (3.1) in that it

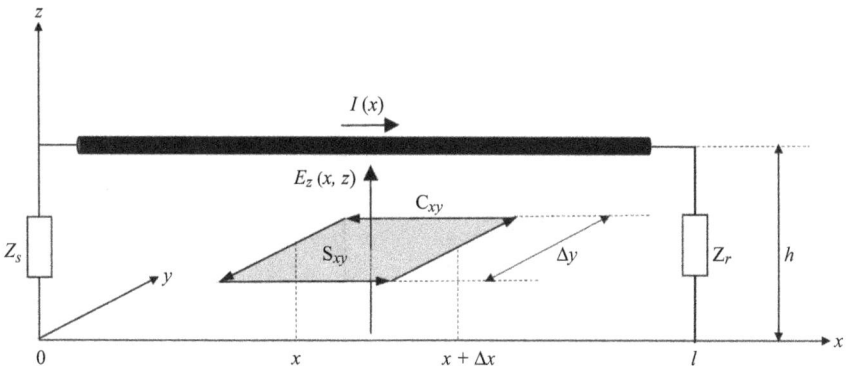

Figure 3.2   Same as Figure 3.1, but showing the closed integration path (contour) C_{xy} bounding surface S_{xy} in the x–y plane used in formulating Ampere's law in integral form. E_z is the vertical electric field and $I(x)$ is the current induced in the horizontal conductor.

corresponds to an active distributed circuit containing distributed series voltage (per unit length) sources specified in terms of the y-component of the incident magnetic field.

Next, we derive the second telegrapher's equation for Taylor *et al.*'s (1965) formulation. Applying Ampere's law to contour C_{xy} enclosing surface S_{xy} in the x–y plane, as shown in Figure 3.2, we can write

$$\frac{1}{\mu_0}\oint_{C_{xy}} \boldsymbol{B}\cdot \mathrm{d}\boldsymbol{l} = j\omega \int_{S_{xy}} \boldsymbol{D}\cdot \mathrm{d}\boldsymbol{S}, \qquad \mathrm{d}\boldsymbol{S} = \mathrm{d}x\,\mathrm{d}y\,\boldsymbol{a}_z$$
$$= j\omega\varepsilon_0 \int_{S_{xy}} E_z\,\mathrm{d}S \tag{3.11}$$

where $\boldsymbol{B}$ is the total magnetic flux density, $\boldsymbol{D}$ is the total electric flux density, E_z is the total electric field in the z-direction, $\mathrm{d}\boldsymbol{l}$ is the differential displacement vector along contour C_{xy} (the direction of integration is shown by arrows), and $\boldsymbol{a}_z$ is the z-directed unit vector that is normal to surface S_{xy}. The right-hand side of (3.11) represents the displacement current between the horizontal conductor and ground, with the conduction current being zero (air can be viewed as a perfect dielectric).

For arbitrarily small Δx and Δy, (3.11) can be written as

$$\frac{1}{\mu_0}\left[B_y(x+\Delta x,0,z)-B_y(x,0,z)\right]\Delta y$$
$$+\frac{1}{\mu_0}\left[B_x(x,-\Delta y/2,z)-B_x(x,\Delta y/2,z)\right]\Delta x \tag{3.12}$$
$$= j\omega\varepsilon_0 E_z(x,0,z)\Delta x\Delta y$$

Dividing both sides of (3.12) by $\varepsilon_0\,\Delta x\Delta y$ and taking the limit as both Δx and Δy approach 0 gives

$$\frac{1}{\varepsilon_0\mu_0}\frac{\partial B_y(x,z)}{\partial x}-\frac{1}{\varepsilon_0\mu_0}\frac{\partial B_x(x,z)}{\partial y}= j\omega E_z(x,z) \tag{3.13}$$

Integration of (3.13) over z from $z=0$ to $z=h$ yields

$$\frac{1}{\varepsilon_0\mu_0}\int_0^h \frac{\partial B_y(x,z)}{\partial x}\,\mathrm{d}z-\frac{1}{\varepsilon_0\mu_0}\int_0^h \frac{\partial B_x(x,z)}{\partial y}\,\mathrm{d}z$$
$$= j\omega\int_0^h E_z(x,z)\,\mathrm{d}z \tag{3.14}$$

Using (3.7) on the right-hand side and decomposing the x- and y-components of the total magnetic flux density each into the incident and scattered components, we get

$$-j\omega V(x)=\frac{1}{\varepsilon_0\mu_0}\int_0^h \left[\frac{\partial B_y{}^i(x,z)}{\partial x}+\frac{\partial B_y{}^s(x,z)}{\partial x}\right]\,\mathrm{d}z$$
$$-\frac{1}{\varepsilon_0\mu_0}\int_0^h \left[\frac{\partial B_x{}^i(x,z)}{\partial y}+\frac{\partial B_x{}^s(x,z)}{\partial y}\right]\,\mathrm{d}z \tag{3.15}$$

Since the scattered field has the TEM structure, there is no scattered magnetic flux density in the x-direction and, hence, the term containing $B_x{}^s$ can be dropped from (3.15). Further, the term containing $B_y{}^s$ can be expressed in terms of current $I(x)$ giving rise to $B_y{}^s$ using (3.9). Thus, (3.15) becomes

$$-j\omega V(x) = \frac{1}{\varepsilon_0\mu_0}\int_0^h \frac{\partial B_y{}^i(x,z)}{\partial x}\,dz + \frac{1}{\varepsilon_0\mu_0}L'\frac{\partial I(x)}{\partial x}$$
$$-\frac{1}{\varepsilon_0\mu_0}\int_0^h \frac{\partial B_x{}^i(x,z)}{\partial y}\,dz \tag{3.16}$$

We now apply Ampere's law to contour C_{xy} enclosing surface S_{xy} in the x–y plane (see Figure 3.2) again, but this time only for the incident field components (just adding superscript "i" to each of the three field components in (3.13)), which yields

$$\frac{1}{\varepsilon_0\mu_0}\frac{\partial B_y{}^i(x,z)}{\partial x} - \frac{1}{\varepsilon_0\mu_0}\frac{\partial B_x{}^i(x,z)}{\partial y} = j\omega E_z{}^i(x,z) \tag{3.17}$$

Integrating (3.17) over z from $z = 0$ to $z = h$, we get

$$\frac{1}{\varepsilon_0\mu_0}\int_0^h \frac{\partial B_y{}^i(x,z)}{\partial x}\,dz - \frac{1}{\varepsilon_0\mu_0}\int_0^h \frac{\partial B_x{}^i(x,z)}{\partial y}\,dz$$
$$= j\omega \int_0^h E_z{}^i(x,z)\,dz \tag{3.18}$$

Using (3.18), we can eliminate the dependence of induced voltage $V(x)$ on magnetic field components $B^i{}_y$ and $B^i{}_x$ in (3.16), by replacing it with the dependence on $E^i{}_z$:

$$-j\omega V(x) = j\omega \int_0^h E_z{}^i(x,z)\,dz + \frac{1}{\varepsilon_0\mu_0}L'\frac{\partial I(x)}{\partial x} \tag{3.19}$$

Noting that $\varepsilon_0\mu_0/L' = C'$ (Paul 1994, p. 24), multiplying all terms in (3.19) by C', and rearranging the terms, we get

$$\frac{dI(x)}{dx} + j\omega C'V(x) = -j\omega C'\int_0^h E_z{}^i(x,z)\,dz \tag{3.20}$$

where C' is the capacitance per unit length of the horizontal conductor above ground ($C' = 2\pi\varepsilon_0/\ln(2h/a)$ for $a \ll h$ (Paul 1994, p. 30)). This is the second telegrapher's equation in Taylor *et al.*'s (1965) coupling model formulation. The term on the right-hand side corresponds to distributed shunt current (per unit length) sources specified in terms of the z-component of the incident electric field.

Equations (3.10) and (3.20) can be solved for $I(x)$ and $V(x)$ with the following boundary conditions:

$$V(0) = -Z_sI(0) \tag{3.21}$$
$$V(l) = Z_rI(l) \tag{3.22}$$

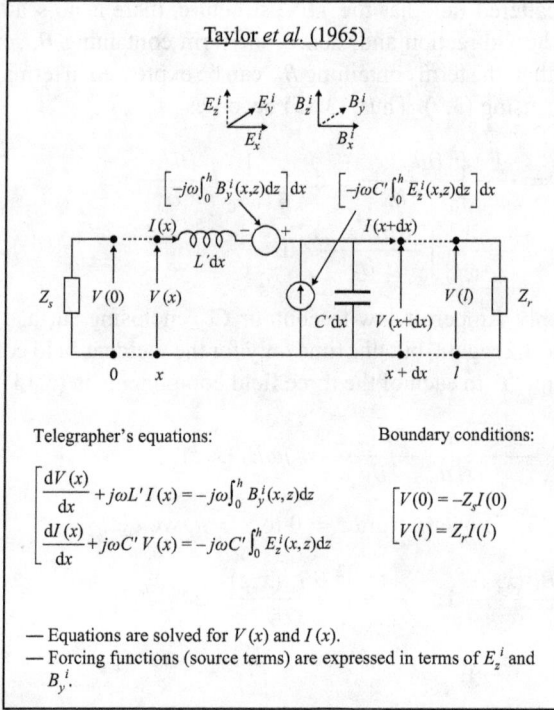

Figure 3.3   *Equivalent circuit (including a differential line segment and terminations) of the electromagnetic coupling model formulation of Taylor et al. (1965) for the case of lossless horizontal conductor above flat perfectly conducting ground. The incident field components utilized in this formulation are shown by broken lines.*

where Z_s and Z_r are the conductor termination impedances (see Figure 3.1). The equivalent circuit corresponding to Taylor *et al.*'s (1965) coupling model formulation (Equations (3.10) and (3.20)) is shown in Figure 3.3. In this model, excitation is represented by vertical electric (E_z^i) and transverse magnetic (B_y^i) fields.

3.3   Model of Agrawal, Price, and Gurbaxani (1980)

In Agrawal *et al.*'s (1980) coupling model formulation, the telegrapher's equations are in terms of $I(x)$ and $V^s(x)$, where $V^s(x)$ is the voltage component related to the scattered component of vertical electric field, E_z^s. In order to find the total voltage, $V(x)$, one needs to separately find the incident voltage component, $V^i(x)$, related to the incident component of vertical electric field, E_z^i, and add it to $V^s(x)$. Detailed derivations are presented below.

Applying Faraday's law to contour C_{xz} enclosing surface S_{xz}, as shown in Figure 3.1, and following the same procedure as in Section 3.2, we can write (see (3.6)):

$$-\frac{\partial}{\partial x}\int_0^h E_z(x,z)\mathrm{d}z = -j\omega\int_0^h B_y(x,z)\,\mathrm{d}z \tag{3.23}$$

Decomposing the total electric and magnetic fields each into their incident and scattered components, we can rewrite (3.23) as

$$-\frac{\partial}{\partial x}\int_0^h \left[E_z^{\,i}(x,z) + E_z^{\,s}(x,z)\right]\mathrm{d}z$$
$$= -j\omega\int_0^h \left[B_y^{\,i}(x,z) + B_y^{\,s}(x,z)\right]\,\mathrm{d}z \tag{3.24}$$

With the purpose to eventually eliminate $E_z^{\,i}$ and $B_y^{\,i}$, we rearrange the terms in (3.24) as follows:

$$-\frac{\partial}{\partial x}\int_0^h E_z^{\,s}(x,z)\mathrm{d}z + j\omega\int_0^h B_y^{\,s}(x,z)\,\mathrm{d}z$$
$$= \frac{\partial}{\partial x}\int_0^h E_z^{\,i}(x,z)\mathrm{d}z - j\omega\int_0^h B_y^{\,i}(x,z)\,\mathrm{d}z \tag{3.25}$$

We now apply Faraday's law again, but this time only to the incident electric and magnetic field components. The result is similar to (3.23), except for the need to add the second term, $E_x^i(x,h)$, on the left-hand side. This modification is needed because of the removal, as per definition of incident field, of the lossless horizontal conductor that had made $E_x(x,\,h) = 0$ in deriving (3.23) or (3.6) (see (3.4) and (3.5)). The result is as follows:

$$-\frac{\partial}{\partial x}\int_0^h E_z^{\,i}(x,z)\mathrm{d}z + E_x^{\,i}(x,h) = -j\omega\int_0^h B_y^{\,i}(x,z)\,\mathrm{d}z \tag{3.26}$$

Comparing this equation with (3.25), one can see that the right-hand side of (3.25) is equal to $E_x^i(x,h)$; thus, (3.25) can be rewritten as

$$-\frac{\partial}{\partial x}\int_0^h E_z^{\,s}(x,z)\mathrm{d}z + j\omega\int_0^h B_y^{\,s}(x,z)\,\mathrm{d}z = E_x^{\,i}(x,h) \tag{3.27}$$

Because of our TEM assumption ($h \ll \lambda$ where λ is the shortest significant wavelength), $E_z^{\,s}$ is uniquely related to the voltage, which we label V^s, between the horizontal conductor and ground (Paul 1994, p. 15), and the first term on the left-hand side of (3.27) can be replaced with $\mathrm{d}V^s(x)/\mathrm{d}x$. Further, the second term on the left-hand side of (3.27) can be expressed in terms of $I(x)$, as per (3.9).

Thus, (3.27) can be written as

$$\frac{\mathrm{d}V^s(x)}{\mathrm{d}x} + j\omega L'I(x) = E_x^{\,i}(x,h) \tag{3.28}$$

where $V^s(x)$ is the so-called scattered voltage defined as

$$V^s(x) = -\int_0^h E_z^s(x,z)\mathrm{d}z \qquad (3.29)$$

The total voltage is obtained as the sum of the scattered $V^s(x)$ and incident $V^i(x)$ voltage components:

$$V(x) = V^s(x) + V^i(x) \qquad (3.30)$$

where

$$V^i(x) = -\int_0^h E_z^i(x,z)\mathrm{d}z \qquad (3.31)$$

Equation (3.28) is the first telegrapher's equation in Agrawal *et al.*'s (1980) formulation. The term on the right-hand side corresponds to distributed series voltage (per unit length) sources specified in terms of the horizontal (axial) component of incident electric field.

The second telegrapher's equation in Agrawal *et al.*'s (1980) formulation can be derived using the continuity equation for a horizontal conductor segment of arbitrarily small length Δx and relating the line charge density on this conductor segment to scattered voltage via capacitance per unit length or applying Ampere's law to contour C_{xy} enclosing surface S_{xy} in the x–y plane (see Figure 3.2) for scattered field components. Here, we will derive the second telegrapher's equation using both approaches.

A. Continuity equation approach

The continuity equation is not a member of the classical set of four Maxwell's equations, but is closely related to it. In fact, the continuity equation is the basis for formulating Ampere's law for time-varying fields. This law in differential form can be written as

$$\frac{1}{\mu_0}\nabla \times \boldsymbol{B} = \boldsymbol{J} + j\omega\boldsymbol{D} \qquad (3.32)$$

If we take the divergence on both sides of (3.32) and note that $\nabla \cdot \nabla \times \boldsymbol{B} = 0$, we get $\nabla \cdot \boldsymbol{J} = -j\omega\nabla \cdot \boldsymbol{D}$. Now, using Gauss's law, $\nabla \cdot \boldsymbol{D} = \rho_v$, on the right-hand side, we obtain the continuity equation in differential form: $\nabla \cdot \boldsymbol{J} = -j\omega\rho_v$, where $\boldsymbol{J}$ is the conduction current density and ρ_v is the volume charge density. Current induced by the incident electromagnetic field in the horizontal conductor (see Figure 3.2) must satisfy this continuity equation,

Let's consider a conductor segment of length Δx with line charge density ρ_L and conduction current density $\boldsymbol{J}$ surrounded by an imaginary closed cylindrical surface S, also of length Δx, as shown in Figure 3.4. Then, the continuity equation in integral form can be written as

$$\int_S \boldsymbol{J}(x) \cdot d\boldsymbol{S} = -j\omega\rho_L(x)\Delta x \qquad (3.33)$$

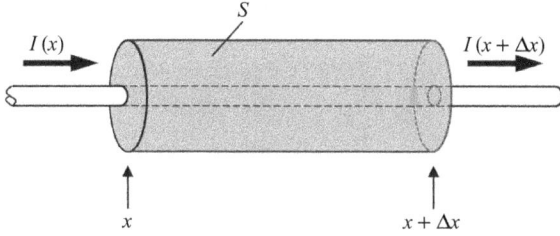

Figure 3.4   A segment of horizontal conductor of length Δx surrounded by an imaginary closed cylindrical surface S, also of length Δx, used for formulating the continuity equation in integral form. I(x) and I(x + Δx) are the conduction currents at points x (into S) and x + Δx (out of S).

where $\rho_L(x)\Delta x$ is the total charge inside S, and the left-hand side represents the net conduction current flowing out of S. Since the conduction current flows only along the horizontal conductor (there is no transverse conduction current through surrounding air and, hence, through the lateral surface of S), we can rewrite (3.33) as

$$I(x + \Delta x) - I(x) = -j\omega\rho_L(x)\Delta x \qquad (3.34)$$

where $I(x)$ and $I(x + \Delta x)$ are conduction currents in and out of the cylindrical surface (see Figure 3.4), respectively. Now, dividing both sides of (3.34) by Δx and taking the limit as Δx approaches 0, we get

$$\frac{\partial I(x)}{\partial x} = -j\omega\rho_L(x) \qquad (3.35)$$

Finally, $\rho_L(x)$ can be related through capacitance per unit length C', to the voltage it produces, which is scattered voltage $V^s(x)$:

$$\rho_L(x) = C'V^s(x) \qquad (3.36)$$

Substituting (3.36) into (3.35), we get

$$\frac{\partial I(x)}{\partial x} + j\omega C'V^s(x) = 0 \qquad (3.37)$$

which is the second telegrapher's equation of Agrawal *et al.*'s (1980) coupling model formulation. It is derived from the continuity equation and is a form of continuity equation by itself.

B.   Ampere's law approach

We will now derive (3.37) applying Ampere's law to contour C_{xy} enclosing surface S_{xy} in the x–y plane (see Figure 3.2) for scattered field components. In doing so, we start with (3.14) which is based on Ampere's law and written for total fields. We rewrite that equation for scattered fields by adding superscript "s" to all field components and dropping the term containing the x-component of magnetic fields, $B_x{}^s$ (because of the TEM structure of scattered field):

$$\frac{1}{\varepsilon_0\mu_0}\int_0^h \frac{\partial B_y{}^s(x,z)}{\partial x}\,dz = j\omega \int_0^h E_z{}^s(x,z)\,dz \tag{3.38}$$

As per (3.9) and (3.29), (3.38) can be written as

$$\frac{L'}{\varepsilon_0\mu_0}\frac{\partial I(x)}{\partial x} = -j\omega V^s(x) \tag{3.39}$$

Noting that $\varepsilon_0\mu_0/L' = C'$ (Paul 1994, p. 24), multiplying all terms in (3.39) by C', and rearranging the terms, we get

$$\frac{\partial I(x)}{\partial x} + j\omega C'V^s(x) = 0 \tag{3.40}$$

which is the same as (3.37).

The second telegrapher's equation of Agrawal *et al.*'s (1980) coupling model formulation (Equation (3.37) or (3.40)) is the same as the second classical telegrapher's equation (3.2), except it contains V^s, not V. In contrast to the second equation of Taylor *et al.* (1965), this one contains no sources.

Equations (3.28) and (3.37) (or (3.40)) are solved for $V^s(x)$ and $I(x)$, and $V(x)$ is found by adding $V^i(x)$ given by (3.31) to $V^s(x)$. The corresponding equivalent circuit is shown in Figure 3.5. This circuit can be viewed as biased at each point x by $V^i(x)$ relative to the reference ground. In many cases, since h is usually relatively small, it is acceptable to use an approximation $V^i(x) = -h\,E_z{}^i(x)$, instead of using (3.31).

The boundary conditions for the total voltages are the same as (3.21) and (3.22), which are reproduced below.

$$V(0) = V^s(0) + V^i(0) = -Z_s I(0) \tag{3.41}$$

$$V(l) = V^s(l) + V^i(l) = Z_r I(l) \tag{3.42}$$

Accordingly, the boundary conditions for the scattered voltage (also represented in the equivalent circuit shown in Figure 3.5) are

$$V^s(0) = -Z_s I(0) + \int_0^h E_z{}^i(0,z)\,dz \tag{3.43}$$

$$V^s(l) = Z_r I(l) + \int_0^h E^i{}_z(l,z)\,dz \tag{3.44}$$

The second term on the right-hand side of (3.43) and (3.44) represents a lumped voltage source at $x = 0$ and $x = l$, respectively (see Figure 3.5).

Among all the coupling models, the model of Agrawal *et al.* (1980) has been most widely used for the evaluation of lightning-induced effects on overhead

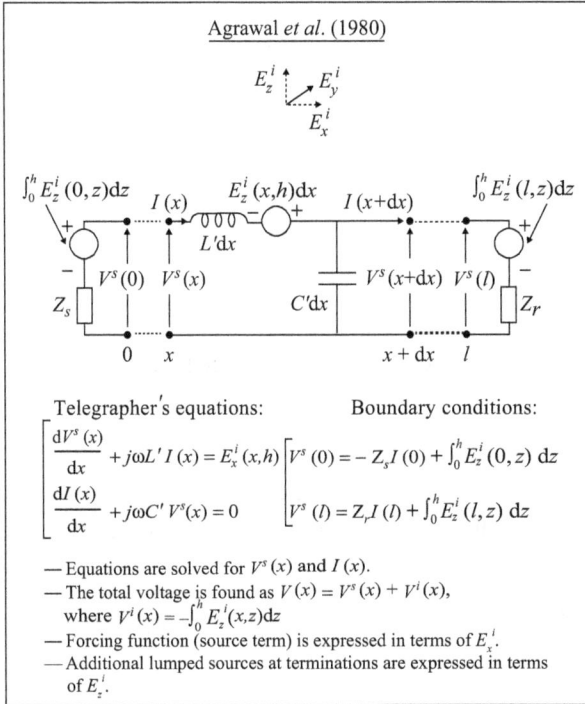

Figure 3.5   *Equivalent circuit (including a differential line segment and terminations) of the electromagnetic coupling model formulation of Agrawal et al. (1980) for the case of lossless horizontal conductor above flat perfectly conducting ground. The incident field components utilized in this formulation are shown by broken lines.*

conductors. It has been also employed in simulations of lightning-induced surges on overhead conductors using the FDTD method (Yee 1966) for solving discretized Maxwell's equations (e.g., Ren *et al.* 2008; Soto *et al.* 2014; Zhang, Q. *et al.* 2014a, 2014b, 2015; Rizk *et al.* 2017, 2020; Zhang, J. *et al.* 2019; Zhang, L. *et al.* 2019). In these works, the FDTD method in the 2D cylindrical coordinate system is used to compute incident (horizontal) electric fields over lossy ground, needed to specify the source term in (3.28), without relying on the approximate Cooray–Rubinstein formula (Rubinstein 1996; Cooray 2002).

3.4   Model of Rachidi (1993)

The first telegrapher's equation in Rachidi's (1993) coupling model formulation can be derived starting with the first telegrapher's equation in Taylor *et al.*'s (1965) model (Equation (3.10)) in which the total current $I(x)$ is postulated to be the sum of the incident $I^i(x)$ and scattered $I^s(x)$ components:

$$\frac{dV(x)}{dx} + j\omega L'\left[I^i(x) + I^s(x)\right] = -j\omega \int_0^h B_y{}^i(x, z)\, dz \qquad (3.45)$$

The incident current is defined as

$$I^i(x) = -\frac{1}{L'}\int_0^h B_y{}^i(x, z)\, dz \qquad (3.46)$$

Substituting (3.46) into (3.45), we get the first telegrapher's equation in Rachidi's (1993) formulation, which does not contain a source term:

$$\frac{dV(x)}{dx} + j\omega L' I^s(x) = 0 \qquad (3.47)$$

To derive the second telegrapher's equation in Rachidi's (1993) coupling model formulation, we start with (3.20) (Taylor *et al.*'s (1965) formulation) in which $I(x)$ is replaced with $I^i(x) + I^s(x)$:

$$\frac{\partial I^i(x)}{\partial x} + \frac{\partial I^s(x)}{\partial x} + j\omega C' V(x) = -j\omega C' \int_0^h E_z{}^i(x, z)\, dz \qquad (3.48)$$

Substituting (3.46) in the first term and rearranging the terms, we get

$$\frac{\partial I^s(x)}{\partial x} + j\omega C' V(x) - \frac{1}{L'}\frac{\partial}{\partial x}\int_0^h B_y{}^i(x, z)\, dz = -j\omega C' \int_0^h E_z{}^i(x, z)\, dz \qquad (3.49)$$

Next, applying Ampere's law to incident fields in air, $\nabla \times \mathbf{B}^i = \mu_0\varepsilon_0 j\omega\,\mathbf{E}^i$ and noting that $\mu_0\varepsilon_0 = L'C'$ (Paul 1994, p. 24), we can write:

$$j\omega E_z{}^i(x, z) = \left(\frac{\partial B_y{}^i(x, z)}{\partial x} - \frac{\partial B_x{}^i(x, z)}{\partial y}\right)\frac{1}{L'C'} \qquad (3.50)$$

Integrating both sides of (3.50) over z from $z = 0$ to $z = h$ and multiplying through by C', we get

$$j\omega C' \int_0^h E_z{}^i(x, z)\, dz = \frac{1}{L'}\int_0^h \frac{\partial B_y{}^i(x, z)}{\partial x}\, dz - \frac{1}{L'}\int_0^h \frac{\partial B_x{}^i(x, z)}{\partial y}\, dz \qquad (3.51)$$

Now we can eliminate the terms containing E_z^i and $B_y{}^i$ from (3.49), replacing them with the term containing $B_x{}^i$, and obtain the second telegrapher's equation in Rachidi's (1993) coupling model formulation:

$$\frac{dI^s(x)}{dx} + j\omega C' V(x) = \frac{1}{L'}\int_0^h \frac{\partial B_x{}^i(x, z)}{\partial y}\, dz \qquad (3.52)$$

The term on the right-hand side corresponds to distributed shunt current (per unit length) sources specified in terms of the x-directed (axial) component of incident magnetic field, $B_x{}^i$.

Equations (3.47) and (3.52) are solved for $I^s(x)$ and $V(x)$. The incident current $I^i(x)$ is found separately using (3.46), which is reproduced below:

$$I^i(x) = -\frac{1}{L'}\int_0^h B_y{}^i(x,z)\,dz \tag{3.53}$$

and the total current is

$$I(x) = I^i(x) + I^s(x) \tag{3.54}$$

The boundary conditions for the total current are

$$I(0) = I^i(0) + I^s(0) = -\frac{V(0)}{Z_s} \tag{3.55}$$

$$I(l) = I^i(l) + I^s(l) = \frac{V(l)}{Z_r} \tag{3.56}$$

Accordingly, the boundary conditions for the scattered current are

$$I^s(0) = -\frac{V(0)}{Z_s} + \frac{1}{L'}\int_0^h B_y{}^i(0,z)dz \tag{3.57}$$

$$I^s(l) = \frac{V(l)}{Z_r} + \frac{1}{L'}\int_0^h B_y{}^i(l,z)dz \tag{3.58}$$

Thus, Rachidi's (1993) formulation requires lumped current sources, expressed in terms of transverse component of incident magnetic field, $B_y{}^i$, at line terminations. The telegrapher's equations (3.47) and (3.52) along with the boundary conditions (3.57) and (3.58) are represented by the equivalent circuit shown in Figure 3.6.

3.5 Equivalence of the three coupling models

Nucci and Rachidi (1995) have compared waveforms of voltages induced at either end of a horizontal lossless conductor of length 1 km and height 10 m above flat, perfectly conducting ground, computed using the three coupling models presented in Sections 3.2–3.4. Figure 3.7 shows those waveforms on an 8-μs time scale. The strike point is located at a distance of 50 m from the midpoint of the horizontal conductor and equidistant from either end of the conductor, each terminated in a matching resistor. The incident lightning electromagnetic fields are calculated using the modified transmission-line model with exponential current decay with height (MTLE) (Nucci *et al.* 1988). The channel-base current has a peak of 12 kA and maximum rate-of-rise of 40 kA/μs, which is typical for subsequent return strokes. Total voltages (identical for all three models) are shown by solid lines and individual voltage components associated with different

Rachidi (1993)

$$B_z^i \uparrow \quad \cdot B_y^i$$
$$B_x^i$$

$$\frac{1}{L'}\int_0^h B_y^i(0,z)\,dz \qquad \left[\frac{1}{L'}\int_0^h \frac{\partial B_x^i(x,z)}{\partial y}\,dz\right]dx \qquad \frac{1}{L'}\int_0^h B_y^i(l,z)\,dz$$

$$I^s(x) \qquad\qquad I^s(x+dx)$$

$$L'dx$$

$$V(0)\ V(x) \qquad\qquad V(x+dx)\ V(l)$$
$$Z_s \qquad\qquad C'dx \qquad\qquad Z_r$$

$$0 \quad x \qquad\qquad\qquad x+dx \quad l$$

Telegrapher's equations: Boundary conditions:

$$\left[\frac{dV(x)}{dx}+j\omega L'\,I^s(x)=0\right. \qquad \left[I^s(0)=-\frac{V(0)}{Z_s}+\frac{1}{L'}\int_0^h B_y^i(0,z)\,dz\right.$$

$$\left.\frac{dI^s(x)}{dx}+j\omega C'\,V(x)=\frac{1}{L'}\int_0^h \frac{\partial B_x^i(x,z)}{\partial y}dz\right| \quad \left.I^s(l)=\frac{V(l)}{Z_r}+\frac{1}{L'}\int_0^h B_y^i(l,z)\,dz\right.$$

— Equations are solved for $V(x)$ and $I^s(x)$
— The total current is found as $I(x)=I^i(x)+I^s(x)$
     where $I^i(x)=-\frac{1}{L'}\int_0^h B_y^i(x,z)\,dz$
— Forcing function (source term) is expressed in terms of B_x^i.
— Additional lumped sources at terminations are expressed in terms
   of B_y^i

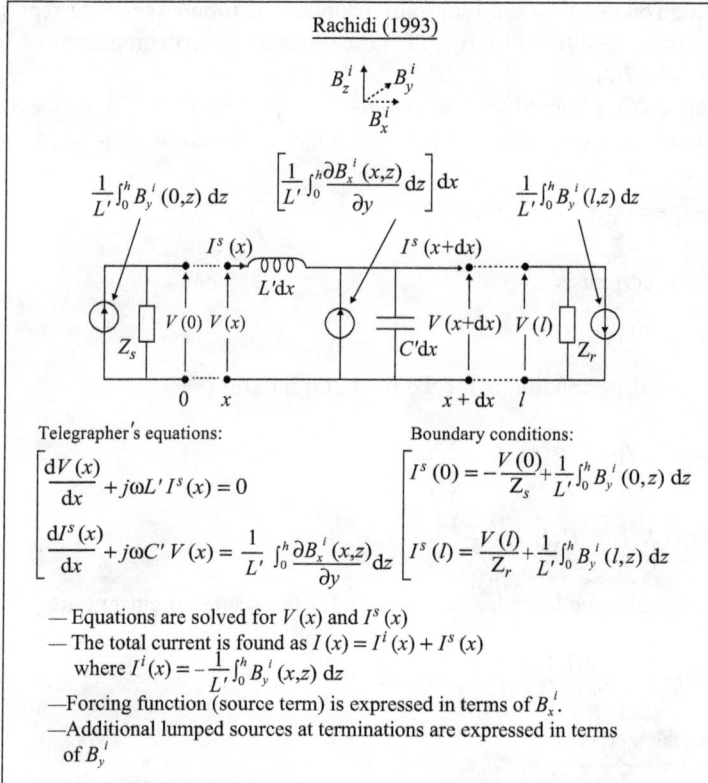

Figure 3.6   Equivalent circuit (including a differential line segment and
terminations) of the electromagnetic coupling model formulation of
Rachidi (1993) for the case of lossless horizontal conductor above flat
perfectly conducting ground. The incident field components utilized in
this formulation are shown by broken lines

incident field components are shown by broken lines. Clearly, the three coupling models
are equivalent to each other.

3.6   Summary

In this chapter, we have introduced three different sets of telegrapher's equations with
source terms and corresponding equivalent circuits that can be used for studying voltage
and current surges induced on an overhead conductor by transient electromagnetic
fields such as those produced by lightning. The source terms incorporated into the
classical telegrapher's equations have been derived, using the electromagnetic theory,
following works of Taylor *et al.* (1965), Agrawal *et al.* (1980), and Rachidi (1993), who
arrived at different formulations. Since all three formulations are based on Maxwell's

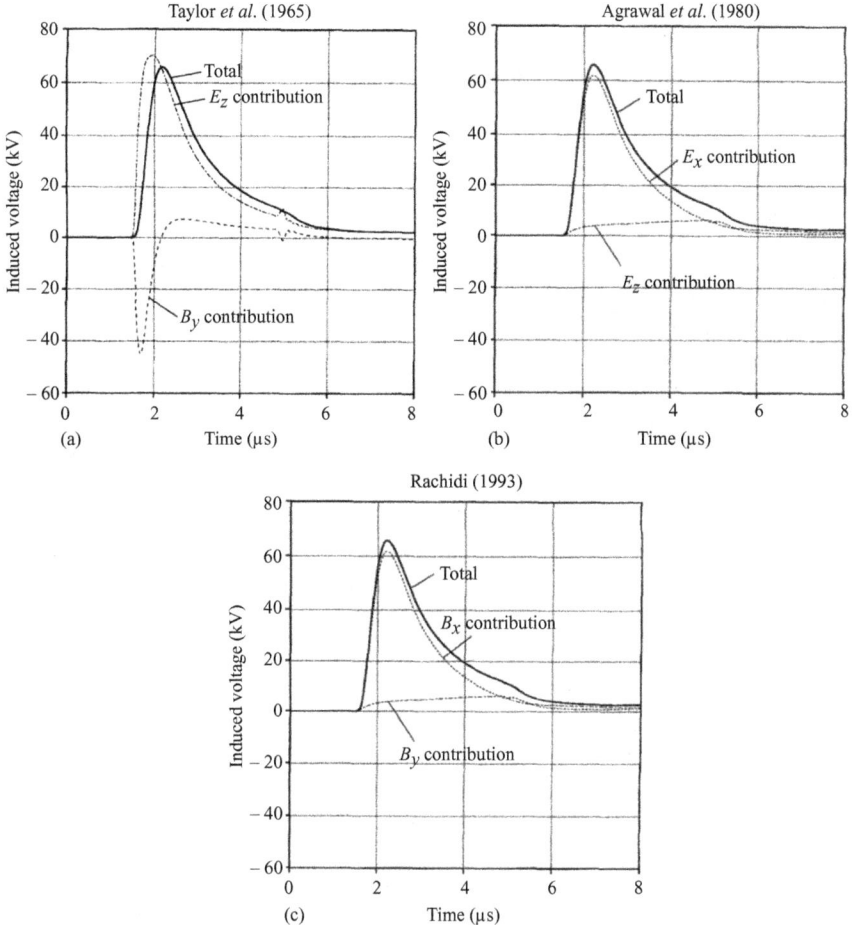

Figure 3.7   *Waveforms of voltages induced at either end of a 1-km long horizontal lossless conductor placed at a height of 10 m above flat, perfectly conducting ground, computed using (a) the model of Taylor et al. (1965), (b) the model of Agrawal et al. (1980), and (c) the model of Rachidi (1993). The strike point is located at a distance of 50 m from the midpoint of the horizontal conductor, and equidistant from either end, each terminated in a matching resistor. Lightning was represented by the modified transmission-line model with exponential current decay with height (MTLE) (Nucci et al. 1988). Lightning current waveform was representative of subsequent strokes. The waveforms are shown on an 8-μs time scale. Total voltages (identical for all three models) are shown by solid lines and individual voltage components associated with different incident field components are shown by broken lines. Reprinted, with permission, from Nucci and Rachidi (1995, Figure 2)*

equations, they yield identical results. As of today, the model of Agrawal *et al.* (1980) has been most widely used for the evaluation of lightning-induced effects on power and telecommunication lines, sometimes in conjunction with the FDTD method which is employed for computing the incident fields above lossy ground.

References

Agrawal, A. K., Price, H. J., and Gurbaxani, S. H. (1980), Transient response of multiconductor transmission lines excited by a nonuniform electromagnetic field, *IEEE Transactions on Electromagnetic Compatibility*, vol. 22, no. 3, pp. 119–129.

Cooray, V. (2002), Some considerations on the "Cooray-Rubinstein" formulation used in deriving the horizontal electric field of lightning return strokes over finitely conducting ground, *IEEE Transactions on Electromagnetic Compatibility*, vol. 44, no. 4, pp. 560–566.

Cooray, V., Rachidi, F., and Rubinstein, M. (2017), Formulation of field-to-transmission line coupling equations in terms of scalar and vector potentials, *IEEE Transactions on Electromagnetic Compatibility*, vol. 59, no. 5, pp. 1586–1591.

Cooray, V., Nucci, C. A., Piantini, A., Rachidi, F., and Rubinstein, M. (2020), Field-to-transmission line coupling models, Chapter 6, *Lightning Interaction with Power Systems*, Volume 1: Fundamentals and Modelling (Edited by Piantini, A.), IET, pp. 217–249.

Nucci, C. A., Mazzetti, C., Rachidi, F., and Ianoz, M. (1988), On lightning return stroke models for LEMP calculations, Paper presented at the 19th International Conference on Lightning Protection, Graz, pp. 463–470.

Nucci, C. A., and Rachidi, F. (1995), On the contribution of the electromagnetic field components in field-to-transmission line interaction, *IEEE Transactions on Electromagnetic Compatibility*, vol. 37, no. 4, pp. 505–508.

Nucci, C. A., and Rachidi, F. (2012), Interaction of electromagnetic fields generated by lightning with overhead electrical networks, Chapter 12, *The Lightning Flash* (2nd edition) (Edited by Cooray, V.), IET, pp. 559–609.

Nucci, C. A., Rachidi, F., and Rubinstein, M. (2012), Interaction of lightning-generated electromagnetic fields with overhead and underground cables, Chapter 18, *Lightning Electromagnetics* (Edited by Cooray, V.), IET, pp. 687–718.

Paul, C. R. (1994), *Analysis of Multiconductor Transmission Lines*, John Wiley & Sons, 584 pages.

Rachidi, F. (1993), Formulation of the field-to-transmission line coupling equations in terms of magnetic excitation field, *IEEE Transactions on Electromagnetic Compatibility*, vol. 35, no. 3, pp. 404–407.

Rachidi, F., and Tkachenko, S. V. (2008), *Electromagnetic Field Interaction with Transmission Lines: From Classical Theory to HF Radiation Effects*, WIT Press, 288 pages.

Ren, H.-M., Zhou, B.-H., Rakov, V. A., Shi, L.-H., Gao, C., and Yang, J.-H. (2008), Analysis of lightning-induced voltages on overhead lines using a 2-D FDTD method and Agrawal coupling model, *IEEE Transactions on Electromagnetic Compatibility*, vol. 50, no. 3, pp. 651–659.

Rizk, M. E. M., Mahmood, F., Lehtonen, M., Badran, E. A., and Abdel-Rahman, M. H. (2017), Computation of peak lightning-induced voltages due to the typical first and subsequent strokes considering high ground resistivity, *IEEE Transactions on Power Delivery*, vol. 32, no. 4, pp. 1861–1871.

Rizk, M. E. M., Lehtonen, M., Baba, Y., and Ghanem, A. (2020), Protection against lightning-induced voltages: Transient model for points of discontinuity on multi-conductor overhead line, *IEEE Transactions on Electromagnetic Compatibility*, vol. 62, no. 4, pp. 1209–1218.

Rubinstein, M. (1996), An approximate formula for the calculation of the horizontal electric field from lightning at close, intermediate, and long range, *IEEE Transactions on Electromagnetic Compatibility*, vol. 38, no. 3, pp. 531–535.

Soto, E., Perez, E., and Younes, C. (2014), Influence of non-flat terrain on lightning induced voltages on distribution networks, *Electric Power Systems Research*, vol. 113, pp. 115–120.

Taylor, C., Satterwhite, R., and Harrison, C. (1965), The response of a terminated two-wire transmission line excited by a nonuniform electromagnetic field, *IEEE Transactions on Antennas and Propagation*, vol. 13, no. 6, pp. 987–989.

Wuyts, I., and De Zutter, D. (1994), Circuit model for plane-wave incidence on multi-conductor transmission lines, *IEEE Transactions on Electromagnetic Compatibility*, vol. 36, no. 3, pp. 206–212.

Yee, K. S. (1966), Numerical solution of initial boundary value problems involving Maxwell's equations in isotropic media, *IEEE Transactions on Antennas and Propagation*, vol. 14, no. 3, pp. 302–307.

Zhang, J., Zhang, Q., Hou, W., Zhang, L., Zhou, F., and Ma, Y. (2019), Evaluation of the lightning-induced voltages of multiconductor lines for striking cone-shaped mountain, *IEEE Transactions on Electromagnetic Compatibility*, vol. 61, no. 5, pp. 1534–1542.

Zhang, L., Wang, L. Yang, J. Jin, X. and Zhang, J. (2019), Effect of overhead shielding wires on the lightning-induced voltages of multiconductor lines above the lossy ground, *IEEE Transactions on Electromagnetic Compatibility*, vol. 61, no. 2, pp. 458–466.

Zhang, Q., Tang, X., Gao, J., Zhang, L., and Li, D. (2014a), The influence of the horizontally stratified conducting ground on the lightning-induced voltages, *IEEE Transactions on Electromagnetic Compatibility*, vol. 56, no. 2, pp. 435–443.

Zhang, Q., Zhang, L., Tang, X., and Gao, J. (2014b), An approximate formula for estimating the peak value of lightning-induced overvoltage considering the stratified conducting ground, *IEEE Transactions on Power Delivery*, vol. 29, no. 2, pp. 884–889.

Zhang, Q., Chen, Y., and Hou, W. (2015), Lightning-induced voltages caused by lightning strike to tall objects considering the effect of frequency dependent soil, *Journal of Atmospheric and Solar-Terrestrial Physics*, vol. 133, pp. 145–156.

Chapter 4

Finite-difference time-domain method

Electromagnetic computation methods (ECMs) have been widely used in analyzing lightning electromagnetic pulses (LEMPs) and lightning-caused surges in various systems. One of the advantages of ECMs, relative to circuit simulation methods, is that they allow a self-consistent, full-wave solution for both the transient current distribution in a 3D conductor system and resultant electromagnetic fields, although these methods are computationally expensive. Among ECMs, the finite-difference time-domain (FDTD) method has been most frequently used in LEMP and surge simulations. In this chapter, update equations for electric and magnetic fields used in the FDTD computation in the 3D Cartesian, 2D cylindrical, and 2D spherical coordinate systems are given. A subgridding technique, which allows one to employ locally finer grids, is described. Representations of lumped sources and lumped circuit elements such as a resistor, an inductor and a capacitor are described. Representations of a thin-wire conductor and the lightning sources are discussed. Also, representations of nonlinear elements, such as surge arrester, and nonlinear phenomena, such as corona on a horizontal conductor, are explained. Absorbing boundary conditions, which are needed for the analysis of electromagnetic fields in an unbounded space, are reviewed.

Key Words: Lightning; lightning return stroke; Maxwell's equations; electromagnetic field; FDTD method; subgridding technique; lumped source; lumped circuit element; thin wire; lightning channel excitation; surge arrester; corona; absorbing boundary conditions

4.1 Introduction

Electromagnetic computation methods (ECMs), which include the method of moments (MoM) (Harrington 1968), the finite-element method (FEM) (Sadiku 1989), the partial-element equivalent-circuit (PEEC) method (Ruehli 1974), the hybrid electromagnetic/circuit model (HEM) (Visacro and Soares 2005), the transmission-line-modeling (TLM) method (Johns and Beurle 1971), and the finite-difference time-domain (FDTD) method (Yee 1966) have been widely used in analyzing lightning electromagnetic pulses (LEMPs) and lightning-caused surges in various systems (e.g., Baba and Rakov 2007a, 2008, 2009, 2014, 2016). One of the advantages of ECMs, relative to circuit simulation methods, is that most of them allow a self-consistent full-wave solution for both the transient current distribution in the lightning channel or in a 3D conductor system and resultant electromagnetic fields, although they are usually computationally expensive.

Among ECMs, the FDTD method has been most frequently used in LEMP calculations. The first peer-reviewed paper, in which the FDTD method was used in a lightning electromagnetic field simulation, was published in 2003 (Baba and Rakov 2003), and it was first applied to the analysis of surge performance of grounding electrodes by Tanabe (2001) in 2001. The amount of published material on applications of the FDTD method to LEMP and surge simulations is quite large. Interest in using the FDTD method continues to grow because of the availability of numerical codes and increased computational capabilities.

The FDTD method uses the central difference approximation to Maxwell's curl equations for time-varying fields, which are Faraday's law and Ampere's law, in the time domain. Gauss's law is also satisfied. The resultant update equations for electric and magnetic fields are solved at each time step and at each discretized space point in the working volume using the leapfrog method. For the analysis of the electromagnetic response of a structure in an unbounded space, an absorbing boundary condition such as Liao's condition (Liao *et al.* 1984) or perfectly matched layers (Berenger 1994) should be applied to suppress unwanted reflections. Advantages of the FDTD method relative to other ECMs can be summarized as follows:

1. It is based on a simple procedure and, therefore, its programming is relatively easy;
2. It is capable of treating complex geometries and inhomogeneities;
3. It is capable of incorporating nonlinear effects and components; and
4. It can handle wideband quantities in one run with a time-to-frequency transforming tool.

Its disadvantages are:

1. It is computationally expensive compared to other methods, such as the MoM;
2. It cannot deal with oblique boundaries that are not aligned with the Cartesian grid when the standard orthogonal grid is used; a staircase approximation for oblique boundaries is usually employed; and
3. It would require a complex procedure for incorporating dispersive materials/media.

Additional details on the FDTD method are given in works of Kunz and Luebbers (1993); Taflove (1995); Uno (1998); Sullivan (2000); Hao and Mittra (2009); Yu *et al.* (2009); Inan and Marshall (2011); and Baba and Rakov (2016).

In this chapter, update equations for electric and magnetic fields used in the FDTD computation in the 3D Cartesian, 2D cylindrical, and 2D spherical coordinate systems are given. A subgridding technique, which allows one to employ locally finer grids, is described. Representations of lumped sources and lumped circuit elements such as a resistor, an inductor, and a capacitor are described. Representations of a thin-wire conductor and the lightning sources are discussed. Also, representations of nonlinear elements, such as surge arrester, and nonlinear phenomena, such as corona on a horizontal conductor, are explained. Absorbing boundary conditions, which are needed for the analysis of electromagnetic fields in an unbounded space, are reviewed.

4.2   Finite-difference expressions of Maxwell's equations

4.2.1   3D Cartesian coordinate system

The FDTD method in the 3D Cartesian coordinate system requires the whole working space, which accommodates a conductor system to be analyzed, to be divided into cubic or rectangular parallelepiped cells with side lengths Δx, Δy, and Δz, as shown in Figure 4.1. The electric field components are placed at the mid-points of the sides of cells: E_x components are placed at the midpoints of sides oriented in the x-direction, E_y components are placed at the midpoints of y-directed sides, and E_z components are placed at the midpoints of z-directed sides. The magnetic field components are placed at the center points of the faces of the cubic or rectangular parallelepiped cells, and are oriented normal to the faces: H_x components are placed at the center points on yz-faces, H_y components are placed at the center points on zx-faces, and H_z components are placed at the center points on xy-faces. The electric field components are computed at integer time steps $n\Delta t$, where n is an integer number and Δt is the time increment, and the magnetic field components are computed at half-integer time steps $(n - 1/2)\,\Delta t$.

Time-update equations for electric field components in x-, y-, and z-directions, E_x, E_y, and E_z, are derived from Ampere's law, and those for magnetic field components, H_x, H_y, and H_z, are derived from Faraday's law. These are shown below.

Ampere's law is given as follows:

$$\nabla \times \boldsymbol{H}^{n-\frac{1}{2}} = \varepsilon \frac{\partial \boldsymbol{E}^{n-\frac{1}{2}}}{\partial t} + \boldsymbol{J}^{n-\frac{1}{2}} = \varepsilon \frac{\partial \boldsymbol{E}^{n-\frac{1}{2}}}{\partial t} + \sigma \boldsymbol{E}^{n-\frac{1}{2}} \tag{4.1}$$

where $\boldsymbol{H}$ is the magnetic field vector, $\boldsymbol{E}$ is the electric field vector, $\boldsymbol{J}(=\sigma\boldsymbol{E})$ is the conduction-current-density vector, ε is the electric permittivity, σ is the electric

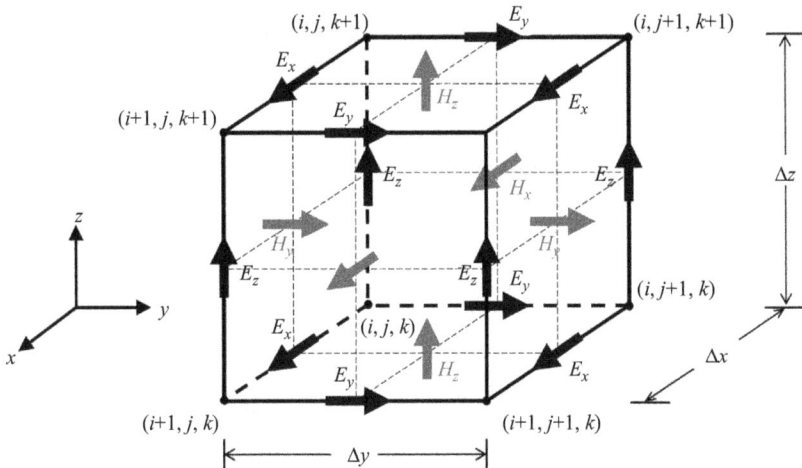

Figure 4.1   *Placement of electric field and magnetic field components on the sides and at the center points of the faces of a cubic cell, respectively*

conductivity, t is the time, and $n - 1/2$ is the time step number for the magnetic field computations. $\varepsilon\,\partial E/\partial t$ is the displacement-current-density vector (due to time variation of the electric field vector). Ampere's law states that the conduction current and/or time-variation of electric field create magnetic field in the direction of right-hand curl. If the time-derivative term in (4.1) is approximated by its central finite difference, (4.1) is expressed as follows:

$$\varepsilon\frac{\partial E^{n-\frac{1}{2}}}{\partial t} + \sigma E^{n-\frac{1}{2}} \approx \varepsilon\frac{E^n - E^{n-1}}{\Delta t} + \sigma\frac{E^n + E^{n-1}}{2} \approx \nabla \times H^{n-\frac{1}{2}} \tag{4.2}$$

Note that $E^{n-1/2}$ in the second term of (4.2) is approximated by its average value, $(E^n + E^{n-1})/2$. If (4.2) is rearranged, the update equation for the electric field vector at a time step number n, E^n, from its one time-step previous value, E^{n-1}, and the half time-step previous magnetic-field curl value, $\nabla \times H^{n-1/2}$, is obtained as follows:

$$E^n = \left(\frac{1 - \frac{\sigma\Delta t}{2\varepsilon}}{1 + \frac{\sigma\Delta t}{2\varepsilon}}\right) E^{n-1} + \left(\frac{\frac{\Delta t}{\varepsilon}}{1 + \frac{\sigma\Delta t}{2\varepsilon}}\right) \nabla \times H^{n-\frac{1}{2}} \tag{4.3}$$

From (4.3), the update equation for $E_x{}^n$ at a location $(i+1/2, j, k)$ (see Figure 4.2(a)), for example, is expressed as follows:

$$E_x{}^n\left(i+\frac{1}{2},j,k\right)$$

$$= \frac{1 - \dfrac{\sigma(i+1/2,j,k)\,\Delta t}{2\varepsilon(i+1/2,j,k)}}{1 + \dfrac{\sigma(i+1/2,j,k)\,\Delta t}{2\varepsilon(i+1/2,j,k)}} E_x{}^{n-1}\left(i+\frac{1}{2},j,k\right) + \frac{\dfrac{\Delta t}{\varepsilon(i+1/2,j,k)}}{1 + \dfrac{\sigma(i+1/2,j,k)\,\Delta t}{2\varepsilon(i+1/2,j,k)}}$$

$$\times \left[\frac{\partial H_z{}^{n-\frac{1}{2}}\left(i+\dfrac{1}{2},j,k\right)}{\partial y} - \frac{\partial H_y{}^{n-\frac{1}{2}}\left(i+\dfrac{1}{2},j,k\right)}{\partial z}\right]$$

$$= \frac{1 - \dfrac{\sigma(i+1/2,j,k)\,\Delta t}{2\varepsilon(i+1/2,j,k)}}{1 + \dfrac{\sigma(i+1/2,j,k)\,\Delta t}{2\varepsilon(i+1/2,j,k)}} E_x{}^{n-1}\left(i+\frac{1}{2},j,k\right) + \frac{\dfrac{\Delta t}{\varepsilon(i+1/2,j,k)}}{1 + \dfrac{\sigma(i+1/2,j,k)\,\Delta t}{2\varepsilon(i+1/2,j,k)}}\frac{1}{\Delta y\Delta z}$$

$$\times \begin{bmatrix} H_z{}^{n-\frac{1}{2}}\left(i+\dfrac{1}{2},j+\dfrac{1}{2},k\right)\Delta z - H_z{}^{n-\frac{1}{2}}\left(i+\dfrac{1}{2},j-\dfrac{1}{2},k\right)\Delta z \\ -H_y{}^{n-\frac{1}{2}}\left(i+\dfrac{1}{2},j,k+\dfrac{1}{2}\right)\Delta y + H_y{}^{n-\frac{1}{2}}\left(i+\dfrac{1}{2},j,k-\dfrac{1}{2}\right)\Delta y \end{bmatrix}$$

$$\tag{4.4}$$

where the spatial derivative terms in (4.4) are approximated by their central finite differences. Update equations for $E_x{}^n$ and $E_y{}^n$, which are derived in the same manner, are given below:

(a)

(b)

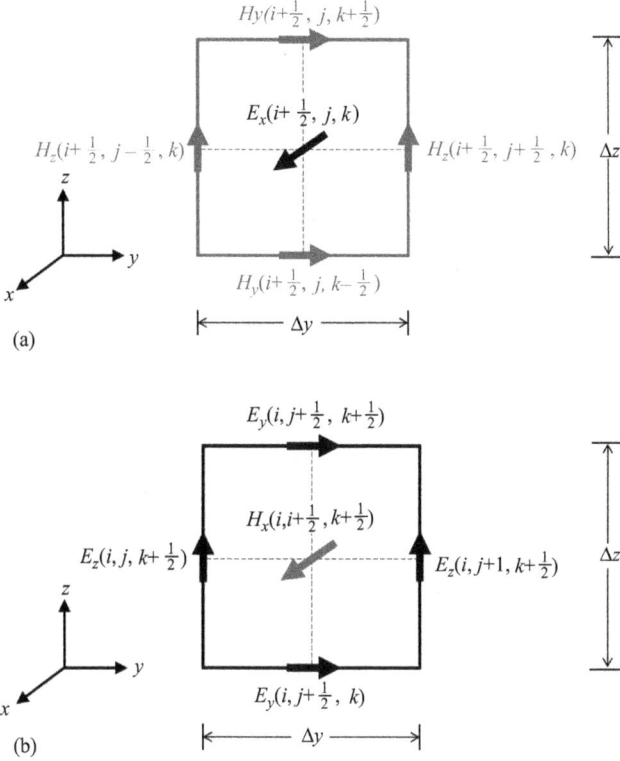

Figure 4.2 (a) Electric field component in the x-direction E_x at location $(i+1/2, j, k)$ and the circulating set of four magnetic field components closest to it, and (b) magnetic field component in the x-direction H_x at location $(i, j+1/2, k+1/2)$ and the circulating set of four electric field components closest to it

$$
E_y^n\left(i,j+\frac{1}{2},k\right) = \frac{1 - \dfrac{\sigma(i,j+1/2,k)\,\Delta t}{2\varepsilon(i,j+1/2,k)}}{1 + \dfrac{\sigma(i,j+1/2,k)\,\Delta t}{2\varepsilon(i,j+1/2,k)}} E_y^{n-1}\left(i,j+\frac{1}{2},k\right)
$$

$$
+ \frac{\dfrac{\Delta t}{\varepsilon(i,j+1/2,k)}}{1 + \dfrac{\sigma(i,j+1/2,k)\,\Delta t}{2\varepsilon(i,j+1/2,k)}} \frac{1}{\Delta z \Delta x}
$$

$$
\times \left[\begin{array}{l} H_x^{n-\frac{1}{2}}\left(i,j+\dfrac{1}{2},k+\dfrac{1}{2}\right)\Delta x - H_x^{n-\frac{1}{2}}\left(i,j+\dfrac{1}{2},k-\dfrac{1}{2}\right)\Delta x \\[2mm] -H_z^{n-\frac{1}{2}}\left(i+\dfrac{1}{2},j+\dfrac{1}{2},k\right)\Delta z + H_z^{n-\frac{1}{2}}\left(i-\dfrac{1}{2},j+\dfrac{1}{2},k\right)\Delta z \end{array} \right]
$$

$$(4.5)$$

$$E_z{}^n\left(i,j,k+\frac{1}{2}\right) = \frac{1 - \dfrac{\sigma(i,j,k+1/2)\,\Delta t}{2\varepsilon(i,j,k+1/2)}}{1 + \dfrac{\sigma(i,j,k+1/2)\,\Delta t}{2\varepsilon(i,j,k+1/2)}} E_z{}^{n-1}\left(i,j,k+\frac{1}{2}\right)$$

$$+\frac{\dfrac{\Delta t}{\varepsilon(i,j,k+1/2)}}{1 + \dfrac{\sigma(i,j,k+1/2)\,\Delta t}{2\varepsilon(i,j,k+1/2)}}\frac{1}{\Delta t\,\Delta x\Delta y}$$

$$\times\left[\begin{array}{l} H_y{}^{n-\frac{1}{2}}\left(i+\dfrac{1}{2},j,k+\dfrac{1}{2}\right)\Delta y - H_y{}^{n-\frac{1}{2}}\left(i-\dfrac{1}{2},j,k-\dfrac{1}{2}\right)\Delta y \\[2mm] -H_x{}^{n-\frac{1}{2}}\left(i,j+\dfrac{1}{2},k+\dfrac{1}{2}\right)\Delta x + H_x{}^{n-\frac{1}{2}}\left(i,j-\dfrac{1}{2},k+\dfrac{1}{2}\right)\Delta x \end{array}\right]$$

$$\tag{4.6}$$

Faraday's law is given as follows:

$$\nabla \times E^n = -\mu\frac{\partial H^n}{\partial t} \tag{4.7}$$

where μ is the magnetic permeability. Faraday's law states that the time variation of magnetic field creates electric field in the negative direction of right-hand curl. If the time-derivative term in (4.7) is approximated by its central finite difference, (4.7) is expressed as follows:

$$\mu\frac{\partial H^n}{\partial t} \approx \mu\frac{H^{n+\frac{1}{2}} - H^{n-\frac{1}{2}}}{\Delta t} \approx -\nabla \times E^n \tag{4.8}$$

If (4.8) is rearranged, the update equation for magnetic field at a time step number $n+1/2$ is obtained from its one-time-step previous value $H^{n-1/2}$ and the half time-step previous electric-field curl value $\nabla \times E^n$ as follows:

$$H^{n+\frac{1}{2}} = H^{n-\frac{1}{2}} - \frac{\Delta t}{\mu}\nabla \times E^n \tag{4.9}$$

From (4.9), the update equation for $H_x^{n+1/2}$ at a location $(i, j + 1/2, k + 1/2)$ (see Figure 4.2(b)), for example, is expressed as follows:

$$
H_x^{n+\frac{1}{2}}\left(i, j+\frac{1}{2}, k+\frac{1}{2}\right) = H_x^{n-\frac{1}{2}}\left(i, j+\frac{1}{2}, k+\frac{1}{2}\right) - \frac{\Delta t}{\mu(i, j+1/2, k+1/2)}
$$
$$
\times \left[\frac{\partial E_z^n\left(i, j+\frac{1}{2}, k+\frac{1}{2}\right)}{\partial y} - \frac{\partial E_y^n\left(i, j+\frac{1}{2}, k+\frac{1}{2}\right)}{\partial z} \right]
$$
$$
= H_x^{n-\frac{1}{2}}\left(i, j+\frac{1}{2}, k+\frac{1}{2}\right) - \frac{\Delta t}{\mu(i, j+1/2, k+1/2)} \frac{1}{\Delta y \Delta z}
$$
$$
\times \left[E_z^n\left(i, j+1, k+\frac{1}{2}\right)\Delta z - E_z^n\left(i, j, k+\frac{1}{2}\right)\Delta z \right.
$$
$$
\left. - E_y^n\left(i, j+\frac{1}{2}, k+1\right)\Delta y + E_y^n\left(i, j+\frac{1}{2}, k\right)\Delta y \right]
$$
$$
\tag{4.10}
$$

where the spatial derivative terms in (4.10) are approximated by their central finite differences. Update equations for $H_y^{n+1/2}$ and $H_z^{n+1/2}$ are derived in the same manner and given below:

$$
H_y^{n+\frac{1}{2}}\left(i+\frac{1}{2}, j, k+\frac{1}{2}\right) = H_y^{n-\frac{1}{2}}\left(i+\frac{1}{2}, j, k+\frac{1}{2}\right) - \frac{\Delta t}{\mu(i+1/2, j, k+1/2)} \frac{1}{\Delta z \Delta x}
$$
$$
\times \left[E_x^n\left(i+\frac{1}{2}, j, k+1\right)\Delta x - E_x^n\left(i+\frac{1}{2}, j, k\right)\Delta x \right.
$$
$$
\left. - E_z^n\left(i+1, j, k+\frac{1}{2}\right)\Delta z + E_z^n\left(i, j, k+\frac{1}{2}\right)\Delta z \right]
$$
$$
\tag{4.11}
$$

$$
H_z^{n+\frac{1}{2}}\left(i+\frac{1}{2}, j+\frac{1}{2}, k\right) = H_z^{n-\frac{1}{2}}\left(i+\frac{1}{2}, j+\frac{1}{2}, k\right) - \frac{\Delta t}{\mu(i+1/2, j+1/2, k)} \frac{1}{\Delta x \Delta y}
$$
$$
\times \left[E_y^n\left(i+1, j+\frac{1}{2}, k\right)\Delta y - E_y^n\left(i, j+\frac{1}{2}, k\right)\Delta y \right.
$$
$$
\left. - E_x^n\left(i+\frac{1}{2}, j+1, k\right)\Delta x + E_x^n\left(i+\frac{1}{2}, j, k\right)\Delta x \right]
$$
$$
\tag{4.12}
$$

By updating $E_x^n, E_y^n, E_z^n, H_x^{n+1/2}, H_y^{n+1/2}$, and $H_z^{n+1/2}$ at every point in the working volume, transient electric and magnetic fields throughout the working volume are calculated.

For the FDTD solution to be stable, the time increment Δt needs to be set to fulfill the Courant stability condition (Courant *et al.* 1928) given as follows:

$$\Delta t \leq \frac{1}{c\sqrt{\frac{1}{(\Delta x)^2} + \frac{1}{(\Delta y)^2} + \frac{1}{(\Delta z)^2}}} \tag{4.13}$$

where c is the speed of light.

Note that the 3D working volume is not necessarily divided uniformly into cubic or rectangular parallelepiped cells. Nonuniform grids or locally finer grids (called subgrids) could be employed for efficiently representing locally fine structures or boundaries (e.g., Thang *et al.* 2012, 2015). The computation procedure for a non-uniform grid is essentially the same as described above. A subgridding technique (e.g., Chevalier *et al.* 1997) is presented in Section 4.3. Nonorthogonal grids could be employed (e.g., Taflove 1995) or differently shaped cells such as tetrahedral or tri-angular prism cells (Hano and Itoh 1996; Tanabe *et al.* 2003; Nakagawa *et al.* 2016) could be employed for representing oblique boundaries without using a staircase approximation. Further, parallel computational approaches could be used in order to accelerate the FDTD computations (e.g., Oliveira and Sobrinho 2009; Oikawa *et al.* 2012; Livesey *et al.* 2012).

4.2.2   2D cylindrical coordinate system

In analyzing electromagnetic pulses, which are radiated from a vertical lightning channel and propagate over a rotationally symmetrical ground, it is more advantageous to use the 2D cylindrical coordinate system (e.g., Yang and Zhou 2004; Ren *et al.* 2008; Taniguchi *et al.* 2008a; Baba and Rakov 2008a, 2009, 2011; Yang *et al.* 2011; Tran *et al.* 2017) since it requires less computation time and memory than the 3D Cartesian coordinate system.

In the 2D cylindrical coordinate system, there exist only radial and vertical components of electric field, E_r and E_z, and azimuthal component of magnetic field, H_φ. The FDTD method in this coordinate system requires the whole 2D working space to be divided into square or rectangular cells. Time-update equations for vertical and radial electric fields, E_r and E_z, are derived from Ampere's law equation (4.1), and that for azimuthal magnetic field, H_φ, is derived from Faraday's law equation (4.7), similar to the derivations for the 3D Cartesian coordinate system. The curl of magnetic-field vector in the 2D cylindrical coordinate system is given by

$$\begin{aligned}
\nabla \times \boldsymbol{H} &= \left[\frac{1}{r}\frac{\partial H_z}{\partial \varphi} - \frac{\partial H_\varphi}{\partial z}, \quad \frac{\partial H_r}{\partial z} - \frac{\partial H_z}{\partial r}, \quad \frac{1}{r}\left(\frac{\partial(rH_\varphi)}{\partial r} - \frac{\partial H_r}{\partial \varphi}\right) \right] \\
&= \left[-\frac{\partial H_\varphi}{\partial z}, \quad 0, \quad \frac{1}{r}\frac{\partial(rH_\varphi)}{\partial r} \right]
\end{aligned} \tag{4.14}$$

From (4.3) and (4.14), update equations for E_r at a location $(i+1/2, j)$ and E_z at a location $(i, j+1/2)$ (see Figure 4.3) are given as follows:

$$
E_r{}^n\left(i+\frac{1}{2}, j\right) = \frac{1 - \dfrac{\sigma(i+1/2, j)\,\Delta t}{2\varepsilon(i+1/2, j)}}{1 + \dfrac{\sigma(i+1/2, j)\,\Delta t}{2\varepsilon(i+1/2, j)}} E_r{}^{n-1}\left(i+\frac{1}{2}, j\right)
$$

$$
+ \frac{\dfrac{\Delta t}{\varepsilon(i+1/2, j)}}{1 + \dfrac{\sigma(i+1/2, j)\,\Delta t}{2\varepsilon(i+1/2, j)}} \left[-\frac{\partial H_\varphi{}^{n-\frac{1}{2}}\left(i+\dfrac{1}{2}, j\right)}{\partial z} \right]
$$

$$
= \frac{1 - \dfrac{\sigma(i+1/2, j)\,\Delta t}{2\varepsilon(i+1/2, j)}}{1 + \dfrac{\sigma(i+1/2, j)\,\Delta t}{2\varepsilon(i+1/2, j)}} E_r{}^{n-1}\left(i+\frac{1}{2}, j\right) + \frac{\dfrac{\Delta t}{\varepsilon(i+1/2, j)}}{1 + \dfrac{\sigma(i+1/2, j)\,\Delta t}{2\varepsilon(i+1/2, j)}} \frac{1}{\Delta z}
$$

$$
\times \left[-H_\varphi{}^{n-\frac{1}{2}}\left(i+\frac{1}{2}, j+\frac{1}{2}\right) + H_\varphi{}^{n-\frac{1}{2}}\left(i+\frac{1}{2}, j-\frac{1}{2}\right) \right]
$$

$$\tag{4.15}$$

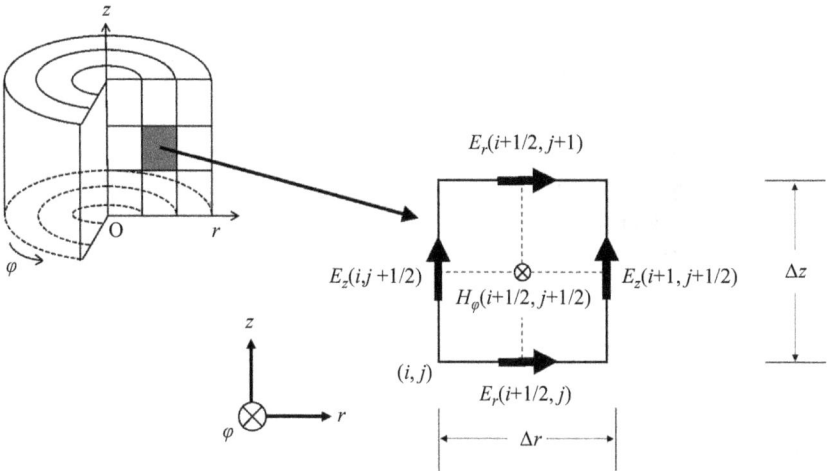

Figure 4.3   *Placement of radial E_r and vertical E_z components of electric field and azimuthal component of magnetic field H_φ in a cell in the 2D cylindrical coordinate system*

$$E_z{}^n\left(i,j+\frac{1}{2}\right) = \frac{1-\dfrac{\sigma(i,j+1/2)\,\Delta t}{2\varepsilon(i,j+1/2)}}{1+\dfrac{\sigma(i,j+1/2)\,\Delta t}{2\varepsilon(i,j+1/2)}}E_z{}^{n-1}\left(i,j+\frac{1}{2}\right)$$

$$+\frac{\dfrac{\Delta t}{\varepsilon(i,j+1/2)}}{1+\dfrac{\sigma(i,j+1/2)\,\Delta t}{2\varepsilon(i,j+1/2)}}\left[\frac{1}{r_i}\frac{\partial r_i\,H_\varphi{}^{n-\frac{1}{2}}\left(i,j+\frac{1}{2}\right)}{\partial z}\right]$$

$$=\frac{1-\dfrac{\sigma(i,j+1/2)\,\Delta t}{2\varepsilon(i,j+1/2)}}{1+\dfrac{\sigma(i,j+1/2)\,\Delta t}{2\varepsilon(i,j+1/2)}}E_z{}^{n-1}\left(i,j+\frac{1}{2}\right)+\frac{\dfrac{\Delta t}{\varepsilon(i,j+1/2)}}{1+\dfrac{\sigma(i,j+1/2)\,\Delta t}{2\varepsilon(i,j+1/2)}}\frac{1}{r_i\,\Delta r}$$

$$\times\left[r_{i+1/2}H_\varphi{}^{n-\frac{1}{2}}\left(i+\frac{1}{2},j+\frac{1}{2}\right)-r_{i-1/2}H_\varphi{}^{n-\frac{1}{2}}\left(i-\frac{1}{2},j+\frac{1}{2}\right)\right]$$

$$(4.16)$$

where r_i is the radial distance from the z-axis to the location of $E_z(i,j+1/2)$, $r_{i-1/2}$ is the distance from the z-axis to the location of $H_\varphi(i-1/2,j+1/2)$, $r_{i+1/2}$ is the distance from the z-axis to the location of $H_\varphi(i+1/2,j+1/2)$, Δr is the cell side length in the radial direction, and Δz is the cell side length in the vertical direction.

The curl of electric-field vector in the 2D cylindrical coordinate system is given by

$$\nabla \times E = \left[\frac{1}{r}\frac{\partial E_z}{\partial \varphi}-\frac{\partial E_\varphi}{\partial z},\ \frac{\partial E_r}{\partial z}-\frac{\partial E_z}{\partial r},\ \frac{1}{r}\left(\frac{\partial(rE_\varphi)}{\partial r}-\frac{\partial E_r}{\partial \varphi}\right)\right]$$

$$=\left[0,\ \frac{\partial E_r}{\partial z}-\frac{\partial E_z}{\partial r},\ 0\right]$$

$$(4.17)$$

From (4.9) and (4.17), the update equation for $H_\varphi^{n+1/2}$ at a location $(i+1/2,j+1/2)$ (see Figure 4.3) is given as follows:

$$H_\varphi{}^{n+\frac{1}{2}}\left(i+\frac{1}{2},j+\frac{1}{2}\right)$$

$$=H_\varphi{}^{n-\frac{1}{2}}\left(i+\frac{1}{2},j+\frac{1}{2}\right)-\frac{\Delta t}{\mu(i+1/2,j+1/2)}$$

$$\times\left[\frac{\partial E_r{}^n\left(i+\frac{1}{2},j+\frac{1}{2}\right)}{\partial z}-\frac{\partial E_z{}^n\left(i+\frac{1}{2},j+\frac{1}{2}\right)}{\partial r}\right]$$

$$=H_\varphi{}^{n-\frac{1}{2}}\left(i+\frac{1}{2},j+\frac{1}{2}\right)-\frac{\Delta t}{\mu(i+1/2,j+1/2)\,\Delta z}\frac{1}{}\left[E_r{}^n\left(i+\frac{1}{2},j+1\right)\right.$$

$$\left.-E_r{}^n\left(i+\frac{1}{2},j\right)\right]+\frac{\Delta t}{\mu(i+1/2,j+1/2)\,\Delta r}\frac{1}{}\left[E_z{}^n\left(i+1,j+\frac{1}{2}\right)-E_z{}^n\left(i,j+\frac{1}{2}\right)\right]$$

$$(4.18)$$

By updating E_r^n, E_z^n, and $H_\varphi^{n+1/2}$ at every point in the working space, transient electric and magnetic fields throughout the working space are obtained.

The time increment Δt needs to be set to fulfill the Courant stability condition given as follows:

$$\Delta t \leq \frac{1}{c\sqrt{\frac{1}{(\Delta r)^2} + \frac{1}{(\Delta z)^2}}} \tag{4.19}$$

4.2.3   2D spherical coordinate system

In analyzing electromagnetic pulses at far distances, which are radiated from a vertical lightning channel and propagate over a rotationally symmetrical curved ground, the 2D spherical coordinate system is often used (e.g., Azadifar *et al.* 2017; Yamamoto *et al.* 2019), since it requires less computation time and memory than the 3D Cartesian coordinate system. The 2D spherical coordinate system is sufficient to account for the curved ground surface.

In the 2D spherical coordinate system, there exist only radial and polar components of electric field, E_r and E_θ, and azimuthal component of magnetic field, H_φ. The FDTD method in this coordinate system requires the whole 2D working space to be divided into annular sectorial cells. Time-update equations for radial and polar electric fields, E_r and E_θ, are derived from Ampere's law equation (4.1), and that for azimuthal magnetic field, H_φ, is derived from Faraday's law equation (4.7), similar to the derivations for the 2D cylindrical coordinate system. The curl of magnetic-field vector in the 2D spherical coordinate system is given by

$$\nabla \times \mathbf{H} = \left[\frac{1}{r\sin\theta}\left(\frac{\partial(H_\varphi\sin\theta)}{\partial\theta} - \frac{\partial H_\theta}{\partial\varphi}\right), \frac{1}{r}\left(\frac{1}{\sin\theta}\frac{\partial H_r}{\partial\varphi} - \frac{\partial(rH_\varphi)}{\partial r}\right), \frac{1}{r}\left(\frac{\partial(rH_\theta)}{\partial r} - \frac{\partial H_r}{\partial\theta}\right) \right]$$

$$= \left[\frac{1}{r\sin\theta}\frac{\partial(H_\varphi\sin\theta)}{\partial\theta}, \quad -\frac{1}{r}\frac{\partial(rH_\varphi)}{\partial r}, 0 \right] \tag{4.20}$$

From (4.3) and (4.20), update equations for E_r at a location $(i+1/2, j)$ and E_θ at a location $(i, j+1/2)$ (see Figure 4.4) are given as follows:

$$E_r^n\left(i+\frac{1}{2},j\right) = \frac{1 - \dfrac{\sigma(i+1/2,j)\Delta t}{2\varepsilon(i+1/2,j)}}{1 + \dfrac{\sigma(i+1/2,j)\Delta t}{2\varepsilon(i+1/2,j)}}\; E_r^{n-1}\left(i+\frac{1}{2},j\right)$$

$$+ \frac{\dfrac{\Delta t}{\varepsilon(i+1/2,j)}}{1 + \dfrac{\sigma(i+1/2,j)\Delta t}{2\varepsilon(i+1/2,j)}} \left[\frac{1}{r_{i+\frac{1}{2}}\sin\theta_j} \frac{\partial H_\varphi^{n-\frac{1}{2}}\left(i+\frac{1}{2},j\right)\sin\theta_j}{\partial\theta} \right]$$

$$= \frac{1 - \dfrac{\sigma(i+1/2,j)\Delta t}{2\varepsilon(i+1/2,j)}}{1 + \dfrac{\sigma(i+1/2,j)\Delta t}{2\varepsilon(i+1/2,j)}} E_r^{\,n-1}\left(i+\frac{1}{2},j\right) + \frac{\dfrac{\Delta t}{\varepsilon(i+1/2,j)}}{1 + \dfrac{\sigma(i+1/2,j)\Delta t}{2\varepsilon(i+1/2,j)}}$$

$$\times \frac{1}{r_{i+\frac{1}{2}}} \frac{1}{\sin\theta_j \, \Delta\theta} \left[\begin{array}{l} H_\varphi^{\,n-\frac{1}{2}}\left(i+\dfrac{1}{2},j+\dfrac{1}{2}\right)\sin\left(\theta_j + \dfrac{\Delta\theta}{2}\right) \\[2mm] -H_\varphi^{\,n-\frac{1}{2}}\left(i+\dfrac{1}{2},j-\dfrac{1}{2}\right)\sin\left(\theta_j - \dfrac{\Delta\theta}{2}\right) \end{array} \right]$$

$$(4.21)$$

$$E_\theta^{\,n}\left(i,j+\frac{1}{2}\right) = \frac{1 - \dfrac{\sigma(i,j+1/2)\Delta t}{2\varepsilon(i,j+1/2)}}{1 + \dfrac{\sigma(i,j+1/2)\Delta t}{2\varepsilon(i,j+1/2)}} E_\theta^{\,n-1}\left(i,j+\frac{1}{2}\right) - \frac{\dfrac{\Delta t}{\varepsilon(i,j+1/2)}}{1 + \dfrac{\sigma(i,j+1/2)\Delta t}{2\varepsilon(i,j+1/2)}}$$

$$\times \left[\frac{1}{r_i} \frac{\partial\, r_i H_\varphi^{\,n-\frac{1}{2}}\left(i,j+\dfrac{1}{2}\right)}{\partial r} \right]$$

$$= \frac{1 - \dfrac{\sigma(i,j+1/2)\Delta t}{2\varepsilon(i,j+1/2)}}{1 + \dfrac{\sigma(i,j+1/2)\Delta t}{2\varepsilon(i,j+1/2)}} E_\theta^{\,n-1}\left(i,j+\frac{1}{2}\right) - \frac{\dfrac{\Delta t}{\varepsilon(i,j+1/2)}}{1 + \dfrac{\sigma(i,j+1/2)\Delta t}{2\varepsilon(i,j+1/2)}} \frac{1}{r_i \Delta r}$$

$$\times \left[r_{i+1/2} H_\varphi^{\,n-\frac{1}{2}}\left(i+\frac{1}{2},j+\frac{1}{2}\right) - r_{i-1/2} H_\varphi^{\,n-\frac{1}{2}}\left(i-\frac{1}{2},j+\frac{1}{2}\right) \right]$$

$$(4.22)$$

where r_i is the radial distance from the coordinate origin to the location of $E_\theta(i,j+1/2)$, $r_{i-1/2}$ is the distance from the coordinate origin to the location of $H_\varphi(i-1/2,j+1/2)$, $r_{i+1/2}$ is the distance from the coordinate origin to the location of $H_\varphi(i+1/2,j+1/2)$, Δr is the cell side length in the radial direction, and $\Delta\theta$ is the angle corresponding to cell side length in the polar direction.

The curl of electric-field vector in the 2D spherical coordinate system is given by

$$\nabla \times \boldsymbol{E} = \left[\frac{1}{r\sin\theta}\left(\frac{\partial(E_\varphi \sin\theta)}{\partial\theta} - \frac{\partial E_\theta}{\partial\varphi}\right), \frac{1}{r}\left(\frac{1}{\sin\theta}\frac{\partial E_r}{\partial\varphi} - \frac{\partial(rE_\varphi)}{\partial r}\right), \frac{1}{r}\left(\frac{\partial(rE_\theta)}{\partial r} - \frac{\partial E_r}{\partial\theta}\right) \right]$$

$$= \left[0,\ 0, \frac{1}{r}\left(\frac{\partial(rE_\theta)}{\partial r} - \frac{\partial E_r}{\partial\theta}\right) \right]$$

$$(4.23)$$

From (4.9) and (4.23), the update equation for $H_\varphi^{n+1/2}$ at a location $(i+1/2, j+1/2)$ (see Figure 4.4) is given as follows:

$$H_\varphi^{n+\frac{1}{2}}\left(i+\frac{1}{2}, j+\frac{1}{2}\right) = H_\varphi^{n-\frac{1}{2}}\left(i+\frac{1}{2}, j+\frac{1}{2}\right)$$

$$-\frac{\Delta t}{\mu(i+1/2, j+1/2)\, r_{i+\frac{1}{2}}} \frac{1}{} \left[\frac{\partial r_{i+\frac{1}{2}} E_\theta^n\left(i+\frac{1}{2}, j+\frac{1}{2}\right)}{\partial r} - \frac{\partial E_r^n\left(i+\frac{1}{2}, j+\frac{1}{2}\right)}{\partial \theta} \right]$$

$$= H_\varphi^{n-\frac{1}{2}}\left(i+\frac{1}{2}, j+\frac{1}{2}\right)$$

$$-\frac{\Delta t}{\mu(i+1/2, j+1/2)\, r_{i+\frac{1}{2}}} \frac{1}{} \left[\frac{r_{i+1} E_\theta^n\left(i+1, j+\frac{1}{2}\right) - r_i E_\theta^n\left(i, j+\frac{1}{2}\right)}{\Delta r} \right.$$
$$\left. - \frac{E_r^n\left(i+\frac{1}{2}, j+1\right) - E_r^n\left(i+\frac{1}{2}, j\right)}{\Delta \theta} \right]$$

$$(4.24)$$

By updating E_r^n, E_θ^n, and $H_\varphi^{n+1/2}$ at every point in the working space, the distributions of electric and magnetic fields throughout the working space are obtained as a function of time.

The time increment Δt needs to fulfill the Courant stability condition given as follows:

$$\Delta t \leq \frac{1}{c\sqrt{\frac{1}{(\Delta r)^2} + \frac{1}{(r_{min}\Delta\theta)^2}}} \qquad (4.25)$$

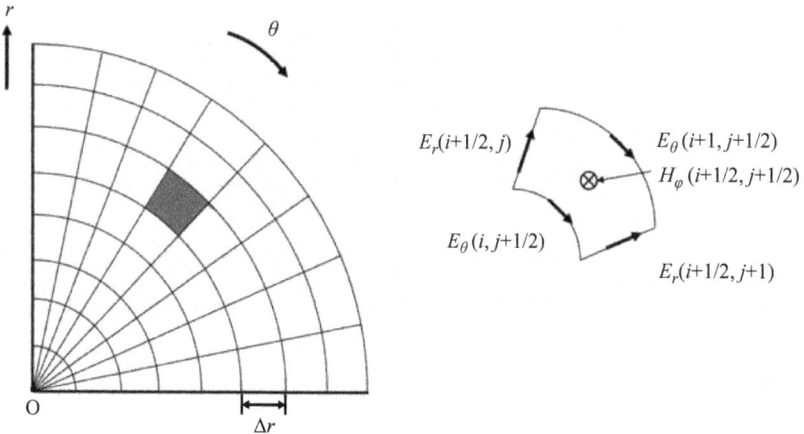

Figure 4.4   Placement of radial E_r and polar E_θ components of electric field and azimuthal component of magnetic field H_φ in a cell in the 2D spherical coordinate system

where r_{min} is the smallest radial distance from the coordinate origin to the working space.

4.3   Subgridding technique

It is computationally efficient to employ a local grid (LG) for representing a relatively small (thin) structure (e.g., power-line conductor) or small element that exists locally in the relatively large working volume. This technique is called subgridding technique. It was employed in lightning electromagnetic field and surge simulations by, for example, Thang *et al.* (2012) and Thang *et al.* (2015).

Figure 4.5 shows a portion of FDTD grid near the boundary between the main grid (MG) and a LG, in which the cell size ratio of the MG to LG is 3 to 1. In Figure 4.5, E and H indicate MG electric and magnetic fields, and e and h indicate LG electric and magnetic fields, respectively. The ratio is required to be an odd integer such as 3 to 1, 5 to 1, and so on, which provides collocated fields in time and space with the MG (Chevalier *et al.* 1997). When the cell size ratio is an odd integer, every MG field value at the MG–LG boundary or within the LG region has a corresponding LG field that is spatially collocated with it.

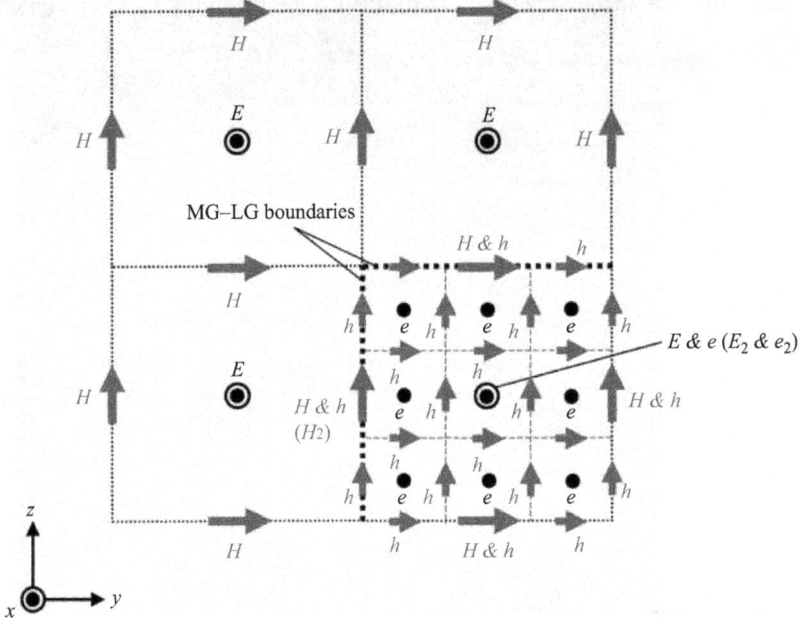

Figure 4.5   Illustration of main and local grids. E and H indicate main-grid electric and magnetic fields, respectively, and e and h indicate local-grid electric and magnetic fields, respectively. Adapted from Chevalier et al. (1997, Figure 1)

When MG fields on or near the MG–LG boundary need to be determined, the required MG fields within the LG region can be obtained from the collocated LG fields without any interpolation. Since the magnetic fields, either normal or tangential, are continuous over the MG–LG boundary when no magnetic material crosses the boundary, it is advantageous to use the tangential magnetic fields on the interface. All fields located on the MG mesh, which include the fields tangential to the MG–LG boundary, are computed using the usual FDTD update equations. When the update equations are used for computing MG-magnetic fields on the MG–LG boundary, electric and magnetic fields at MG locations but inside the LG region are needed. The electric field inside the LG, which are located near the MG–LG boundary and collocated with MG fields, are obtained by a weighted average of collocated MG and LG equation results. The magnetic fields tangential to the MG–LG boundary located on either this boundary or one LG cell inside the LG are determined using combinations of interpolation and weighted sums. All other fields located on the LG mesh, which include those collocated with MG fields, are computed using the usual update equations.

Collocated magnetic fields on the MG–LG boundary are computed as follows. For example, H_1 on the MG–LG boundary at LG time steps $t = (n - 1/2 + 1/3)\Delta t$, $(n - 1/2 + 2/3)\Delta t$, and $(n - 1/2 + 1)\Delta t$ is evaluated by assuming a quadratic function in time from $H_1^{n-3/2}$, $H_1^{n-1/2}$, and $H_1^{n+1/2}$ (Chevalier *et al.* 1997).

$$H_1^{n-\frac{1}{2}+m} = H_1^{n-\frac{1}{2}} + A\,m + \frac{B\,m^2}{2}$$

with   (4.26)

$$m = \frac{1}{3}, \frac{2}{3}, 1, \quad A = \frac{H_1^{n+\frac{1}{2}} - H_1^{n-\frac{3}{2}}}{2}, \quad B = H_1^{n+\frac{1}{2}} + H_1^{n-\frac{3}{2}} - 2H_1^{n-\frac{1}{2}}$$

All other collocated tangential magnetic fields on the MG–LG interface are computed in the same manner.

Magnetic fields, which are located on the MG–LG boundary but not collocated with the MG, are evaluated using a spatial interpolation. For example, z-directed h_{pq} located on the z–x plane of the MG–LG boundary, which is shown in Figure 4.6, is evaluated as follows:

$$h_{pq} = (1 - p)(1 - q)H_1 + p(1 - q)H_2 + q(1 - p)H_3 + pqH_4$$

with   (4.27)

$$p = 0, \frac{1}{3}, \frac{2}{3}, 1, \qquad q = 0, \frac{1}{3}, \frac{2}{3}, 1$$

where H_1, H_2, H_3, and H_4 are z-directed MG magnetic fields that surround h_{pq}.

When a perfect conductor crosses the MG–LG interface, (4.27) needs to be modified. For example, when a perfectly conducting plate parallel to the x–y plane and a perfectly conducting plate parallel to the z–x plane cross the MG–LG boundary parallel to the y–z plane as shown in Figure 4.7, nearby z-directed magnetic fields are

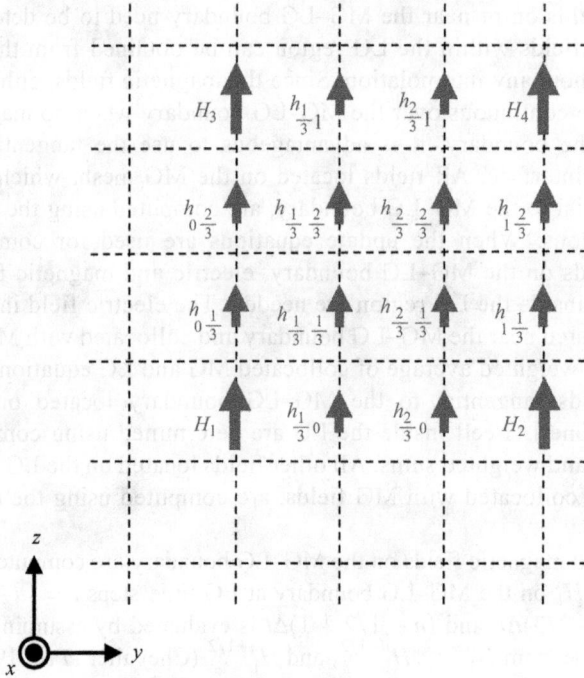

Figure 4.6   Illustration of z-directed magnetic fields on the y–z surface of the MG–LG boundary. Adapted from Chevalier et al. (1997, Figure 3)

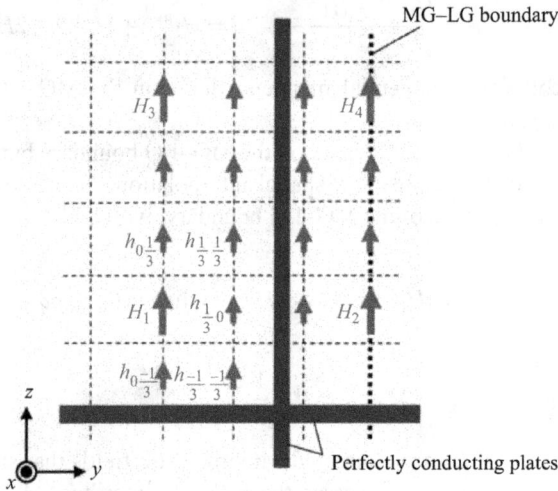

Figure 4.7   Illustration of z-directed magnetic field in the presence of a perfectly conducting plate parallel to the x–y plane and a perfectly conducting plate parallel to the z–x plane crossing the MG–LG boundary parallel to the y–z plane. Adapted from Chevalier et al. (1997, Figure 4)

approximated as follows (Chevalier *et al.* 1997):

$$h_{0\,\frac{1}{3}} = \frac{2}{3}H_1 + \frac{1}{3}H_3, \quad h_{\frac{1}{3}\,0} = H_1, \quad h_{\frac{1}{3}\,\frac{1}{3}} = h_{0\,\frac{1}{3}}, \quad h_{0\,\frac{-1}{3}} = \frac{1}{3}H_1, \quad h_{\frac{-1}{3}\,\frac{-1}{3}} = h_{0\,\frac{-1}{3}},$$
(4.28)

where H_1, H_2, H_3, and H_4 are z-directed MG magnetic fields nearby.

When the subgridding described above is employed, numerical instability sometimes occurs. In order to suppress the numerical instability, the following measures have been proposed (Chevalier *et al.* 1997). Figure 4.8 shows a region near an MG–LG boundary with collocated magnetic field H_1 on the MG–LG boundary and parallel LG magnetic fields h_2 and h_3. The value of h_2 is averaged as follows:

$$h_2' = 0.95h_2 + 0.05\frac{H_1 + h_3}{2}$$
(4.29)

where h_2' is the modified magnetic field. This modification needs to be applied to all magnetic fields located one cell inside the MG–LG boundary. Electric fields, which are collocated within the LG and closest to the MG–LG boundary, also need to be modified. For example, E_2 and e_2 in Figure 4.5, are modified as follows:

$$\begin{aligned} E_2' &= 0.8E_2 + 0.2\,e_2, \\ e_2' &= 0.2E_2 + 0.8\,e_2, \end{aligned}$$
(4.30)

where E_2' and e_2' are modified electric fields.

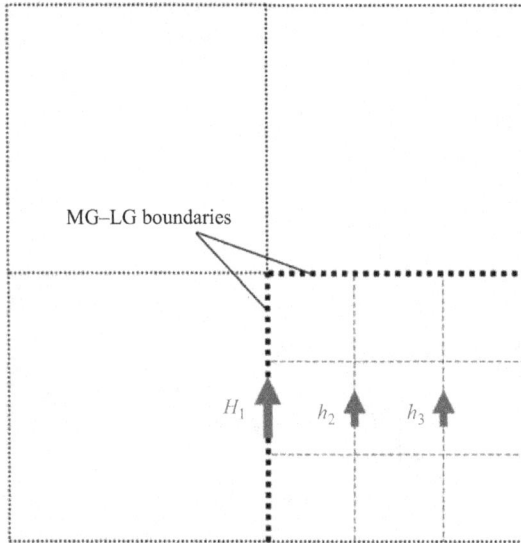

Figure 4.8   *A region near an MG–LG boundary with collocated magnetic field H_1 on the MG–LG boundary and parallel LG magnetic fields h_2 and h_3. Adapted from Chevalier et al. (1997, Figure 5)*

4.4   Representation of lumped sources and lumped circuit elements

4.4.1   Lumped voltage source

A lumped voltage source V_s^n, along the z-axis the z-direction, at a location $(i, j, k + 1/2)$ is represented by specifying vertical electric field E_z^n at the source location as follows:

$$E_z^n\left(i, j, k + \frac{1}{2}\right) = -\frac{V_z^n\left(i, j, k + \frac{1}{2}\right)}{\Delta z} \tag{4.31}$$

A lumped voltage source V_s^n along the x- or y-axis is represented similarly. Also, a lumped voltage source at point $(0, j+1/2)$ along the z-axis in the 2D cylindrical coordinate system is represented in the same manner and the update equation is given below:

$$E_z^n\left(i, j + \frac{1}{2}\right) = -\frac{V_s^n\left(i, j + \frac{1}{2}\right)}{\Delta z} \tag{4.32}$$

Note that a lumped voltage source at point $(i+1/2, 0)$ along the r-axis (zenith direction) in the 2D spherical coordinate system is represented essentially by the same expression as (4.32).

4.4.2   Lumped current source

A lumped current source $I_s^{n-1/2}$, along the z-axis, at location $(i, j, k+1/2)$ is represented by specifying the z-component $J_z^{n-1/2}$ of conduction-current density $\boldsymbol{J}^{n-1/2}$ in (4.1) at the source location as follows:

$$J_z^{n-\frac{1}{2}}\left(i, j, k + \frac{1}{2}\right) = \frac{1}{\Delta x \Delta y} I_s^{n-\frac{1}{2}}\left(i, j, k + \frac{1}{2}\right) \tag{4.33}$$

Therefore, the update equation for E_z at $(i, j, k+1/2)$ is given by

$$E_z^n\left(i, j, k + \frac{1}{2}\right) = \frac{1 - \dfrac{\sigma(i, j, k+1/2)\Delta t}{2\varepsilon(i, j, k+1/2)}}{1 + \dfrac{\sigma(i, j, k+1/2)\Delta t}{2\varepsilon(i, j, k+1/2)}} E_z^{n-1}\left(i, j, k + \frac{1}{2}\right) + \frac{\dfrac{\Delta t}{\varepsilon(i, j, k+1/2)}}{1 + \dfrac{\sigma(i, j, k+1/2)\Delta t}{2\varepsilon(i, j, k+1/2)}}$$

$$\times \frac{1}{\Delta x \Delta y} \begin{bmatrix} H_y^{n-\frac{1}{2}}\left(i+\frac{1}{2}, j, k+\frac{1}{2}\right)\Delta y - H_y^{n-\frac{1}{2}}\left(i-\frac{1}{2}, j, k+\frac{1}{2}\right)\Delta y \\ -H_x^{n-\frac{1}{2}}\left(i, j+\frac{1}{2}, k+\frac{1}{2}\right)\Delta x + H_x^{n-\frac{1}{2}}\left(i, j-\frac{1}{2}, k+\frac{1}{2}\right)\Delta x \end{bmatrix}$$

$$- \frac{\dfrac{\Delta t}{\varepsilon(i, j, k+1/2)}}{1 + \dfrac{\sigma(i, j, k+1/2)\Delta t}{2\varepsilon(i, j, k+1/2)}} \frac{1}{\Delta x \Delta y} I_s^{n-\frac{1}{2}}\left(i, j, k + \frac{1}{2}\right) \tag{4.34}$$

Note that a lumped current source $I_s^{n-1/2}$, along the z-axis, at location $(i, j, k+1/2)$ for the case of $\Delta x = \Delta y$ can be represented in a simpler way by specifying the circulating set of four magnetic fields closest to the source as follows (Baba and Rakov 2003):

$$H_x^{n-\frac{1}{2}}\left(i,j+\frac{1}{2},k+\frac{1}{2}\right) = -\frac{1}{4\Delta x}I_s^{n-\frac{1}{2}}\left(i,j,k+\frac{1}{2}\right)$$

$$H_x^{n-\frac{1}{2}}\left(i,j-\frac{1}{2},k+\frac{1}{2}\right) = \frac{1}{4\Delta x}I_s^{n-\frac{1}{2}}\left(i,j,k+\frac{1}{2}\right)$$

$$H_y^{n-\frac{1}{2}}\left(i+\frac{1}{2},j,k+\frac{1}{2}\right) = \frac{1}{4\Delta y}I_s^{n-\frac{1}{2}}\left(i,j,k+\frac{1}{2}\right)$$

$$H_y^{n-\frac{1}{2}}\left(i-\frac{1}{2},j,k+\frac{1}{2}\right) = -\frac{1}{4\Delta y}I_s^{n-\frac{1}{2}}\left(i,j,k+\frac{1}{2}\right)$$

(4.35)

Representation of a lumped current source along the x- or y-axis is similar to (4.34) or (4.35).

A lumped current source $I_s^{n-1/2}$ at point $(0, j+1/2)$ along the z-axis in the 2D cylindrical coordinate system is represented in the same manner as in the 3D Cartesian coordinate system and the update equation is given below:

$$E_z^n\left(0,j+\frac{1}{2}\right) = \frac{1 - \frac{\sigma(0,j+1/2)\Delta t}{2\varepsilon(0,j+1/2)}}{1 + \frac{\sigma(0,j+1/2)\Delta t}{2\varepsilon(0,j+1/2)}} E_z^{n-1}\left(0,j+\frac{1}{2}\right)$$

$$+ \frac{\frac{\varepsilon(0,j+1/2)}{\sigma(0,j+1/2)\Delta t}}{1 + \frac{\sigma(0,j+1/2)\Delta t}{2\varepsilon(0,j+1/2)}} \frac{4}{\Delta r} H_\varphi^{n-\frac{1}{2}}\left(\frac{1}{2},j+\frac{1}{2}\right)$$

(4.36)

$$- \frac{\frac{\varepsilon(0,j+1/2)}{\sigma(0,j+1/2)\Delta t}}{1 + \frac{\sigma(0,j+1/2)\Delta t}{2\varepsilon(0,j+1/2)}} \frac{1}{\pi\left(\frac{\Delta r}{2}\right)^2} I_s^{n-\frac{1}{2}}\left(0,j+\frac{1}{2}\right)$$

Note that a simpler representation in the 2D cylindrical coordinate system can be given as follows:

$$H_\varphi^{n-\frac{1}{2}}\left(\frac{1}{2},j+\frac{1}{2}\right) = \frac{1}{2\pi\left(\frac{\Delta r}{2}\right)} I_s^{n-\frac{1}{2}}\left(0,j+\frac{1}{2}\right)$$

(4.37)

Also note that a lumped current source $I_s^{n-1/2}$ at point $(i+1/2, 0)$ along the r-axis (zenith direction) in the 2D spherical coordinate system is represented in the same manner as in the 2D cylindrical coordinate system and the update equation is given as follows:

$$H_\varphi^{n-\frac{1}{2}}\left(i+\frac{1}{2},0\right) = \frac{1}{2\pi\left(\frac{r_{i+1/2}\Delta\theta}{2}\right)} I_s^{n-\frac{1}{2}}\left(i+\frac{1}{2},0\right)$$

(4.38)

4.4.3   Lumped resistance

A lumped resistance R, along the z-axis, at location $(i, j, k+1/2)$ in a lossless medium $(\sigma = 0)$ is represented by specifying the z-component $J_z^{n-1/2}$ of conduction-current density $J^{n-1/2}$ in (4.1) at the lumped-resistance location as follows:

$$J_z^{n-\frac{1}{2}}\left(i,j,k+\frac{1}{2}\right) = \frac{1}{\Delta x \Delta y}I_z^{n-\frac{1}{2}}\left(i,j,k+\frac{1}{2}\right) = \frac{1}{\Delta x \Delta y}\frac{E_z^{n-\frac{1}{2}}\left(i,j,k+\frac{1}{2}\right)\Delta z}{R}$$

$$\approx \frac{\Delta z}{\Delta x \Delta y}\frac{1}{R}\frac{E_z^n\left(i,j,k+\frac{1}{2}\right) + E_z^{n-1}\left(i,j,k+\frac{1}{2}\right)}{2} \tag{4.39}$$

Therefore, the update equation for E_z at $(i, j, k+1/2)$ is given by

$$E_z^n\left(i,j,k+\frac{1}{2}\right) = \frac{1 - \dfrac{\Delta t \Delta z}{2R\varepsilon(i,j,k+1/2)\Delta x \Delta y}}{1 + \dfrac{\Delta t \Delta z}{2R\varepsilon(i,j,k+1/2)\Delta x \Delta y}}E_z^{n-1}\left(i,j,k+\frac{1}{2}\right)$$

$$+ \frac{\dfrac{\Delta t}{\varepsilon(i,j,k+1/2)}}{1 + \dfrac{\Delta t \Delta z}{2R\varepsilon(i,j,k+1/2)\Delta x \Delta y}}\frac{1}{\Delta x \Delta y}$$

$$\times \left[\begin{array}{l} H_y^{n-\frac{1}{2}}\left(i+\frac{1}{2},j,k+\frac{1}{2}\right)\Delta y - H_y^{n-\frac{1}{2}}\left(i-\frac{1}{2},j,k+\frac{1}{2}\right)\Delta y \\ -H_x^{n-\frac{1}{2}}\left(i,j+\frac{1}{2},k+\frac{1}{2}\right)\Delta x + H_x^{n-\frac{1}{2}}\left(i,j-\frac{1}{2},k+\frac{1}{2}\right)\Delta x \end{array} \right]$$

$$\tag{4.40}$$

A lumped resistance along the x- or y-axis is represented in the same manner.

A lumped resistance R at point $(0, j+1/2)$ along the z-axis in the 2D cylindrical coordinate system is represented similarly and the update equation is given below:

$$E_z^n\left(0,j+\frac{1}{2}\right) = \frac{1 - \dfrac{\Delta t \Delta z}{2R\varepsilon(0,j+1/2)\pi\left(\dfrac{\Delta r}{2}\right)^2}}{1 + \dfrac{\Delta t \Delta z}{2R\varepsilon(0,j+1/2)\pi\left(\dfrac{\Delta r}{2}\right)^2}}E_z^{n-1}\left(0,j+\frac{1}{2}\right)$$

$$+ \frac{\dfrac{\Delta t}{\varepsilon(0,j+1/2)}}{1 + \dfrac{\Delta t \Delta z}{2R\varepsilon(0,j+1/2)\pi\left(\dfrac{\Delta r}{2}\right)^2}}\frac{1}{\Delta r}\frac{2}{}H_\varphi^{n-\frac{1}{2}}\left(\frac{1}{2},j+\frac{1}{2}\right)$$

$$\tag{4.41}$$

A lumped resistance R at point $(i+1/2, 0)$ along the r-axis (zenith direction) in the 2D spherical coordinate system is represented similarly and the update equation is given below:

$$E_r{}^n\left(i+\frac{1}{2},0\right) = \frac{1 - \dfrac{\Delta t \Delta r}{2R\varepsilon(i+1/2,0)\pi\left(r_{i+1/2}\Delta\theta/2\right)^2}}{1 + \dfrac{\Delta t \Delta r}{2R\varepsilon(i+1/2,0)}\pi\left(r_{i+1/2}\Delta\theta/2\right)^2} E_r{}^{n-1}\left(i+\frac{1}{2},0\right)$$

$$+ \frac{\dfrac{\Delta t}{\varepsilon(i+1/2,0)}}{1 + \dfrac{\Delta t \Delta r}{2R\varepsilon(i+1/2,0)\pi\left(r_{i+1/2}\Delta\theta/2\right)^2}} \frac{1}{r_{i+1/2}\Delta\theta} H_\varphi{}^{n-\frac{1}{2}}\left(i+\frac{1}{2},\frac{1}{2}\right)$$

$$(4.42)$$

4.4.4 Lumped inductance

A lumped inductance L, along the z-axis, at location $(i,j,k+1/2)$ in a lossless medium $(\sigma = 0)$ is represented by specifying the z-component $J_z^{n-1/2}$ of conduction-current density $\boldsymbol{J}^{n-1/2}$ in (4.1) at the lumped-inductance location as follows:

$$J_z^{n-\frac{1}{2}}\left(i,j,k+\frac{1}{2}\right) = \frac{1}{\Delta x \Delta y} I_z^{n-\frac{1}{2}}\left(i,j,k+\frac{1}{2}\right)$$

$$= \frac{1}{\Delta x \Delta y}\frac{1}{L}\int_0^{(n-\frac{1}{2})\Delta t} E_z\left(i,j,k+\frac{1}{2}\right)\Delta z \; dt \qquad (4.43)$$

$$\approx \frac{1}{\Delta x \Delta y}\frac{\Delta z \Delta t}{L}\sum_{m=1}^{n-1} E_z{}^m\left(i,j,k+\frac{1}{2}\right)$$

Therefore, the update equation for E_z at $(i, j, k+1/2)$ is given by

$$E_z{}^n\left(i,j,k+\frac{1}{2}\right) = E_z{}^{n-1}\left(i,j,k+\frac{1}{2}\right) + \frac{\Delta t}{\varepsilon(i,j,k+1/2)}\frac{1}{\Delta x \Delta y}$$

$$\times \left[\begin{array}{l} H_y{}^{n-\frac{1}{2}}\left(i+\frac{1}{2},j,k+\frac{1}{2}\right)\Delta y - H_y{}^{n-\frac{1}{2}}\left(i-\frac{1}{2},j,k+\frac{1}{2}\right)\Delta y \\ -H_x{}^{n-\frac{1}{2}}\left(i,j+\frac{1}{2},k+\frac{1}{2}\right)\Delta x + H_x{}^{n-\frac{1}{2}}\left(i,j-\frac{1}{2},k+\frac{1}{2}\right)\Delta x \end{array} \right]$$

$$- \frac{\Delta z(\Delta t)^2}{L\varepsilon(i,j,k+1/2)}\frac{1}{\Delta x \Delta y}\sum_{m=1}^{n-1} E_z{}^m\left(i,j,k+\frac{1}{2}\right)$$

$$(4.44)$$

A lumped inductance along the x- or y-axis is represented in the same manner.

A lumped inductance L at point $(0, j+1/2)$ along the z-axis in a lossless medium $(\sigma = 0)$ in the 2D cylindrical coordinate system is represented similarly and the update equation is given by:

$$E_z{}^n\left(0,j+\frac{1}{2}\right) = E_z{}^{n-1}\left(0,j+\frac{1}{2}\right) + \frac{\Delta t}{\varepsilon(0,j+1/2)\,\Delta r} \frac{1}{2} \, H_\varphi{}^{n-\frac{1}{2}}\left(\frac{1}{2},j+\frac{1}{2}\right)$$

$$- \frac{\Delta z(\Delta t)^2}{L\varepsilon(i,j,k+1/2)} \frac{1}{\pi\left(\frac{\Delta r}{2}\right)^2} \sum_{m=1}^{n-1} E_z{}^m\left(0,j+\frac{1}{2}\right)$$

$$(4.45)$$

A lumped inductance L at point $(i+1/2, 0)$ along the r-axis (zenith direction) in a lossless medium ($\sigma = 0$) in the 2D spherical coordinate system is represented similarly and the update equation is given below:

$$E_r{}^n\left(i+\frac{1}{2},0\right) = E_r{}^{n-1}\left(i+\frac{1}{2},0\right) + \frac{\Delta t}{\varepsilon(i+1/2,0)\,r_{i+1/2}\Delta\theta} \frac{1}{2} \, H_\varphi{}^{n-\frac{1}{2}}\left(i+\frac{1}{2},\frac{1}{2}\right)$$

$$+ \frac{\Delta r(\Delta t)^2}{L\,\varepsilon(i+1/2,0)\pi\left(\frac{r_{i+1/2}\Delta\theta}{2}\right)^2} \sum_{m=1}^{n-1} E_r{}^m\left(i+\frac{1}{2},0\right)$$

$$(4.46)$$

4.4.5 Lumped capacitance

A lumped capacitance C, along the z-axis, at location $(i, j, k+1/2)$ in a lossless medium ($\sigma = 0$) is represented by specifying the z-component $J_z^{n-1/2}$ of conduction-current density $J^{n-1/2}$ in (4.1) at the lumped-capacitance location as follows:

$$J_z{}^{n-\frac{1}{2}}\left(i,j,k+\frac{1}{2}\right) = \frac{1}{\Delta x\Delta y}I_z{}^{n-\frac{1}{2}}\left(i,j,k+\frac{1}{2}\right) = \frac{1}{\Delta x\Delta y}C\frac{dE_z{}^{n-\frac{1}{2}}\left(i,j,k+\frac{1}{2}\right)\Delta z}{dt}$$

$$\approx \frac{1}{\Delta x\Delta y}\frac{C\Delta z}{\Delta t}\left[E_z{}^n\left(i,j,k+\frac{1}{2}\right) - E_z{}^{n-1}\left(i,j,k+\frac{1}{2}\right)\right]$$

$$(4.47)$$

Therefore, the update equation for E_z at $(i, j, k+1/2)$ is given by

$$E_z{}^n\left(i,j,k+\frac{1}{2}\right) = E_z{}^{n-1}\left(i,j,k+\frac{1}{2}\right) + \frac{\dfrac{\Delta t}{\varepsilon(i,j,k+1/2)}}{1+\dfrac{C\Delta z}{\varepsilon(i,j,k+1/2)\Delta x\Delta y}}\frac{1}{\Delta x\Delta y}$$

$$\times \begin{bmatrix} H_y{}^{n-\frac{1}{2}}\left(i+\frac{1}{2},j,k+\frac{1}{2}\right)\Delta y - H_y{}^{n-\frac{1}{2}}\left(i-\frac{1}{2},j,k+\frac{1}{2}\right)\Delta y \\ -H_x{}^{n-\frac{1}{2}}\left(i,j+\frac{1}{2},k+\frac{1}{2}\right)\Delta x + H_x{}^{n-\frac{1}{2}}\left(i,j-\frac{1}{2},k+\frac{1}{2}\right)\Delta x \end{bmatrix}$$

$$(4.48)$$

A lumped capacitance along the x- or y-axis is represented in the same manner.

A lumped capacitance C at point $(0, j+1/2)$ along the z-axis in a lossless medium $(\sigma = 0)$ in the 2D cylindrical coordinate system is represented similarly and the update equation is given below:

$$E_z{}^n\left(0, j + \frac{1}{2}\right) = E_z{}^{n-1}\left(0, j + \frac{1}{2}\right) + \frac{\frac{\Delta t}{\varepsilon(0, j+1/2)}}{1 + \frac{C\Delta z}{\varepsilon(0, j+1/2)\pi\left(\frac{\Delta r}{2}\right)^2}} \frac{1}{\frac{\Delta r}{2}} H_\varphi{}^{n-\frac{1}{2}}\left(\frac{1}{2}, j + \frac{1}{2}\right)$$

(4.49)

A lumped capacitance C at point $(i+1/2, 0)$ along the r-axis (zenith direction) in a lossless medium $(\sigma = 0)$ in the 2D spherical coordinate system is represented similarly and the update equation is given below:

$$E_r{}^n\left(i + \frac{1}{2}, 0\right) = E_r{}^{n-1}\left(i + \frac{1}{2}, 0\right)$$

$$+ \frac{\frac{\Delta t}{\varepsilon(i+1/2, 0)}}{\varepsilon(i+1/2, 0)\pi\left(r_{i+1/2}\Delta\theta/2\right)^2} \frac{1}{r_{i+1/2}\Delta\theta} H_\varphi{}^{n-\frac{1}{2}}\left(i + \frac{1}{2}, \frac{1}{2}\right)$$

(4.50)

4.4.6 Lumped series resistance and inductance

Since a lumped series-connected resistance and inductance (*RL* element) is frequently used, for example, for representing a lightning return-stroke channel, its representation is described here. A series *RL* element along the z-axis in a cell at location $(i, j, k + 1/2)$ in a lossless medium $(\sigma = 0)$ is also represented by modifying the update equation for $E_z^n(i, j, k + 1/2)$, which is explained below.

The following relation is fulfilled along a z-directed cell side, along which a series *RL* element is located, on the basis of Kirchhoff's voltage law:

$$L\frac{d\, I_z{}^{n-\frac{1}{2}}(i, j, k + \frac{1}{2})}{d\, t} + R\, I_z{}^{n-\frac{1}{2}}\left(i, j, k + \frac{1}{2}\right) = E_z{}^{n-\frac{1}{2}}\left(i, j, k + \frac{1}{2}\right)\Delta z$$

(4.51)

It is approximated by the following finite-difference expression:

$$L\frac{J_z{}^n\left(i, j, k + \frac{1}{2}\right) - J_z{}^{n-1}\left(i, j, k + \frac{1}{2}\right)}{\Delta t}\Delta x\, \Delta y$$

$$+ R\frac{J_z{}^n\left(i, j, k + \frac{1}{2}\right) + J_z{}^{n-1}\left(i, j, k + \frac{1}{2}\right)}{2}\Delta x\, \Delta y$$

$$= \frac{E_z{}^n\left(i, j, k + \frac{1}{2}\right) + E_z{}^{n-1}\left(i, j, k + \frac{1}{2}\right)}{2}\Delta z$$

(4.52)

If (4.52) is rearranged, the z-component $J_z^{n-1/2}$ of conduction-current density $J^{n-1/2}$ in (4.1) at the location of lumped series RL element is given as follows:

$$
J_z^n\left(i,j,k+\frac{1}{2}\right) = \frac{\left(\dfrac{L}{\Delta t}-\dfrac{R}{2}\right)}{\left(\dfrac{L}{\Delta t}+\dfrac{R}{2}\right)}J_z^{n-\frac{1}{2}}\left(i,j,k+\frac{1}{2}\right)
$$

$$
+\frac{\dfrac{\Delta z}{2}}{\left(\dfrac{L}{\Delta t}+\dfrac{R}{2}\right)\Delta x\,\Delta y}\left[E_z^n\left(i,j,k+\frac{1}{2}\right)+E_z^{n-1}\left(i,j,k+\frac{1}{2}\right)\right]
$$

(4.53)

Therefore, the update equation for E_z at $(i, j, k+1/2)$ is given by

$$
E_z^n\left(i,j,k+\frac{1}{2}\right)=\frac{\left[\dfrac{\varepsilon(i,j,k+1/2)}{\Delta t}-\frac{1}{2}\dfrac{\frac{\Delta z}{2}}{\left(\frac{L}{\Delta t}+\frac{R}{2}\right)\Delta x\Delta y}\right]}{\left[\dfrac{\varepsilon(i,j,k+1/2)}{\Delta t}+\frac{1}{2}\dfrac{\frac{\Delta z}{2}}{\left(\frac{L}{\Delta t}+\frac{R}{2}\right)\Delta x\Delta y}\right]}E_z^{n-1}(i,j,k+1/2)
$$

$$
+\frac{1}{\left[\dfrac{\varepsilon(i,j,k+1/2)}{\Delta t}+\frac{1}{2}\dfrac{\frac{\Delta z}{2}}{\left(\frac{L}{\Delta t}+\frac{R}{2}\right)\Delta x\Delta y}\right]}\frac{1}{\Delta x\,\Delta y}
\begin{bmatrix}
H_y^{n-\frac{1}{2}}\left(i+\frac{1}{2},j,k+\frac{1}{2}\right)\Delta y\\[6pt]
-H_y^{n-\frac{1}{2}}\left(i-\frac{1}{2},j,k+\frac{1}{2}\right)\Delta y\\[6pt]
-H_x^{n-\frac{1}{2}}\left(i,j+\frac{1}{2},k+\frac{1}{2}\right)\Delta x\\[6pt]
+H_x^{n-\frac{1}{2}}\left(i,j-\frac{1}{2},k+\frac{1}{2}\right)\Delta x
\end{bmatrix}
$$

$$
-\frac{\frac{1}{2}\left[\dfrac{\left(\frac{L}{\Delta t}-\frac{R}{2}\right)}{\left(\frac{L}{\Delta t}+\frac{R}{2}\right)}+1\right]}{\left[\dfrac{\varepsilon(i,j,k+1/2)}{\Delta t}+\frac{1}{2}\dfrac{\frac{\Delta z}{2}}{\left(\frac{L}{\Delta t}+\frac{R}{2}\right)\Delta x\Delta y}\right]}J_z^{n-\frac{1}{2}}\left(i,j,k+\frac{1}{2}\right)
$$

(4.54)

A lumped series RL element along the x- or y-axis is represented in the same manner.

A lumped series RL element at point $(0, j+1/2)$ along the z-axis in the 2D cylindrical coordinate system is represented similarly, and the update equation is given below:

$$E_z^n\left(0,j+\frac{1}{2}\right) = \frac{\left[\dfrac{\varepsilon(0,j+1/2)}{\Delta t} - \dfrac{1}{2}\dfrac{\frac{\Delta z}{2}}{\left(\frac{L}{\Delta t}+\frac{R}{2}\right)\pi\left(\frac{\Delta r}{2}\right)^2}\right]}{\left[\dfrac{\varepsilon(0,j+1/2)}{\Delta t} + \dfrac{1}{2}\dfrac{\frac{\Delta z}{2}}{\left(\frac{L}{\Delta t}+\frac{R}{2}\right)\pi\left(\frac{\Delta r}{2}\right)^2}\right]} E_z^{n-1}\left(0,j+\frac{1}{2}\right)$$

$$+ \frac{1}{\left[\dfrac{\varepsilon(0,j+1/2)}{\Delta t} + \dfrac{1}{2}\dfrac{\frac{\Delta z}{2}}{\left(\frac{L}{\Delta t}+\frac{R}{2}\right)\pi\left(\frac{\Delta r}{2}\right)^2}\right]} \frac{1}{\frac{\Delta r}{2}} H_\varphi^{\,n-\frac{1}{2}}\left(\frac{1}{2},j+\frac{1}{2}\right)$$

$$- \frac{\dfrac{1}{2}\left[\dfrac{\left(\frac{L}{\Delta t}-\frac{R}{2}\right)}{\left(\frac{L}{\Delta t}+\frac{R}{2}\right)}+1\right]}{\left[\dfrac{\varepsilon(0,j+1/2)}{\Delta t} + \dfrac{1}{2}\dfrac{\frac{\Delta z}{2}}{\left(\frac{L}{\Delta t}+\frac{R}{2}\right)\pi\left(\frac{\Delta r}{2}\right)^2}\right]} J_z^{\,n-\frac{1}{2}}\left(0,j+\frac{1}{2}\right)$$

$$(4.55)$$

The z-component $J_z^{n-1/2}$ of conduction-current density at the location of lumped series RL element is given as follows:

$$J_z^n\left(0,j+\frac{1}{2}\right) = \frac{\left(\frac{L}{\Delta t}-\frac{R}{2}\right)}{\left(\frac{L}{\Delta t}+\frac{R}{2}\right)} J_z^{\,n-\frac{1}{2}}\left(0,j+\frac{1}{2}\right)$$

$$+ \frac{\frac{\Delta z}{2}}{\left(\frac{L}{\Delta t}+\frac{R}{2}\right)\pi\left(\frac{\Delta r}{2}\right)^2}\left[E_z^n\left(0,j+\frac{1}{2}\right) + E_z^{n-1}\left(0,j+\frac{1}{2}\right)\right]$$

$$(4.56)$$

A lumped series RL element at point $(i+1/2, 0)$ along the r-axis (zenith direction) in the 2D spherical coordinate system is represented similarly, and the update equation is given below:

$$E_z^n\left(i+\frac{1}{2},0\right) = \frac{\left[\dfrac{\varepsilon(i+1/2,0)}{\Delta t} - \dfrac{1}{2}\dfrac{\frac{\Delta z}{2}}{\left(\frac{L}{\Delta t}+\frac{R}{2}\right)\pi\left(\frac{r_{i+1/2}\Delta\theta}{2}\right)^2}\right]}{\left[\dfrac{\varepsilon(i+1/2,0)}{\Delta t} + \dfrac{1}{2}\dfrac{\frac{\Delta z}{2}}{\left(\frac{L}{\Delta t}+\frac{R}{2}\right)\pi\left(\frac{r_{i+1/2}\Delta\theta}{2}\right)^2}\right]}E_z^{n-1}\left(i+\frac{1}{2},0\right)$$

$$+\frac{1}{\left[\dfrac{\varepsilon(i+1/2,0)}{\Delta t} + \dfrac{1}{2}\dfrac{\frac{\Delta z}{2}}{\left(\frac{L}{\Delta t}+\frac{R}{2}\right)\pi\left(\frac{r_{i+1/2}\Delta\theta}{2}\right)^2}\right]}\frac{1}{\dfrac{r_{i+1/2}\Delta\theta}{2}}H_\varphi^{n-\frac{1}{2}}\left(i+\frac{1}{2},\frac{1}{2}\right)$$

$$-\frac{\dfrac{1}{2}\left[\dfrac{\left(\frac{L}{\Delta t}-\frac{R}{2}\right)}{\left(\frac{L}{\Delta t}+\frac{R}{2}\right)}+1\right]}{\left[\dfrac{\varepsilon(i+1/2,0)}{\Delta t} + \dfrac{1}{2}\dfrac{\frac{\Delta z}{2}}{\left(\frac{L}{\Delta t}+\frac{R}{2}\right)\pi\left(\frac{r_{i+1/2}\Delta\theta}{2}\right)^2}\right]}J_z^{n-\frac{1}{2}}\left(i+\frac{1}{2},0\right) \qquad (4.57)$$

The z-component $J_z^{n-1/2}$ of conduction-current density at the location of lumped series RL element is given as follows:

$$J_z^n\left(i+\frac{1}{2},0\right) = \frac{\left(\frac{L}{\Delta t}-\frac{R}{2}\right)}{\left(\frac{L}{\Delta t}+\frac{R}{2}\right)}J_z^{n-\frac{1}{2}}\left(i+\frac{1}{2},0\right)$$

$$+\frac{\frac{\Delta z}{2}}{\left(\frac{L}{\Delta t}+\frac{R}{2}\right)\pi\left(\frac{r_{i+1/2}\Delta\theta}{2}\right)^2}\left[E_z^n\left(i+\frac{1}{2},0\right)+E_z^{n-1}\left(i+\frac{1}{2},0\right)\right]$$

$$(4.58)$$

4.5   Representation of thin wire

Several representations of thin wire for 3D FDTD simulation have been proposed (e.g., Umashankar *et al.* 1987; Noda and Yokoyama 2002; Baba *et al.* 2005; Railton *et al.* 2006; Taniguchi *et al.* 2008b; Asada *et al.* 2015a; Du *et al.* 2017; Tatematsu 2018; Chen *et al.* 2018; Li *et al.* 2018). Here, the thin wire representation proposed by Noda and Yokoyama (2002), which has been most frequently used in surge simulations, is explained.

Noda and Yokoyama (2002) have shown that a straight, perfectly conducting wire in a lossless medium, represented by forcing the tangential components of electric field along the wire axis to be zero, in 3D FDTD simulations has an equivalent radius $a_0 = 0.23\Delta s$, where Δs is the lateral side length of cells employed. Further, they have represented a wire having radius a other than a_0 by embedding the wire of $a_0 = 0.23\Delta s$ in an artificial-medium parallelepiped. In order to represent a wire thinner than the wire having the corresponding equivalent radius, the relative permeability for calculating the circulating set of four magnetic field components closest to the wire needs to be increased and the relative permittivity for calculating the radial electric field components closest to the wire decreased. In a lossy medium, the conductivity also needs to be modified, similarly to the relative permittivity (Baba *et al.* 2005). The modified conductivity σ', modified relative permittivity ε'_r, and modified relative permeability

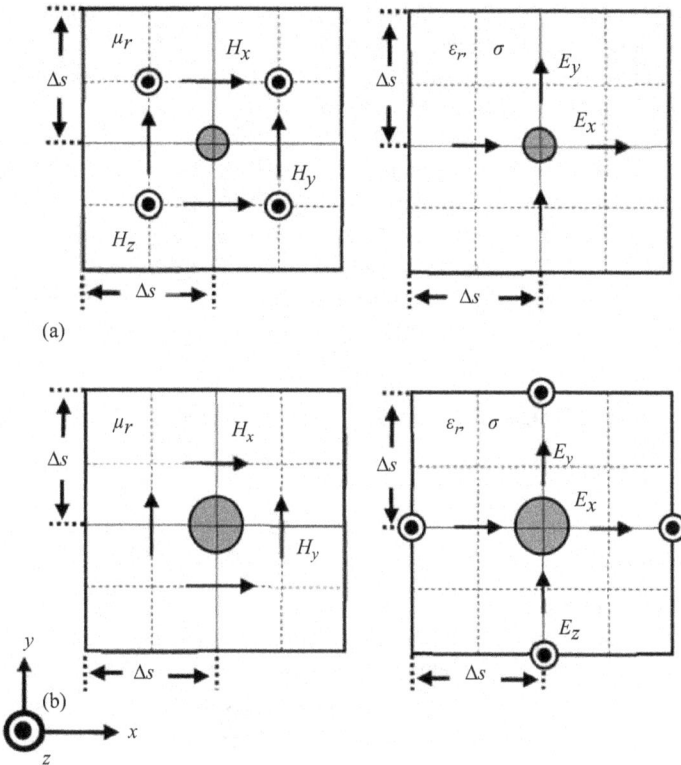

Figure 4.9 (a) Cross-sectional views of a z-directed wire (central shaded circle) having radius a, and the configuration of electric and magnetic field components closest to the wire: (a) $a < a_0 (m < 1)$ and (b) $a > a_0 (m > 1)$. © 2008 IEEE. Reprinted, with permission, from Taniguchi et al. (2008b, Figure 3)

μ_r' are given as follows:

$$\sigma' = m\ \sigma, \qquad \varepsilon_r' = m\ \varepsilon_r, \qquad \mu_r' = \frac{\mu_r}{m}$$

$$m = \frac{\ln\left(\dfrac{\Delta s}{a_0}\right)}{\ln\left(\dfrac{\Delta s}{a}\right)}, \qquad a_0 \approx 0.23\ \Delta s \tag{4.59}$$

where σ, ε_r, and μ_r are the conductivity, relative permittivity, and relative permeability of the original medium, and m is the modification coefficient.

Note that in representing a wire whose radius a is smaller than the equivalent radius a_0 the modified relative permeability μ_r' is also employed in computing axial magnetic field components closest to the wire in addition to the circulating set of four closest magnetic field components, in order to avoid numerical instability (Taniguchi *et al.* 2008b), as shown (for a z-directed wire) in Figure 4.9(a). Also, in representing a wire whose radius a is larger than the equivalent radius a_0, the modified relative permittivity ε_r' is employed in computing axial electric field components closest to the wire, in addition to the closest radial electric field components (Taniguchi *et al.* 2008b), as shown in Figure 4.9(b).

Also note that representations of a lossy thin wire, which can account for the frequency-dependent internal impedance, have been developed recently by Du *et al.* (2017), Tatematsu (2018), Chen *et al.* (2018), and Li *et al.* (2018). Further, representations of a thin coaxial cable have been proposed by Tatematsu (2015, 2018) and Li *et al.* (2018). In these representations, the internal electromagnetic field of the metal sheath conductor of the coaxial cable is described by applying the distributed-circuit theory based on the transverse electromagnetic (TEM) field structure.

4.6   Representation of lightning channel and excitation

4.6.1   *Lightning return-stroke channel*

There are seven types of representation of lightning return-stroke channel used in LEMP and surge computations (Baba and Rakov 2007a, 2008b, 2009, 2014):

1. a perfectly conducting/slightly resistive wire in air above ground;
2. a wire loaded by additional distributed series inductance in air above ground;
3. a wire surrounded by a dielectric medium (other than air) that occupies the entire half space above ground (this fictitious configuration is used only for finding current distribution, which is then applied to a vertical wire in air above ground for calculating electromagnetic fields);
4. a wire coated by a dielectric material in air above ground;
5. a wire coated by a fictitious material having high relative permittivity and high relative permeability in air above ground;
6. two parallel wires having additional distributed shunt capacitance in air (this fictitious configuration is used only for finding current distribution, which is

then applied to a vertical wire in air above ground for calculating electromagnetic fields); and
7. a phased-current-source array in air above ground, each current source being
activated by the arrival of lightning return-stroke front propagating upward at a
specified speed.

These seven channel representations are illustrated in Figure 4.10.

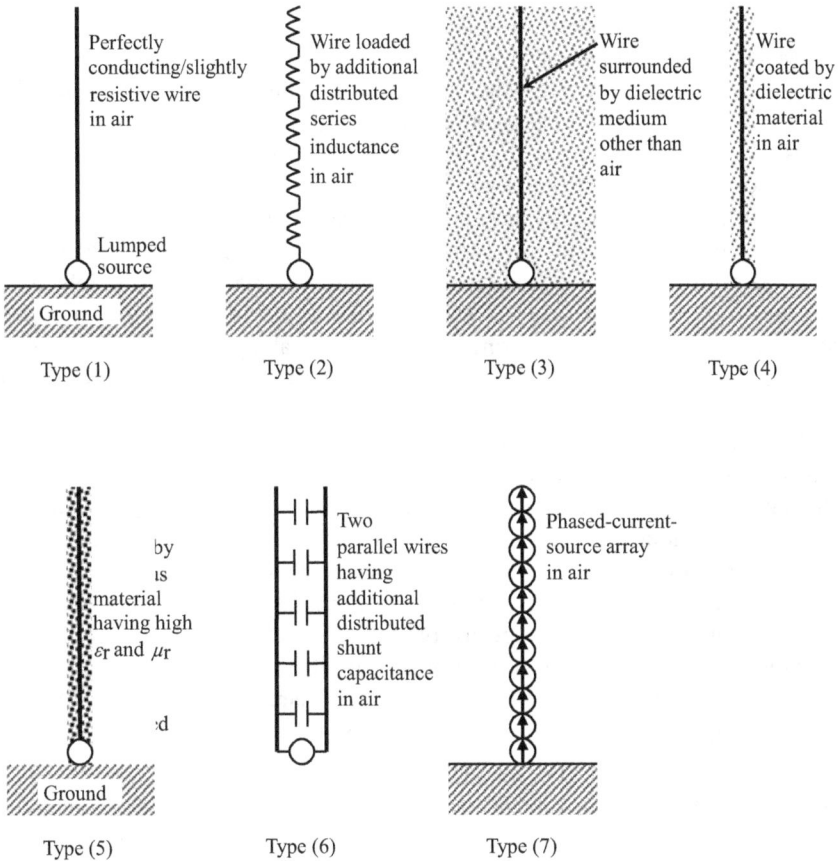

Figure 4.10   Different representations of lightning return-stroke channel. © 2009
IEEE. Reprinted, with permission, from Baba and Rakov (2009,
Figure 12)

The return-stroke speed, along with the current peak, largely determines the
radiation field initial peak (e.g., Rakov and Dulzon 1987). The characteristic
impedance of the lightning channel influences the magnitude of lightning current
and/or the current reflection coefficient at the top of the strike object when a
lumped voltage source is employed. It is therefore desirable that the return-stroke

speed and the characteristic impedance of simulated lightning channel agree with observations that can be summarized as follows:

(i) typical values of return-stroke speed are in the range from $c/3$ to $c/2$, , where c is the speed of light (Rakov 2007), as observed using optical techniques;

(ii) the equivalent impedance of the lightning channel is expected to be in the range from 0.6 to 2.5 kΩ (Gorin and Shkilev 1984).

Type (1) of lightning-channel representation was used, for example, by Baba and Rakov (2003) in their FDTD simulation of electromagnetic fields due to a lightning strike to flat ground. Note that this lightning-channel representation was first used by Podgorski and Landt (1987) in their simulation of lightning current in a tall structure using the MoM in the time domain (Miller *et al.* 1973). The speed of the current wave propagating along a vertical perfectly conducting/slightly resistive wire is nearly equal to the speed of light, which is two to three times higher than typical measured values of return-stroke front speed ($c/3$ to $c/2$). This discrepancy is the main deficiency of this representation. The characteristic impedance of the channel-representing vertical wire varies with height above ground and for a radius of 3 cm is estimated to be around 0.6 kΩ at a height of 500 m. This is right at the lower end of its expected range of variation (0.6 to 2.5 kΩ). Note that a current wave suffers attenuation (distortion) as it propagates along a vertical wire even if that wire has no ohmic losses (Baba and Rakov 2005a). Further attenuation can be achieved by loading the wire by distributed series resistance.

Type (2) was used, for example, by Baba and Rakov (2007a) in their FDTD simulation of current distribution along a vertical lightning channel. Note that this lightning-channel representation was first used by Kato *et al.* (1999) in their simulation of lightning current in a tall structure and its associated electromagnetic fields with the MoM in the time domain. The speed of the current wave propagating along a vertical wire loaded by additional distributed series inductance of 17 and 6.3 µH/m in air is $c/3$ and $c/2$, respectively, if the natural inductance of vertical wire is assumed to be $L_0 = 2.1$ µH/m (as estimated by Rakov (1998) for a 3-cm-radius conductor at a height of 500 m above ground). The characteristic impedance ranges from 1.2 to 1.8 kΩ (0.6 k$\Omega \times [(17 + 2.1)/2.1]^{1/2} = 1.8$ kΩ, and 0.6 k$\Omega \times [(6.3 + 2.1)/2.1]^{1/2} = 1.2$ kΩ) for the speed ranging from $c/3$ to $c/2$. The characteristic impedance of the inductance-loaded wire is within the range of values of the expected equivalent impedance of the lightning channel. Note that additional inductance has no physical meaning and is invoked only to reduce the speed of current wave propagating along the wire to a value lower than the speed of light. The use of this representation allows one to calculate both the distribution of current along the channel-representing wire and remote electromagnetic fields in a single, self-consistent procedure. Bonyadi-Ram *et al.* (2008) have incorporated additional distributed series inductance that increases with increasing height in order to simulate the optically observed reduction in return-stroke speed with increasing height (e.g., Idone and Orville 1982).

Type (3) was used, for example, by Baba and Rakov (2007a) in their FDTD simulation of current along a vertical lightning channel. Note that this lightning-channel representation was first used by Moini *et al.* (2000) in their simulation on lightning electromagnetic fields with the MoM in the time domain. The artificial dielectric medium was used only for finding current distribution along the lightning channel, which was then removed for calculating electromagnetic fields in air. When the relative permittivity is 9 or 4, the speed is $c/3$ or $c/2$, respectively. The corresponding characteristic impedance ranges from 0.2 to 0.3 kΩ (0.6 k$\Omega/\sqrt{9} = 0.2$ kΩ and 0.6 k$\Omega/\sqrt{4} = 0.3$ kΩ). These characteristic impedance values are smaller than the expected ones (0.6 to 2.5 kΩ).

Type (4) was used, for example, by Baba and Rakov (2007a) in their FDTD simulation of current along a vertical lightning channel. Note that this lightning-channel representation was first used by Kato *et al.* (2001) in their simulation of lightning electromagnetic fields with the MoM in the frequency domain (Harrington 1968). Baba and Rakov (2007a) represented the lightning channel by a vertical perfectly conducting wire, which had a radius of 0.23 m and was placed along the axis of a dielectric rectangular parallelepiped of relative permittivity 9 and cross-section 4 m × 4 m. This dielectric parallelepiped was surrounded by air. The speed of the current wave propagating along the wire was about $0.74c$. Such a representation allows one to calculate both the distribution of current along the wire and the remote electromagnetic fields in a single, self-consistent procedure, while that of a vertical wire surrounded by an artificial dielectric medium occupying the entire upper half space (type (3) described above) requires two steps to achieve the same objective. However, the electromagnetic fields produced by a dielectric-coated wire in air might be influenced by the presence of coating.

Type (5) was first used by Miyazaki and Ishii (2004) in their FDTD simulation of electromagnetic fields due to a lightning strike to a tall structure. The speed of the current wave propagating along the wire was about $0.5c$, although the exact values of relative permittivity and relative permeability of the coating are not given by Miyazaki and Ishii (2004). Similar to type (4), this representation allows one to calculate both the distribution of current along the wire and the remote electromagnetic fields in a single, self-consistent procedure. For the same speed of current wave, the characteristic impedance value for this channel representation is higher than that for type (4), since both relative permittivity and permeability are set at higher values in the type (5) representation.

Type (6) has not been used in LEMP and surge simulations with the FDTD method to date. It was, however, used by Bonyadi-Ram *et al.* (2005) in their simulation based on the MoM in the time domain. In their model, each of the wires has a radius of 2 cm, and the separation between the wires is 30 m. The speed of the current wave propagating along two parallel wires having additional distributed shunt capacitance in air is $0.43c$ when the additional capacitance is 50 pF/m. Similar to type (3) described above, this representation employs a fictitious configuration only for finding a reasonable distribution of current along the lightning channel, and then this current distribution is applied to the actual configuration (vertical wire in air above ground).

Type (7) was used by Baba and Rakov (2003) in their FDTD calculations of lightning electromagnetic fields. This representation can be employed for simulation of "engineering" lightning return-stroke models. Each current source of the phased-current-source array is activated successively by the arrival of lightning return-stroke front that progresses upward at a specified speed. Although the impedance of this channel model is equal to infinity, appropriate reflection coefficients at the top and bottom of the structure and at the lightning attachment point can be implemented to account for the presence of tall strike object and upward-connecting leader (e.g. Baba and Rakov 2005b, 2007b).

Among the above seven types, types (2) and (5) appear to be best in terms of the resultant return-stroke front speed, the characteristic impedance, and the procedure for current and field computations. Type (7) is also useful since the return-stroke front speed and the current attenuation with height are controlled easily with a simple mathematical expression representing "engineering" return-stroke models such as the transmission-line (TL) model (Uman *et al*. 1975) and its modifications (e.g., Rakov and Dulzon 1987; Nucci *et al*. 1988).

Practical aspects of the implementation of various electromagnetic models of the lightning return stroke are reviewed by Karami *et al*. (2016).

4.6.2   Excitation methods

Methods of excitation of the lightning channel used in electromagnetic pulse and surge computations include

1. closing a charged vertical wire at its bottom end with a specified impedance (or circuit);
2. a lumped voltage source (equivalent to a delta-gap electric-field source);
3. a lumped current source; and
4. a phased-current-source array.

Type (1) excitation method was used, for example, by Baba and Rakov (2007a) in their FDTD simulation of currents along a vertical lightning channel. Note that this method was first used by Podgorski and Landt (1987) in their simulation of lightning currents with the MoM in the time domain. Baba and Rakov (2007a) represented a leader/return-stroke sequence by a precharged vertical perfectly conducting wire connected via a nonlinear resistor to flat ground. In their model, closing a charged vertical wire in a specified circuit simulates the lightning return-stroke process.

Type (2) was also used by Baba and Rakov (2007a), but it was first employed by Moini *et al*. (1998) in their simulation of lightning-induced voltages with the MoM in the time domain. This type of source generates a specified electric field, which is independent of current flowing through the source. Since it has zero internal impedance, its presence in series with the lightning channel and strike object does not disturb any transient processes in them. If necessary, one could insert a lumped resistor in series with the voltage source to adjust the impedance seen by waves entering the channel from the strike object to a value consistent with the expected equivalent impedance of the lightning channel.

Type (3) was used, for example, by Noda (2007). However, in contrast with a lumped voltage source, a lumped current source inserted at the attachment point is justified only when there are no reflected waves returning to the source. This is the case for a branchless subsequent lightning stroke terminating on flat ground without (or with a very short) upward connecting leader. The primary reason for the use of a lumped current source at the channel base is a desire to use directly the channel-base current, known from measurements for both natural and triggered lightning, as an input parameter of the model. When one employs a lumped ideal current source at the attachment point in analyzing lightning strikes to a tall grounded object, the lightning channel, owing to the infinitely large impedance of the ideal current source, is electrically isolated from the strike object, so that current waves reflected from ground cannot be directly transmitted to the lightning channel (only electromagnetic coupling is possible). Since this is physically unreasonable, a series ideal current source is not suitable for the modeling of lightning strikes to tall grounded objects (Baba and Rakov 2005).

Features of type (4) excitation are described in Section 4.6.1 for type (7) representation of lightning channel.

4.7   Representation of surge arrester

Tatematsu and Noda (2014) have proposed a technique to represent a surge arrester, the physical size of which is much smaller than the wavelength of interest, by a lumped nonlinear resistor. The voltage versus current (V–I) characteristics of nonlinear resistors are represented by piecewise linear approximation, as shown in Figure 4.11. The specific points on the V–I characteristic are obtained from measured voltage versus current curve. In Figure 4.11, I_m and V_m represent the current and voltage at the mth point, respectively, and the total number of points is denoted by M. The voltage versus current characteristic of the nonlinear resistor shown in Figure 4.11 is approximated as follows:

$$V^{n+\frac{1}{2}} = R_0\left(I^{n+\frac{1}{2}} - I_0\right) + V_0 \qquad \text{for } V^{n+\frac{1}{2}} < V_1$$

$$V^{n+\frac{1}{2}} = R_m\left(I^{n+\frac{1}{2}} - I_m\right) + V_m \qquad \text{for } V_m \leq V^{n+\frac{1}{2}} < V_{m+1} \quad (1 \leq m \leq M - 3)$$

$$V^{n+\frac{1}{2}} = R_{M-2}\left(I^{n+\frac{1}{2}} - I_{M-2}\right) + V_{M-2} \qquad \text{for } V_{M-2} \leq V^{n+\frac{1}{2}}$$

$$R_m = \frac{V_{m+1} - V_m}{I_{m+1} - I_m} \tag{4.60}$$

The V–I characteristic for voltages smaller than V_0 and larger than V_{M-1} are represented by linear extrapolation of (I_0, V_0) and (I_1, V_1), and (I_{M-2}, V_{M-2}) and (I_{M-1}, V_{M-1}), respectively.

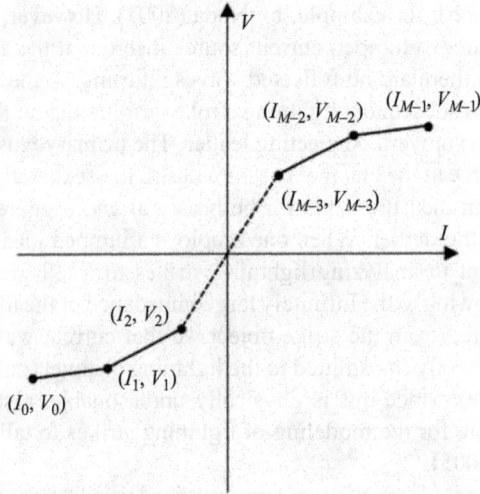

Figure 4.11   Piecewise-linear representation of voltage vs. current characteristic of a surge arrester. Reprinted, with permission, from Tatematsu and Noda (2014, Figure 1)

The current through the nonlinear resistor is obtained from (4.60) as follows:

$$I^{n+\frac{1}{2}} = \frac{V^{n+\frac{1}{2}}}{R_m} - \left(\frac{V_m}{R_m} - I_m\right) \tag{4.61}$$

When the nonlinear resistor is along the z-axis and located at point $(i, j, k+1/2)$ in a lossless medium ($\sigma = 0$), (4.61) becomes:

$$I_z^{n+\frac{1}{2}}\left(i,j,k+\frac{1}{2}\right) = J_z^{n+\frac{1}{2}}\left(i,j,k+\frac{1}{2}\right)\Delta x \Delta y$$

$$= \frac{V_z^{n+\frac{1}{2}}\left(i,j,k+\frac{1}{2}\right)}{R_m} - \left(\frac{V_m}{R_m} - I_m\right)$$

$$= \frac{\left[E_z^{n+1}\left(i,j,k+\frac{1}{2}\right) + E_z^{n-1}\left(i,j,k+\frac{1}{2}\right)\right]\Delta z}{2}\frac{1}{R_m} - \left(\frac{V_m}{R_m} - I_m\right) \tag{4.62}$$

From (4.1) and (4.62), the update equation for E_z at $(i, j, k +1/2)$ is given by

$$E_z^n\left(i,j,k+\frac{1}{2}\right) = \frac{1-\dfrac{\Delta t}{2R_m\varepsilon(i,j,k+1/2)\Delta z}}{1+\dfrac{\Delta t}{2R_{m,}\varepsilon(i,j,k+1/2)\Delta z}} E_z^{n-1}\left(i,j,k+\frac{1}{2}\right)$$

$$+ \frac{\dfrac{\Delta t}{\varepsilon(i,j,k+1/2)}}{1+\dfrac{\Delta t}{2R_m\varepsilon(i,j,k+1/2)\Delta z}}\frac{1}{\Delta x\Delta y}$$

$$\times \left[\begin{array}{l} H_y^{n-\frac{1}{2}}\left(i+\dfrac{1}{2},j,k+\dfrac{1}{2}\right)\Delta y - H_y^{n-\frac{1}{2}}\left(i-\dfrac{1}{2},j,k+\dfrac{1}{2}\right)\Delta y \\ -H_x^{n-\frac{1}{2}}\left(i,j+\dfrac{1}{2},k+\dfrac{1}{2}\right)\Delta x + H_x^{n-\frac{1}{2}}\left(i,j-\dfrac{1}{2},k+\dfrac{1}{2}\right)\Delta x \end{array}\right]$$

$$+ \frac{\dfrac{\Delta t}{\varepsilon(i,j,k+1/2)}}{1+\dfrac{\Delta t}{2R_m\varepsilon(i,j,k+1/2)\Delta z}}\frac{1}{\Delta x\Delta y}\left(\frac{V_m}{R_m}-I_m\right)$$

(4.63)

Since the voltage versus current relation of the lumped nonlinear resistor is given by a piecewise linear function, the electric field along the nonlinear resistor is updated using (4.63) in the following simple procedure:

1. Update each electric field using (4.63) with $m = 0$ with the assumption that $V^{n+1/2}$ satisfies the condition $V^{n+1/2} < V_1$, then go to Step 2.
2. If the computed $V^{n+1/2}$ satisfies the assumption in Step 1, the computed electric field is correct. Otherwise, go to Step 3 with $m = 1$.
3. If $m = M-2$, go to Step 5. Otherwise, update each electric field using (4.63), with m from Step 2 or from Step 4 and with the assumption that $V^{n+1/2}$ satisfies the condition $V_m \leq V^{n+1/2} < V_{m+1}$, then go to Step 4.
4. If $V^{n+1/2}$ satisfies the assumption in Step 3, the computed electric field is correct. Otherwise, add one to m and go back to Step 3.
5. Update each electric field using (4.63) with $m = M - 2$ with the assumption that $V^{n+1/2}$ satisfies the condition $V_{M-2} \leq V^{n+1/2}$, then go to Step 6.
6. If $V^{n+1/2}$ satisfies the assumption in Step 5, the computed electric field is correct.

In this procedure, the electric field along the surge arrester represented by nonlinear resistor can be obtained using (4.63) $M-1$ times at the most.

Imato et al. (2016) have proposed a different technique to represent a surge arrester. The arrester is represented as a combination of small cells, each of which has a nonlinear resistivity (or conductivity) in the x-, y-, and z-directions, depending on the electric field in each direction. The resistivity or conductivity is given as a function of electric field, $\rho(E)$ or $\sigma(E) = 1/\rho(E)$, based on the measured relationship between the voltage across a small ZnO element and the current flowing in it (*V–I* characteristic). In FDTD calculations, $\sigma(E) = 1/\rho(E)$ is incorporated in electric field update equations (4.4), (4.5), and (4.6). This simple representation requires no iterative procedure, and is suitable for FDTD surge simulations. Note that Tanaka *et al.* (2020) have proposed a simple mathematical expression for $\rho(E)$ with three adjustable constants, c_0, c_1, and c_2, which is given below:

$$\rho(E) = 10^{c_0 + c_1 \ (\log E)^{c_2}} \tag{4.64}$$

where log is the base-10 logarithm, E is in V/m, and ρ is in Ωm.

Tsuge *et al.* (2020) has applied this representation to an electromagnetic and thermal simulation of surge arrester subjected to a lightning impulse.

4.8 Representation of corona on a horizontal conductor

Thang *et al.* (2012) have proposed a simplified corona-discharge model for computing surges propagating on overhead wires using the FDTD method. The radial progression of corona streamers from an overhead wire was represented as the radial expansion of a cylindrical conducting region whose conductivity is several tens of microsiemens per meter.

The critical electric field E_0 on the surface of a cylindrical wire of radius r_0 for initiation of corona discharge is given by equation of Hartmann (1984), which is reproduced below.

$$E_0 = m \cdot 2.594 \times 10^6 \left(1 + \frac{0.1269}{r_0^{0.4346}}\right) [\text{V/m}] \tag{4.65}$$

where m is a coefficient depending on the wire surface conditions.

The critical electric field necessary for streamer propagation (Cooray 2003) (which determines the maximum extent of the radially expanding corona region) for positive, E_{cp}, and negative, E_{cn}, polarity is set as follows (Waters *et al.* 1987):

$$\begin{cases} E_{cp} = 0.5 & [\text{MV/m}] \\ E_{cn} = 1.5 & [\text{MV/m}] \end{cases} \tag{4.66}$$

It is shown by Noda (1996) that the statistical inception delay, streamer development process, and ionization process, all of which are microsecond-scale phenomena, should be considered in developing a corona-discharge model for lightning surge computations. In the FDTD computations, the ionization process is roughly approximated by increasing the conductivity of the corona-discharge region from zero to $\sigma_{cor} = 20$ or $40\ \mu\text{S/m}$, and the statistical inception delay and streamer development process are ignored. The corresponding time constants, $RC = \varepsilon_0/\sigma_{cor}$ (R and C are the resistance and capacitance of cylindrical corona discharge region, respectively), are equal to about 0.5 or 0.25 μs. The corona radius

r_c was obtained, using analytical expression (4.67) below, based on E_c (0.5 or 1.5 MV/m, depending on polarity; see (4.66) above) and the FDTD-computed charge per unit length (q). Then, the conductivity of the cells located within r_c was set to $\sigma_{cor} = 20$ or $40\ \mu S/m$.

$$E_c = \frac{q}{2\pi\varepsilon_0 r_c} + \frac{q}{2\pi\varepsilon_0(2h - r_c)}\ [\text{V/m}] \tag{4.67}$$

Equation (4.67), which is an approximation valid for $r_c \ll 2h$, gives the electric field at distance r_c below an infinitely long, horizontal uniform line charge, $+q[\text{C/m}]$, located at height h above flat perfectly conducting ground. A more general equation, not requiring that $r_c \ll 2h$, but assuming that corona region is a good conductor, yields similar results.

Simulation of corona discharge implemented in the FDTD procedure is summarized below:

(a) If the FDTD-computed electric-field, E_{zb}^n, at time step n and at a point located below and closest to the wire (at $0.5\Delta z$ from the wire axis shown in Figure 4.12(a)), exceeds $0.46E_0$, where E_0 is given by (4.65), the conductivity of $\sigma_{cor} = 20$ or $40\ \mu S/m$ is assigned to x- and z-directed sides of the four cells closest

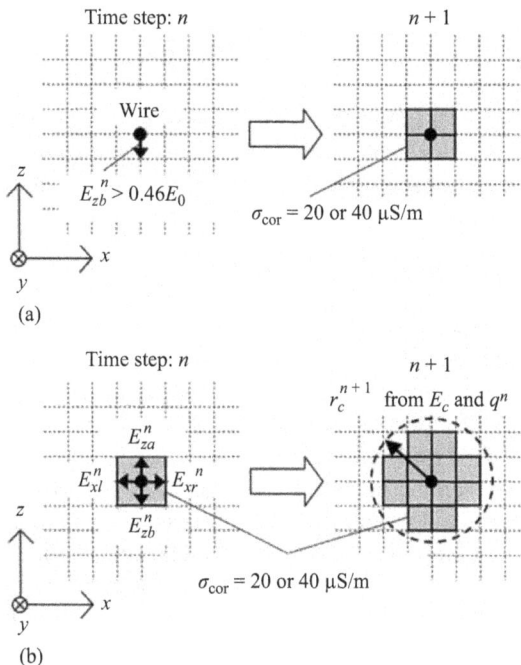

Figure 4.12   FDTD representations of corona on a horizontal conductor.
(a) Inception of corona discharge at the wire surface.
(b) Radial expansion of corona discharge. © 2012 IEEE. Reprinted, with permission, from Thang et al. (2012, Figure 3)

to the wire. Note that E_{zb}^n is almost the same as E_{xl}^n and E_{xr}^n (see Figure 4.12(b)) at points located on the left- and right-hand sides of the wire, respectively, and closest to the wire (at $0.5\Delta x$ from the wire axis). Therefore, only E_{zb}^n is monitored for determining initiation of corona discharge. Also note that neither computed radial current nor q–V curves change if the same conductivity is also assigned to y-directed (axial direction) sides of the four cells.

(b) The radial current I^n per unit length of the wire at $y = j\Delta y$ from the excitation point at time step n is evaluated by numerically integrating radial conduction and displacement current densities as follows:

$$I^n(j\Delta y) = \sigma_{\text{cor}}\left[\left(E_{xl}^n + E_{xr}^n\right)\Delta z + \left(E_{za}^n + E_{zb}^n\right)\Delta x\right]\Delta y$$
$$+ \varepsilon_0\left[\left(\frac{E_{xl}^n - E_{xl}^{n-1}}{\Delta t} + \frac{E_{xr}^n - E_{xr}^{n-1}}{\Delta t}\right)\Delta z + \left(\frac{E_{za}^n - E_{za}^{n-1}}{\Delta t} + \frac{E_{zb}^n - E_{zb}^{n-1}}{\Delta t}\right)\Delta x\right]\Delta y$$
$$(4.68)$$

where E_{xl}, E_{xr}, E_{za}, and E_{zb} are radial electric fields closest to the wire shown in Figure 4.12(b). The total charge (the sum of charge deposited on the wire and corona charge in the surrounding air) per unit length of the wire at $y = j\Delta y$ from the excitation point at time step n is calculated as follows:

$$q^n(j\Delta y) = q^{n-1}(j\Delta y) + \frac{I^{n-1}(j\Delta y) + I^n(j\Delta y)}{2}\Delta t \qquad (4.69)$$

From q^n yielded by (4.69) and E_c given by (4.66), the corona radius r_c^{n+1} at time step $n + 1$ is calculated using (4.67). The conductivity of $\sigma_{\text{cor}} = 20$ or $40\,\mu\text{S/m}$ is assigned to x- and z-directed sides of all cells located within r_c^{n+1}.

4.9   Absorbing boundary conditions

For the analysis of the electromagnetic response of a structure in an unbounded space, an absorbing boundary condition, which suppresses unwanted reflections, needs to be applied to planes that truncate the open space and accommodate the working volume. There are two types of absorbing boundary conditions. One is a differential-based absorbing boundary condition such as Liao's condition (Liao *et al.* 1984), and the other is a material-based absorbing boundary condition such as perfectly matched layers (Berenger 1994). Here, Liao's absorbing boundary condition is explained since it is often used in lightning surge simulations with the FDTD method.

Figure 4.13(a) shows the conceptual picture of a z-directed electric field E_z, which propagates in the negative x-direction with the speed of light c, and crosses the absorbing boundary located at $x = x_1$. The z-directed electric field at x_1 at time step number n, $E_z^n(x_1)$, could be estimated from $E_z^{n-2}(x_1 + 2c\Delta t)$ and $E_z^{n-1}(x_1 + c\Delta t)$ using a linear approximation, which is given below:

$$E_x{}^n(x_1) = 2E_z{}^{n-1}(x_1 + c\Delta t) - E_z{}^{n-2}(x_1 + 2c\Delta t) \qquad (4.70)$$

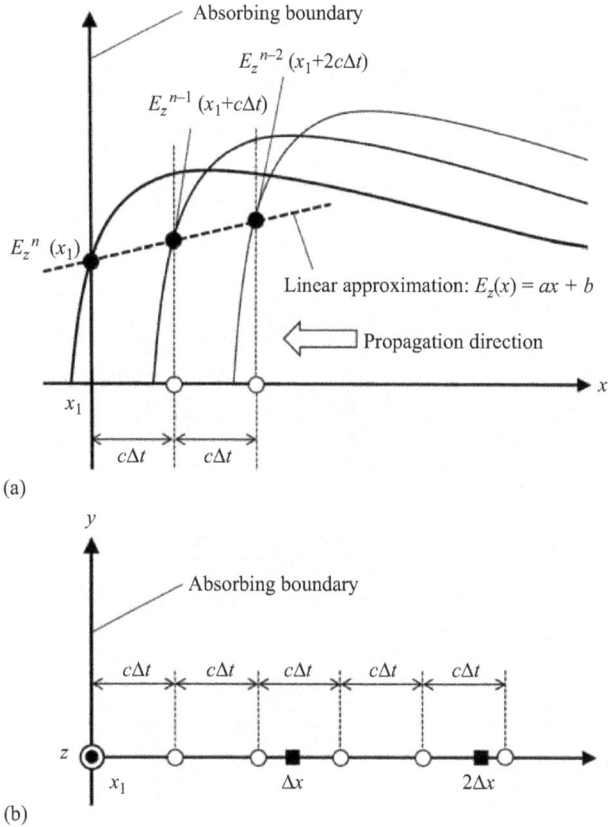

Figure 4.13 (a) Conceptual picture of a z-directed electric field E_z propagating in the negative x-direction (from right to left) with the speed of light c and crossing the absorbing boundary located at $x = x_1$, and (b) electric-field computation points near the absorbing boundary

Since locations of $x_1 + 2c\Delta t$ and $x_1 + c\Delta t$ do not coincide with the electric-field computation points: $x_1 + 2\Delta t, x_1 + \Delta t$, and so on, as shown in Figure 4.13(b), $E_z^{n-2}(x_1 + 2c\Delta t)$ and $E_z^{n-1}(x_1 + c\Delta t)$ are estimated using a quadratic interpolation formula, which is given below:

$$E_z^n(x_1) = 2T_{11}E_z^{n-1}(x_1) + 2T_{12}E_z^{n-1}(x_1 + \Delta x) + 2T_{13}E_z^{n-1}(x_1 + 2\Delta x)$$
$$- T_{11}^2 E_z^{n-2}(x_1) - 2T_{11}T_{12}E_z^{n-2}(x_1 + \Delta x)$$
$$- (2T_{11}T_{13} + T_{12}^2)E_z^{n-2}(x_1 + 2\Delta x) - 2T_{12}T_{13}E_z^{n-2}(x_1 + 3\Delta x)$$
$$- T_{13}^2 E_z^{n-2}(x_1 + 4\Delta x)$$

$$(4.71)$$

where

$$T_{11} = \frac{(2-s)(1-s)}{2}, \quad T_{12} = s(2-s), \quad T_{13} = \frac{s(s-1)}{2}, \quad s = \frac{c\Delta t}{\Delta x}$$

(4.72)

Expression (4.71) is Liao's second-order absorbing boundary condition. Note that the use of (4.71) in a single-precision floating-point computation often causes numerical instability (e.g., Asada *et al.* 2015b). In order to avoid this numerical instability, the following expression for T_{11} is used.

$$T_{11} = \frac{(2-2d-s)(1-s)}{2}$$

(4.73)

Uno (1998) suggests that $d = 0.0075$ should be effective to suppress numerical instability.

4.10   Summary

In this chapter, update equations for electric and magnetic fields in the 3D Cartesian, 2D cylindrical, and 2D spherical coordinate systems have been given. A subgridding technique has been explained. Representations of lumped sources and lumped circuit elements have been described. Representations of lightning channel and its excitation have been discussed. Also, representation of surge arresters and corona on horizontal conductors have been explained. Finally, Liao's absorbing boundary condition, which is needed for the analysis of electromagnetic fields in an unbounded space, has been presented.

References

Asada, T., Baba, Y., Nagaoka, N., and Ametani, A. (2015a), An improved thin wire representation for FDTD transient simulations, *IEEE Transactions on Electromagnetic Compatibility*, vol. 57, no. 3, pp. 484–487.

Asada, T., Baba, Y., Nagaoka, N., and Ametani, A. (2015b), A study of absorbing boundary condition for surge simulations with the FDTD method (in Japanese), *IEEJ Transactions on Power and Energy*, vol. 135, no. 6, pp. 408–416.

Azadifar, M., Li, D., Rachidi, F., *et al.* (2017), Analysis of lightning-ionosphere interaction using simultaneous records of source current and 380 km distant electric field, *Journal of Atmospheric and Solar-Terrestrial Physics*, vol. 159, pp. 48–56.

Baba, Y., and Rakov, V. A. (2003), On the transmission line model for lightning return stroke representation, *Geophysical Research Letters*, vol. 30, no. 24, 4 pages.

Baba, Y., and Rakov, V. A. (2005a), On the mechanism of attenuation of current waves propagating along a vertical perfectly conducting wire above ground: Application to lightning, *IEEE Transactions on Electromagnetic Compatibility*, vol. 47, no. 3, pp. 521–532.

Baba, Y., and Rakov, V. A. (2005b), On the use of lumped sources in lightning return stroke models, *Journal of Geophysical Research*, vol. 110 (D03101), doi:10.1029/2004JD005202.

Baba, Y., and Rakov, V. A. (2007a), Electromagnetic models of the lightning return stroke, *Journal of Geophysical Research*, vol. 112, no. D4, DOI: 10.1029/2006JD007222.

Baba, Y., and Rakov, V. A. (2007b), Influences of the presence of a tall grounded strike object and an upward connecting leader on lightning currents and electromagnetic fields, *IEEE Transactions on Electromagnetic Compatibility*, vol. 49, no. 4, pp. 886–892.

Baba, Y., and Rakov, V. A. (2008a), Influence of strike object grounding on close lightning electric fields, *Journal of Geophysical Research*, vol. 113, no. D12, doi: 10.1029/2008JD009811.

Baba, Y., and Rakov, V. A. (2008b), Applications of electromagnetic models of the lightning return stroke, *IEEE Transactions on Power Delivery*, vol. 23, no. 2, pp. 800–811.

Baba, Y., and Rakov, V. A. (2009), Electric and magnetic fields predicted by different electromagnetic models of the lightning return stroke versus measured fields, *IEEE Transactions on Electromagnetic Compatibility*, vol. 51, no. 3, pp. 479–483.

Baba, Y., and Rakov, V. A. (2011), Simulation of corona at lightning-triggering wire: Current, charge transfer, and the field-reduction effect, *Journal of Geophysical Research*, vol. 116, no. D21, doi: 10.1029/2011JD016341.

Baba, Y., and Rakov, V. A. (2014), Applications of the FDTD method to lightning electromagnetic pulse and surge simulations, *IEEE Transactions on Electromagnetic Compatibility*, vol. 56, no. 6, pp. 1506–1521.

Baba, Y., and Rakov, V. A. (2016), *Electromagnetic Computation Methods for Lightning Surge Protection Studies*, Wiley-IEEE, Singapore, 315 pages.

Baba, Y., Nagaoka, N., and Ametani, A. (2005), Modeling of thin wires in a lossy medium for FDTD simulations, *IEEE Transactions on Electromagnetic Compatibility*, vol. 47, no. 1, pp. 54–60.

Berenger, J. P. (1994), A perfectly matched layer for the absorption of electromagnetic waves, *Journal of Computational Physics*, vol. 114, pp. 185–200.

Bonyadi-Ram, S., Moini, R, Sadeghi, S. H. H., and Rakov, V. A. (2005), Incorporation of distributed capacitive loads in the antenna theory model of lightning return stroke, Paper presented at 16th International Zurich Symposium on Electromagnetic Compatibility, pp. 213–218, Zurich.

Bonyadi-Ram, S., Moini, R., Sadeghi, S. H. H. and Rakov, V. A. (2008), On representation of lightning return stroke as a lossy monopole antenna with inductive loading, *IEEE Transactions on Electromagnetic Compatibility*, vol. 50, no. 1, pp. 118–127.

Chen, H., Du., Y, Yuan, M., and Liu, Q. H. (2018), Analysis of the grounding for the substation under very fast transient using improved lossy thin-wire model for FDTD, *IEEE Transactions on Electromagnetic Compatibility*, vol. 60, no. 6, pp. 1833–1841.

Chevalier, M. W., Luebbers, R. J., and Cable, V. P. (1997), FDTD local grid with material traverse, *IEEE Transactions on Antennas and Propagation*, vol. 45, no. 3, pp. 411–421.

Cooray, V. (2003), *The Lightning Flash*, p. 79, The Institution of Electrical Engineers, UK.

Courant, R., Friedrichs, K., and Lewy, H. (1928), Über die partiellen Differenzengleichungen der mathematischen Physik (in German), *Mathematiche Annalen*, vol. 100, no. 1, pp. 32–74.

Du, Y., Li, B., and Chen, M. (2017), The extended thin-wire model of lossy round wire structures for FDTD simulations, *IEEE Transactions on Power Delivery*, vol. 32, no. 6, pp. 2472–2480.

Gorin, B. N., and Shkilev, A. V. (1984), Measurements of lightning currents at the Ostankino tower (in Russian), *Electrichestrvo*, vol. 8, pp. 64–65.

Hano, M., and Itoh, T. (1996), Three-dimensional time-domain method for solving Maxwell's equations based on circumcenters of elements, *IEEE Transactions on Magnetics*, vol. 32, no. 3, pp. 946–949.

Hao, Y., and Mittra, R. (2009), *FDTD Modeling of Metamaterials*, Artech House Publishers, Boston, USA.

Harrington, R. F. (1968), *Field Computation by Moment Methods*, Macmillan Co., New York.

Hartmann, G. (1984), Theoretical evaluation of Peek's law, *IEEE Transactions on Industry Applications*, vol. 20, no. 6, pp. 1647–1651.

Idone, V. P., and Orville, R. E. (1982), Lightning return stroke velocities in the Thunderstorm Research International Program (TRIP), *Journal of Geophysical Research*, vol. 87, pp. 4903–4915.

Imato, S., Baba, Y., Nagaoka, N., and Itamoto N. (2016), FDTD analysis of the electric field of a substation arrester under a lightning overvoltage, *IEEE Transactions on Electromagnetic Compatibility*, vol. 58, no. 2, pp. 615–618.

Inan, U. S., and Marshall, R. A. (2011), *Numerical Electromagnetics: The FDTD Method*, Cambridge University Press, UK.

Johns, P. B., and Beurle, R. B. (1971), Numerical solutions of 2-dimensional scattering problems using a transmission-line matrix, *Proceeding of the IEE*, vol. 118, no. 9, pp. 1203–1208.

Karami, H., Rachidi, F., and Rubinstein, M. (2016), On practical implementation of electromagnetic models of lightning return-strokes, *Atmosphere*, vol. 7, no. 135, doi:10.3390/atmos7100135.

Kato, S., Narita, T., Yamada, T., and Zaima, E. (1999), Simulation of electromagnetic field in lightning to tall tower, In Proceedings of the 11th International Symposium on High Voltage Engineering, no. 467, London, UK.

Kato, S., Takinami, T., Hirai, T., and Okabe, S. (2001), A study of lightning channel model in numerical electromagnetic field computation (in Japanese), In Proceedings of 2001 IEEJ National Convention, no. 7–140, Nagoya, Japan.

Kunz, K. S., and Luebbers, R. J. (1993), *The Finite Difference Time Domain Method for Electromagnetics*, CRC Press, Boca Raton, USA.

Li, B., Du., Y., and Chen, M. (2018), An FDTD thin-wire model for lossy wire structures with noncircular cross section, *IEEE Transactions on Power Delivery*, vol. 33, no. 6, pp. 3055–3064.

Liao, Z. P., Wong, H. L., Yang, B.-P., and Yuan, Y.-F. (1984), A transmitting boundary for transient wave analysis, *Scientia Sinica*, vol. A27, no. 10, pp. 1063–1076.

Livesey, M., Stack, J. F., Costen, F., Nanri, T., Nakashima, N., and Fujino, S. (2012), Development of a CUDA implementation of the 3D FDTD method, *IEEE Antennas and Propagation Magazine*, vol. 54, no. 5, pp. 186–195.

Miller, E. K., Poggio, A. J., and Burke, G. J. (1973), An integro-differential equation technique for the time domain analysis of thin wire structure: Part I. The numerical method, *Journal of Computational Physics*, vol. 12, pp. 24–28.

Miyazaki, S., and Ishii, M. (2004), Reproduction of electromagnetic fields associated with lightning return stroke to a high structure using FDTD method (in Japanese), In Proceedings of 2004 IEEJ National Convention, no. 7–065, p. 98, Kanagawa, Japan.

Moini, R., Kordi, B., and Abedi, M. (1998), Evaluation of LEMP effects on complex wire structures located above a perfectly conducting ground using electric field integral equation in time domain, *IEEE Transactions on Electromagnetic Compatibility*, vol. 40, no. 2, pp. 154–162.

Moini, R., Kordi, B., Rafi, G. Z., and Rakov, V. A. (2000), A new lightning return stroke model based on antenna theory, *Journal of Geophysical Research*, vol. 105 (D24), pp. 29,693–29,702.

Nakagawa, M., Baba, Y., Tsubata, H., Nishi, T., and Fujisawa, H. (2016), FDTD simulation of lightning current in a multilayer CFRP panel with triangular-prism cells, *IEEE Transactions on Electromagnetic Compatibility*, vol. 58, no. 1, pp. 327–330.

Noda, T. (1996), Development of a transmission-line model considering the skin and corona effects for power system transient analysis, Ph.D. Thesis, Doshisha University.

Noda, T. (2007), A tower model for lightning overvoltage studies based on the result of an FDTD simulation (in Japanese), *IEEJ Transactions on Power and Energy*, vol. 127, no. 2, pp. 379–388.

Noda, T., and Yokoyama, S. (2002), Thin wire representation in finite difference time domain surge simulation, *IEEE Transactions on Power Delivery*, vol. 17, no. 3, pp. 840–847.

Nucci, C. A., Mazzetti, C., Rachidi, F., and Ianoz, M. (1988), On lightning return stroke models for LEMP calculations, In Proceedings of 19th International Conference on Lightning Protection, Graz, Austria, pp. 463–469.

Oikawa, T., Sonoda, J., Sato, M., Honma, N., and Ikegawa, Y. (2012), Analysis of lightning electromagnetic field on large-scale terrain model using three-dimensional MW-FDTD parallel computation (in Japanese), *IEEJ Transactions on Fundamentals and Materials*, vol. 132, no. 1, pp. 44–50.

Oliveira, R. M. e. S. d., and Sobrinho C. L. d. S. S. (2009), Computational environment for simulating lightning strokes in a power substation by finite-difference time-domain method, *IEEE Transactions on Electromagnetic Compatibility*, vol. 51, no. 4, pp. 995–1000.

Podgorski, A. S., and Landt, J. A. (1987), Three dimensional time domain modeling of lightning, *IEEE Transactions on Power Delivery*, vol. 2, no. 3, pp. 931–938.

Railton, C. J., Paul, D. L., and Dumanli, S. (2006), The treatment of thin wire and coaxial structures in lossless and lossy media in FDTD by the modification of assigned material parameters, *IEEE Transactions on Electromagnetic Compatibility*, vol. 48, no. 4, pp. 654–660.

Rakov, V. A. (1998), Some inferences on the propagation mechanisms of dart leaders and return strokes, *Journal of Geophysical Research*, vol. 103 (D2), pp. 1879–1887.

Rakov, V. A. (2007), Lightning return stroke speed, *Journal of Lightning Research*, vol. 1, pp. 80–89.

Rakov, V. A., and Dulzon, A. A. (1987), Calculated electromagnetic fields of lightning return stroke (in Russian), *Tekh. Elektrodinam.*, vol. 1, pp. 87–89.

Ren, H.-M., Zhou, B.-H., Rakov, V. A., Shi, L.-H., Gao, C., and Yang, J.-H. (2008), Analysis of lightning-induced voltages on overhead lines using a 2-D FDTD method and Agrawal coupling model, *IEEE Transactions on Electromagnetic Compatibility*, vol. 50, no. 3, pp. 651–659.

Ruehli, A. (1974), Equivalent circuit models for three-dimensional multiconductor systems, *IEEE Transactions on Microwave Theory and Techniques*, vol. 22, no. 3, pp. 216–221.

Sadiku, M. N. O. (1989), A simple introduction to finite element analysis of electromagnetic problems, *IEEE Transactions on Education*, vol. 32, no. 2, pp. 85–93.

Sullivan, D. M. (2000), *Electromagnetic Simulation using the FDTD Method*, IEEE Press, New Jersey, USA.

Taflove, A. (1995), *Computational Electrodynamics: The Finite-Difference Time-Domain Method*, Artech House Publishers, Boston, USA.

Tanabe, K. (2001), Novel method for analyzing dynamic behavior of grounding systems based on the finite-difference time-domain method, *IEEE Power Engineering Review*, pp. 55–57.

Tanabe, N., Baba, Y., Nagaoka, N., and Ametani, A. (2003), High-accuracy analysis of surges on a slanting conductor and a cylindrical conductor by an FDTD method (in Japanese), *IEEJ Transactions on Power and Energy*, vol. 123, no. 6, pp. 725–733.

Tanaka, T., Tsuge, R., Baba, Tsujimoto, Y., and Tsukamoto, N. (2020), An approximate mathematical expression for nonlinear resistive properties of metal oxide varistor elements for FDTD simulations, *IEEE Transactions on Electromagnetic Compatibility*, doi:10.1109/TEMC.2020.2983200.

Taniguchi, Y., Baba, Y., Nagaoka, N., and Ametani, A. (2008a), Representation of an arbitrary-radius wire for FDTD calculations in the 2D cylindrical coordinate system, *IEEE Transactions on Electromagnetic Compatibility*, vol. 50, no. 4, pp. 1014–1018.

Taniguchi, Y., Baba, Y., Nagaoka, N., and Ametani, A. (2008b), An improved thin wire representation for FDTD computations, *IEEE Transactions on Antennas and Propagation*, vol. 56, no. 10, pp. 3248–3252.

Tatematsu, A. (2015), A technique for representing coaxial cables for FDTD-based surge simulations, *IEEE Transactions on Electromagnetic Compatibility*, vol. 57, no. 3, pp. 488–495.

Tatematsu, A. (2018), Technique for representing lossy thin wires and coaxial cables for FDTD-based surge simulations, *IEEE Transactions on Electromagnetic Compatibility*, vol. 60, no. 3, pp. 705–715.

Tatematsu, A., and Noda, T. (2014), Three-dimensional FDTD calculation of lightning-induced voltages on a multiphase distribution line with the lightning arresters and an overhead shielding wire, *IEEE Transactions on Electromagnetic Compatibility*, vol. 56, no. 1, pp. 159–167.

Thang, T. H., Baba, Y., Nagaoka, N., *et al.* (2012), A simplified model of corona discharge on overhead wire for FDTD computations, *IEEE Transactions on Electromagnetic Compatibility*, vol. 54, no. 3, pp. 585–593.

Thang, T. H., Baba, Y., Rakov, V. A. and Piantini, A. (2015), FDTD computation of lightning-induced voltages on multi-conductor lines with surge arresters and pole transformers, *IEEE Transactions on Electromagnetic Compatibility*, vol. 57, no. 3, pp. 442–447.

Tran, T. H., Baba, Y., Somu, V. B., and Rakov, V. A. (2017), FDTD modeling of LEMP propagation in the earth-ionosphere waveguide with emphasis on realistic representation of lightning source, *Journal of Geophysical Research*, vol. 122, no. 23, pp. 12,918–12,937.

Tsuge, R., Saito, H., Baba, Y., Kawamura, H., and Itamoto, N. (2020), Finite-difference time-domain simulation of a lightning-impulse-applied ZnO element, *IEEE Transactions on Electromagnetic Compatibility*, doi: 10.1109/TEMC.2019.2944390.

Uman, M. A., McLain, D. K., and Krider, E. P. (1975), The electromagnetic radiation from a finite antenna, *American Journal of Physics*, vol. 43, pp. 33–38.

Umashankar, K. R., Taflove, A., and Beker, B. (1987), Calculation and experimental validation of induced currents on coupled wires in an arbitrary shaped cavity, *IEEE Transactions on Antennas and Propagation*, vol. 35, no. 11, pp. 1248–1257.

Uno, T. (1998), *Finite Difference Time Domain Method for Electromagnetic Field and Antennas* (in Japanese), Corona Publishing Co., Ltd., Tokyo, Japan.

Visacro, S., and Soares, A. Jr. (2005), HEM: A model for simulation of lightning related engineering problems, *IEEE Transactions on Power Delivery*, vol. 20, no. 2, pp.1206–1207.

Waters, R. T., German, D. M., Davies, A. E., Harid, N., and Eloyyan, H. S. B. (1987), Twin conductor surge corona, Paper presented at the 5th Int. Symp. High Voltage Engineering, Braunschweig, Federal Republic of Germany.

Yamamoto, J., Baba, Y., Tran, T. H., and Rakov, V. A. (2020), Simulation of LEMP propagation in the earth-ionosphere waveguide using the FDTD method in the 2-D spherical coordinate system, *IEEJ Transactions on Electrical and Electronic Engineering*, vol. 15, no. 3, pp. 335–339.

Yang, B., Zhou, B.-H., Gao, C., Shi, L.-H., Chen, B., and Chen, H.-L. (2011), Using a two-step finite-difference time-domain method to analyze lightning-induced voltages on transmission lines, *IEEE Transactions on Electromagnetic Compatibility*, vol. 53, no. 1, pp. 256–260.

Yang, C., and Zhou, B. (2004), Calculation methods of electromagnetic fields very close to lightning, *IEEE Transactions on Electromagnetic Compatibility*, vol. 46, no. 1, pp. 133–141.

Yee, K. S. (1966), Numerical solution of initial boundary value problems involving Maxwell's equations in isotropic media, *IEEE Transactions on Antennas and Propagation*, vol. 14, no. 3, pp. 302–307.

Yu, W., Yang, X., Liu, Y., and Mittra, R. (2009), *Electromagnetic Simulation Techniques Based on the FDTD Method*, John Wiley & Sons, New Jersey, USA.

Chapter 5

Applications of the FDTD method

The first peer-reviewed paper, in which the finite-difference time-domain (FDTD) method for solving discretized Maxwell's equations was used in a simulation of lightning-induced surges, was published in 2006. About 30 journal papers and a large number of conference papers, which use the FDTD method in simulations of lightning-induced surges, have been published during the last 15 years. FDTD procedures used in simulations of lightning-induced surges are classified into two types in terms of spatial dimension: two-dimensional (2D) and three-dimensional (3D). About 30% of the simulations as of today have employed the 2D-FDTD method in the cylindrical coordinate system, and about 70% of the simulations have used the 3D-FDTD method. In the 2D case, the FDTD method is employed to express the distributed sources in terms of incident electromagnetic fields illuminating overhead conductors, to be incorporated in the equivalent distributed-parameter circuits of those conductors. The equivalent circuit depends on the field-to-conductor coupling model. There are three major coupling models that yield the same results, the most popular of which is the model proposed by Agrawal *et al.* (1980). One of the reasons for using the two-step or hybrid approach is that horizontal closely spaced thin conductors can be represented easily by a distributed circuit. In some 3D-FDTD simulations, the subgridding technique has been used to represent horizontal closely spaced thin conductors. The subgridding technique employs locally fine grids for representing closely spaced thin wires or other small structures that exist in the working volume. In other 3D-FDTD simulations, a nonuniform gridding technique has been used. The transmission-line (TL) engineering model has been most frequently used for representing the lightning return stroke. The modified TL model with linear current decay with height (MTLL) and the modified TL model with exponential current decay with height (MTLE) have also been employed. Further, the TL model extended to include a tall strike object has been developed. As of today, about 70% of the FDTD simulations have been concerned with induced surges associated with lightning strikes to flat ground, and about 30% with lightning strikes to the top of mountain, building, or tall object. In this chapter, journal papers on the FDTD-based studies of lightning-induced surges are classified in terms of spatial dimension, lightning channel representation, and application. An overview of these works is given and six representative works are described in detail.

Key Words: 2D-FDTD method; 3D-FDTD method; electromagnetic-field-to-conductor coupling model; lightning return stroke; lightning-induced voltage; nonuniform grid; subgrid; surge arrester; thin wire; overhead conductor; buried conductor

5.1   Introduction

5.1.1   Classification of applications of the FDTD method

The first peer-reviewed paper, in which the finite-difference time-domain (FDTD) method for solving discretized Maxwell's equations (Yee 1966) was used for a simulation of lightning-induced surges, was published in 2006 (Baba and Rakov 2006). About 30 journal papers and a large number of conference papers, in which the FDTD method is employed in simulations of lightning-induced surges, have been published during the last 15 years.

FDTD methods used in simulations of lightning-induced surges are classified into two types in terms of spatial dimension: two-dimensional (2D) and three-dimensional (3D). As of today, about 30% of the studies (Ren *et al.* 2008; Soto *et al.* 2014; Zhang, Q. *et al.* 2014a, 2014b, 2015a; Rizk *et al.* 2017, 2020; Zhang, J. *et al.* 2019a; Zhang, L. *et al.* 2019) are based on the 2D-FDTD method in the cylindrical coordinate system, and about 70% (Baba and Rakov 2006; Tatematsu and Noda 2010, 2014; Ishii *et al.* 2012; Sumitani *et al.* 2012; Thang *et al.* 2014, 2015a, 2015b; Namdari *et al.* 2015; Zhang, Q. *et al.* 2015b; Du *et al.* 2016; Rizk *et al.* 2016a, 2016b; Tanaka *et al.* 2016; Diaz *et al.* 2017; Chen *et al.* 2018; Natsui *et al.* 2018, 2020; Zhang, J. *et al.* 2019b) on the 3D-FDTD method (see Table 5.1).

In simulations based on the 2D-FDTD method, the method is employed to find the distributed sources, which represent incident lightning electromagnetic pulses (LEMPs) illuminating overhead power distribution or telecommunication lines and are incorporated in their equivalent distributed-parameter circuits corresponding to the field-to-conductor coupling model of Agrawal *et al.* (1980). One of the reasons for using the two-step or hybrid approach is that horizontal closely spaced thin conductors can be conveniently represented by a distributed-parameter circuit (Ren *et al.* 2008; Zhang, J. *et al.* 2019a; Zhang, L. *et al.* 2019; Rizk *et al.* 2020). Note that Yang *et al.* (2011) have employed a different two-step approach for evaluating lightning-induced voltages on an overhead single wire above lossy ground. The first step was 2D-cylindrical FDTD computation of electric and magnetic fields, generated by a nearby lightning strike that illuminated the 3D volume accommodating the overhead wire, in a uniform grid. The second step was 3D-FDTD computation of lightning-induced voltages on the overhead wire illuminated by incident electromagnetic fields originating from the boundary of the 3D working volume. Yang *et al.* (2012) have used the same two-step approach for evaluating lightning-induced currents in a buried insulated conductor.

In some 3D-FDTD simulations (Sumitani *et al.* 2012; Thang *et al.* 2015a, 2015b), the subgridding technique (Chevalier *et al.* 1997) described in Section 4.3 has been used to represent horizontal closely spaced thin conductors. The subgridding technique employs locally fine grids for representing closely

Table 5.1   List of journal papers on FDTD computations of lightning-induced surges, which are classified in terms of spatial dimension (2D, 3D, or hybrid) and gridding (uniform, nonuniform, or subgridding)

Dimension	Spatial gridding	Paper(s)
2D	Uniform	• Ren *et al.* 2008 • Soto *et al.* 2014 • Zhang, Q. *et al.* 2014a, 2014b, 2015a • Rizk *et al.* 2017, 2020 • Zhang, J. *et al.* 2019a • Zhang, L. *et al.* 2019
3D	Uniform	• Baba and Rakov 2006 • Ishii *et al.* 2012 • Zhang, Q. *et al.* 2015b • Diaz *et al.* 2017 • Zhang, J. *et al.* 2019b
	Nonuniform	• Tatematsu and Noda 2010, 2014 • Thang *et al.* 2014 • Namdari *et al.* 2015 • Du *et al.* 2016 • Rizk *et al.* 2016a, 2016b • Tanaka *et al.* 2016 • Chen *et al.* 2018 • Natsui *et al.* 2018, 2020
	Subgridding	• Sumitani *et al.* 2012 • Thang *et al.* 2015a, 2015b
2D-3D hybrid	Uniform	• Yang *et al.* 2011, 2012

spaced thin wires or other small structures that exist in the working volume. In other 3D-FDTD simulations (Tatematsu and Noda 2010, 2014; Thang *et al.* 2014; Namdari *et al.* 2015; Du *et al.* 2016; Rizk *et al.* 2016a, 2016b; Tanaka *et al.* 2016; Chen *et al.* 2018; Natsui *et al.* 2018, 2020), a nonuniform gridding technique has been used. The nonuniform gridding technique changes cell lengths only in one direction (while cell side lengths in the other two directions are fixed). Therefore, there is no boundary at which interpolation of fields of locally fine grid from fields of the main grid is required. Note that in 3D-FDTD simulations for a thin-conductor line (e.g., Baba and Rakov 2006; Yang *et al.* 2012; Namdari *et al.* 2015; Zhang, Q. *et al.* 2015b; Rizk *et al.* 2016a; Chen *et al.* 2018; Zhang, J. *et al.* 2019b) coarse grids have been used with an equivalent thin-wire representation (e.g., Noda and Yokoyama 2002). Also note that the field-to-conductor coupling model of Agrawal *et al.* is employed by Zhang, J. *et al.* (2019b) in their 3D-FDTD simulation in a uniform grid. Table 5.1 gives a list of journal papers on lightning-induced surges, which are classified in terms of spatial dimension and gridding.

Table 5.2 gives a list of journal papers on lightning-induced surges, which are classified in terms of lightning channel representation. The transmission-line (TL) model (Uman and McLain 1969) has been most frequently used, the modified TL model with linear current decay with height (MTLL) (Rakov and Dulzon 1987), and the modified TL model with exponential current decay with height (MTLE) (Nucci *et al.* 1988) have also been employed. The TL model extended to include a tall strike object (Baba and Rakov 2005a) has been employed by Baba and Rakov (2006), Ren *et al.* (2008), and Zhang, Q. *et al.* (2015a). These models are classified as engineering models (Rakov and Uman 1998). In contrast, the return-stroke channel has been represented as a vertical lossy conductor by Ishii *et al.* (2012) and Diaz *et al.* (2017) or as a vertical perfect conductor by Du *et al.* (2016). These

Table 5.2   *List of journal papers on FDTD computations of lightning-induced surges, which are classified in terms of lightning representation*

Model	Paper(s)
TL	• Ren *et al.* 2008 • Tatematsu and Noda 2010, 2014 • Yang *et al.* 2011 • Soto *et al.* 2014 • Thang *et al.* 2014, 2015a, 2015b • Zhang, Q. *et al.* 2014b • Rizk *et al.* 2016a, 2016b, 2017 • Natsui *et al.* 2018, 2020 • Zhang, J. 2019b
MTLL	• Ren *et al.* 2008 • Sumitani *et al.* 2012 • Zhang, Q. *et al.* 2014a, 2015b • Tanaka *et al.* 2016 • Chen *et al.* 2018 • Rizk *et al.* 2020
MTLE	• Ren *et al.* 2008 • Yang *et al.* 2012 • Namdari *et al.* 2015 • Zhang, J. *et al.* 2019a • Zhang, L. *et al.* 2019
Extended TL to include a tall strike object	• Baba and Rakov 2006 • Ren *et al.* 2008 • Zhang, Q. *et al.* 2015a
Vertical conductor	• Ishii *et al.* 2012 (1 $\Omega\cdot$m) • Diaz *et al.* 2017(1 $\Omega\cdot$m) • Du *et al.* 2016 (perfect conductor)

models are classified as antenna-theory or electromagnetic models (Rakov and Uman 1998).

Table 5.3 gives a list of journal papers on lightning-induced surges, which are classified in terms of application. As of today, about 70% of the FDTD simulations have been concerned with induced surges associated with lightning strikes to flat ground, and about 30% with induced surges associated with lightning strikes to the top of mountain, building, or tall object. In Section 5.1.2, each of the works is briefly reviewed.

5.1.2   General overview of applications

Ren *et al.* (2008) have analyzed voltages, induced by rocket-triggered lightning strikes to flat lossy ground, on a two-conductor overhead test distribution line (part of an experimental setup at Camp Blanding, Florida, lightning-triggering facility (Barker *et al.* 1996)). They have used the 2D-FDTD method in the cylindrical coordinate system with the field-to-conductor coupling model of Agrawal *et al.* They have shown that the peak of the induced voltage is insensitive to the choice of model (they considered three engineering return-stroke models: TL, MTLL, and MTLE). Also, they have analyzed voltages, induced by lightning strikes to a tall object, on a single overhead conductor above flat lossy ground. In the simulation, the TL model extended to include a tall strike object (Baba and Rakov 2005a) was employed.

Sumitani *et al.* (2012) have computed voltages, induced by rocket-triggered lightning strikes to flat lossy ground, on a two-conductor overhead test distribution line (the same one as in the study of Ren *et al.* (2008)). 3D subgridding was employed: the spatial discretization is fine in the vicinity of overhead conductors and coarse in the rest of the computational domain. The lightning channel was represented by the MTLL model. They have shown that peak values of FDTD-computed lightning-induced voltages agree fairly well with corresponding measured ones. Also, they have shown influences of ground conductivity, return-stroke current-propagation speed, and other factors. This study is further discussed in Section 5.3.

Natsui *et al.* (2018) have computed voltages, induced by inclined lightning strikes to flat lossy ground, on a single overhead conductor. A 3D-nonuniform grid was employed. Lightning was represented by the TL model, and the distance was 50 m from the midpoint of the overhead conductor. They have shown that peak values of FDTD-computed lightning-induced voltages are higher for a lightning channel inclined toward the overhead conductor and lower for a lightning channel inclined away from the overhead conductor. Using the same approach, Natsui *et al.* (2020) have computed voltages, induced by inclined lightning strikes to flat lossy ground, on a multiconductor overhead line. They have shown how peak values of FDTD-computed lightning-induced voltages are influenced by the channel inclination. Also, they have shown that lightning-induced voltages are significantly reduced by installing two neutral conductors, one above and the other below the phase conductors.

Rizk *et al.* (2016a) have computed voltages, induced by lightning strikes to flat lossy ground of resistivity ranging from 0 to 2 k$\Omega\cdot$m, on a single overhead conductor. A 3D-nonuniform grid is employed. Lightning was represented by the TL model and

Table 5.3 List of journal papers on FDTD computations of lightning-induced surges, which are classified in terms of application

Strike termination	Subject of study	Paper(s)
Flat ground	Sensitivity analysis, including comparison of return-stroke models	Ren et al. 2008
	Influence of ground conductivity and return-stroke current-propagation speed	Sumitani et al. 2012
	Influence of inclined channel	Natsui et al. 2018, 2020
	Influence of low-conductivity soil	Rizk et al. 2016a, 2017
	Influence of stratified ground (two horizontal layers of different conductivity)	Zhang, Q. et al. 2014a, 2014b
	Influence of mixed propagation path	Zhang, Q. et al. 2015b
	Effect of surge arresters	Tatematsu and Noda 2010, 2014
		Namdari et al. 2015
		Thang et al. 2015a, 2015b
		Chen et al. 2018
		Rizk et al. 2020
	Effect of shield or neutral wires	Tatematsu and Noda 2010, 2014
		Namdari et al. 2015
		Zhang, L. et al. 2019
		Natsui et al. 2020
		Rizk et al. 2020
	Influence of pole transformers	Thang et al. 2015a, 2015b
	Influence of nearby buildings	Thang et al. 2015b
	Influence of corona discharge	Thang et al. 2014
	Evaluation of approximate expressions	Zhang, Q. et al. 2014b
		Rizk et al. 2016a, 2017
	Lightning-induced surges in underground telecommunication cables	Yang et al. 2012
		Tanaka et al. 2016
Mountain	Influence of mountain	Soto et al. 2014
		Zhang, J. et al. 2019a, 2019b
	Effect of shield wires	Zhang, J. et al. 2019a
		(Continues)

Building	Induced surges in electrical wiring or coaxial cable	Ishii et al. 2012
		Du et al. 2016
		Diaz et al. 2017
Tall object	Influence of the presence of tall strike object	Baba and Rakov 2006
		Ren et al. 2008
		Zhang, Q. et al. 2015a
	Influence of frequency-dependent soil parameters	Zhang, Q. et al. 2015a
	Influence of stratified ground (two horizontal layers of different conductivity)	Rizk et al. 2016b

was located 51 m from the midpoint of the overhead conductor. They have shown that the peak of lightning-induced voltage increases with increasing ground resistivity but is insensitive to the ground permittivity and the time derivative of lightning current. Also, they have shown that an expression for estimating the peak value of lightning-induced voltage, based on the formula proposed by Rusck (1957) and extended by Darveniza (2007) to consider the effect of lossy ground, yields values smaller than corresponding FDTD-computed ones for high-resistivity ground, and proposed correction to that expression.

Rizk *et al.* (2017) have computed voltages, induced by lightning strikes to flat lossy ground of resistivity ranging from 0.25 to 2 kΩ·m, on a single overhead conductor. They have used the 2D-FDTD method in the cylindrical coordinate system with the field-to-conductor coupling model of Agrawal *et al.* Lightning was represented by the TL model. Both first- and subsequent-stroke currents were considered. They have proposed an expression for estimating the peak value of lightning-induced voltage for first and subsequent strokes, based on Rusck's model. The proposed expression considers the ground resistivity and the risetime of current, and yields values that agree well with corresponding FDTD-computed ones.

Zhang, Q. *et al.* (2014a) have computed voltages induced on a single overhead conductor by lightning strikes to flat stratified ground (two horizontal layers of different conductivity). They have used the 2D-FDTD method and the field-to-conductor coupling model of Agrawal *et al.* Lightning was represented by the MTLL model. They have shown that the peak value of lightning-induced voltage increases with increasing the thickness of the upper layer when its conductivity is lower than that of the lower layer. Zhang, Q. *et al.* (2014b) have proposed an expression for estimating the peak value of lightning-induced voltage on a single overhead conductor above flat stratified (two-layer) ground and tested its validity against the FDTD calculations similar to those of Zhang, Q. *et al.* (2014a), but with the TL model for lightning return-stroke representation.

Zhang, Q *et al.* (2015b) have studied the influence of mixed propagation path on voltages induced on a single overhead conductor by lightning strikes to flat ground. A 3D-uniform grid was employed. Lightning was represented by the MTLL model and was located 60 or 200 m from the midpoint of the overhead conductor. Two lossy-ground regions were considered, whose conductivity values were set to 0.1 and 0.001 S/m. They have shown that the peak of lightning-induced voltage depends on the location of the vertical interface (parallel to the overhead conductor) between the different conductivity regions relative to the overhead conductor and the strike point and on whether lightning strikes the higher or lower conductivity region.

Tatematsu and Noda (2010, 2014) have analyzed voltages, induced by lightning strikes to flat lossy ground, on an overhead three-phase power distribution line with an overhead shield (ground) wire and surge arresters. The nonlinear voltage–current (V–I) relation of the arrester is represented by a piecewise linear approximation, which is based on the measured V–I curve. Lightning was represented by the TL model. Note that the conductor system was accommodated in a 3D-computational domain of 1400 m × 650 m × 700 m, which is divided nonuniformly into cubic cells: 2 cm in the vicinity of the ground wires and phase conductors and increasing gradually to 200 cm beyond that region.

Namdari *et al.* (2015) have analyzed voltages, induced by lightning strikes to flat lossy ground, on an overhead three-phase power distribution line with a neutral wire and surge arresters. They have derived an updating equation of electric field of one side of a cell, along which the arrester is located. At each time step, the Newton-Raphson method was used to solve the electric-field-update equation for the arrester. Lightning was represented by the MTLE model. The conductor system was accommodated in a 3D-computational domain of about 75 m × 850 m × 1400 m, which was divided nonuniformly into rectangular cells.

Thang *et al.* (2015a) have analyzed voltages, induced by lightning strikes to flat perfectly conducting ground, on an overhead three-phase power distribution line with a neutral wire, pole transformers, and surge arresters. A 3D-subgrid model, in which spatial discretization was fine in the vicinity of overhead wires and coarse in the rest of the computational domain, was employed. Lightning was represented by the TL model. They have shown that the computed lightning-induced voltage waveforms agree reasonably well with the corresponding ones measured in the small-scale experiment carried out by Piantini *et al.* (2007). This study is further discussed in Section 5.5.

Thang *et al.* (2015b) have computed voltages, induced by lightning strikes to flat perfectly conducting ground, on an overhead three-phase power distribution line with a neutral wire, pole transformers, and surge arresters in the presence of nearby buildings. A 3D-subgrid model was employed. Lightning was represented by the TL model. As expected, the presence of nearby buildings caused reduction of lightning-induced voltages. The observed trend is in general agreement with that reported from the small-scale experiment carried out by Piantini *et al.* (2000). This study is further discussed in Section 5.6.

Chen *et al.* (2018) have computed voltages, induced by lightning strikes to flat lossy ground, on a single overhead conductor with a neutral wire and surge arresters. A 3D-nonuniform grid was employed. Lightning was represented by the MTLL model. Each arrester was represented by a lumped circuit composed of two diodes, a resistor, a capacitor, and a controlled voltage source. They have examined influences of different arrester spacing arrangements on the suppression of lightning-induced voltages.

Rizk *et al.* (2020) have studied effects of a shield wire, surge arresters, and grounding electrodes on lightning-induced voltages on an overhead three-phase power distribution line. They have used the 2D-FDTD method and the field-to-conductor coupling model of Agrawal *et al.* Lightning was represented by the MTLL model.

Zhang, L. *et al.* (2019a) have studied effects of one or two shield wires on the reduction of lightning-induced voltages on overhead three-phase power distribution lines with vertical and horizontal conductor arrangements. They have examined the accuracy of the shielding factor, defined by Rusck (1957) as the ratio of voltage in the presence of shield wire to that in its absence, and shown that its accuracy decreases with decreasing ground conductivity. They have used the 2D-FDTD method and the field-to-conductor coupling model of Agrawal *et al.* Lightning was represented by the MTLE model.

Thang *et al.* (2014) have computed lightning-induced voltages at different points along a 5-mm-radius, 1-km-long single overhead wire, taking into account corona space charge around the wire. A 3D-nonuniform grid is employed. Lightning was represented by the MTLL model. The progression of corona streamers from the wire was represented as the radial expansion of cylindrical weakly conducting (40 μS/m) region around the wire. The magnitudes of lightning-induced voltages in the presence of corona discharge are larger than those computed without considering corona. Induced voltage risetimes, in the presence of corona discharge, are longer than those computed without considering corona. It appears that the distributed impedance discontinuity, associated with the corona development on the wire, is the primary reason for higher induced-voltage peaks and longer voltage risetimes, compared to the case without corona. This study is further discussed in Section 5.4.

Yang *et al.* (2011) have employed a two-step approach for evaluating lightning-induced voltages on a single overhead conductor above flat lossy ground. The first step is a 2D-cylindrical FDTD computation of electric and magnetic fields generated by a nearby lightning strike that illuminate the 3D domain accommodating the overhead conductor. The second step is a 3D-FDTD computation of lightning-induced voltages on the overhead conductor illuminated by incident electromagnetic fields originating from the boundary of the 3D-computational domain. Lightning was represented by the TL model. Using the same two-step approach, Yang *et al.* (2012) have computed currents, induced by lightning strikes to flat lossy ground, on a 100-m-long conductor in an insulating pipe buried in the ground. In this case, lightning was represented by the MTLE model. They have shown that lightning-induced currents in the conductor in the buried insulating pipe are reduced by installing shield wires on the surface of the ground.

Tanaka *et al.* (2016) have computed currents, induced by lightning strikes to flat ground, in the insulated metallic sheath (represented, along with the insulated wires inside it, by an equivalent solid conductor) of 1-km-long telecommunication cable buried in the ground and its buried shield (bare) wire installed above the cable. A 3D-nonuniform grid was employed and lightning was represented by the MTLL model. They have shown that computed waveforms of currents induced in the cable metallic sheath and in the shield wire agree well with the corresponding ones measured by Barbosa *et al.* (2008). This study is further discussed in Section 5.7.

Soto *et al.* (2014) have studied voltages, induced by lightning strikes to the top of a mountain, on a single overhead conductor located above a slope of the mountain and other non-flat terrain. They have used the 2D-FDTD method and the field-to-conductor coupling model of Agrawal *et al.* Lightning was represented by the TL model. They have shown that lightning-induced voltages are significantly influenced by topography.

Zhang, J. *et al.* (2019a) have analyzed voltages, induced by lightning strikes to the top of a cone-shaped mountain, on an overhead three-phase power distribution line with or without a shield wire with vertical arrangement of conductors and on a three-phase line with or without two shield wires with horizontal arrangement of conductors.

The distribution lines were located on flat lossy ground near the base of the mountain. They have used the 2D-FDTD method and the field-to-conductor coupling model of Agrawal *et al.* Lightning was represented by the MTLE model. They have shown that lightning-induced voltages are enhanced for strikes to the mountain compared to strikes to flat ground, particularly for low-conductivity ground. Also, they have shown that the presence of shield wires can reduce lightning-induced voltages.

Zhang, J. *et al.* (2019b) have analyzed voltages, induced by lightning strikes to the top of an oblique cone-shaped mountain, on a single overhead conductor over the surface of the mountain or over the nearby flat lossy ground. They have used the 3D-FDTD method, uniform grid, and the field-to-conductor coupling model of Agrawal *et al.* Lightning was represented by the TL model. They have shown that lightning-induced voltages are influenced by the steepness of the mountain and the location of the line relative to the mountain.

Ishii *et al.* (2012) have computed currents and voltages on electrical wiring in a building whose lightning rod is struck by lightning. A 3D-uniform grid is employed. Lightning channel was represented by a vertical conductor having a series distributed resistance of 1 Ω/m. They have shown that induced currents and voltages are higher on the top and bottom floors.

Du *et al.* (2016) have computed currents and voltages on electrical wiring in a building struck by lightning. A 3D-nonuniform grid was employed. Lightning channel was represented by a vertical perfect conductor. The building was ignored, and only a single downconductor of its lightning protective system was represented by a simple vertical conductor and electrical wiring was represented by parallel perfect conductors whose terminals were open or connected via lumped elements (or surge protective devices). They also have shown that induced currents and voltages are higher on the top and bottom floors.

Diaz *et al.* (2017) have analyzed voltages and currents in a buried coaxial cable entering building struck by lightning. A 3D-nonuniform grid was employed. Lightning channel was represented by a vertical conductor having a series distributed resistance of 1 Ω/m. They have shown that FDTD-computed results agree well with corresponding ones computed with the method of moments (MoM).

Baba and Rakov (2006) have analyzed voltages, induced by lightning strikes to a tall object and to flat ground, on an overhead single conductor above flat lossy ground. A 3D-uniform grid was employed. Lightning was represented by the TL model extended to include a tall strike object. They have shown that, in a realistic case (the return-stroke wavefront speed is one-third of the speed of light, $v = c/3$, the grounding impedance Z_{gr} is zero (much smaller than the equivalent impedance of lightning channel Z_{ch} or characteristic impedance of tall object Z_{ob}), and Z_{ch} is three times higher than Z_{ob}), the ratio of magnitudes of lightning-induced voltage at a distance of 100 m from the lightning channel due to a lightning strike to the tall object to that due to the strike to flat ground increases with increasing the object height from 0 to 100 m and decreases with increasing the object height from 100 to 300 m. This study is further discussed in Section 5.2.

Zhang, Q. *et al.* (2015a) have analyzed voltages, induced by lightning strikes to a tall object, on a single overhead conductor above flat lossy ground. They have used the 2D-FDTD method and the field-to-conductor coupling model of Agrawal *et al*. The TL model extended to include a tall strike object was employed and the frequency dependence of ground parameters was considered (both the ground resistivity and permittivity decrease with increasing frequency). They have shown that the influence of frequency-dependent resistivity is more significant for higher-resistivity soil.

Rizk *et al*. (2016b) have computed voltages, induced by lightning strikes to the tip of a blade of a wind-turbine-generator tower, on an overhead three-phase power distribution line above a stratified ground. A 3D-nonuniform grid was employed. Lightning was represented by the TL model. The bottom of the TL-represented lightning channel was connected to the tip of a blade of a wind-turbine-generator tower, which was represented by a 109-m-high perfectly conducting structure. They have shown that the lower layer of the stratified ground is more influential for first strokes and for the upper layer with higher resistivity.

5.2   Voltages induced on an overhead single conductor by lightning strikes to a nearby tall grounded object

5.2.1   Introduction

In order to optimize lightning protection means of telecommunication and power distribution lines, one needs to know voltages that can be induced on line conductors by lightning strikes to ground or to nearby grounded objects. The presence of tall strike object can serve to either increase or decrease lightning electric fields and lightning-induced voltages, as discussed below.

Fisher and Schnetzer (1994) examined the dependence of triggered-lightning electric fields on the height of strike object at Fort McClellan, Alabama. The fields were measured on the ground surface at distances of 9.3 and 19.3 m from the base of grounded metallic strike rod whose height was either 4.5 or 11 m. They observed that the leader electric fields (approximately equal in magnitude to the corresponding return-stroke fields at such close distances) tended to be reduced as the strike object height increased. Thus, it appears that the presence of strike object served to reduce electric fields in its vicinity relative to the case of lightning strike to flat ground.

Miyazaki and Ishii (2004), using the Numerical Electromagnetic Code (NEC-2) (Burke and Poggio 1980), examined the influence of the presence of tall strike object (60 to 240 m in height) on the associated electromagnetic fields at ground level 100 m to 500 km away from the base of the strike object. They represented the lightning channel by a vertical wire having distributed resistance ($1~\Omega/m$) and additional distributed inductance ($3~\mu H/m$), energized by a voltage source connected between the channel and the strike object represented by a vertical perfectly conducting wire. The voltage source had internal resistance of $300~\Omega$. Grounding resistance of the strike object was assumed to be $30~\Omega$ and ground conductivity was set to 0.003 S/m. The ratio of the calculated vertical electric field due to a lightning strike to the tall object to that due to the same strike to flat ground was found to be smaller than unity at horizontal distances of 100 to 600 m from the lightning channel and larger than unity beyond 600 m. The ratio reached its peak around several kilometers from the channel and then exhibited a decrease with increasing horizontal distance. Miyazaki and Ishii noted that the latter decrease was due to the propagation effects (attenuation of electromagnetic waves as they propagate over lossy ground).

Baba and Rakov (2005b) compared the distance dependences of vertical electric and azimuthal magnetic fields due to a lightning strike to a tall object with those due to the same strike to flat ground, using the TL model extended to include a tall strike object (Baba and Rakov 2005a). In this model, any grounding impedance can be directly specified and the total charge transfer to ground is the same regardless of the presence of strike object. Their findings can be summarized as follows. The electric field for the strike-object case is reduced relative to the flat-ground case at closer distances from the object. In an idealized case that is characterized by the return stroke front speed equal to the speed of light, $v = c$, the current reflection coefficient at the bottom of the strike object $\rho_{bot} = 1$ (grounding impedance $Z_{gr} = 0$), and that at the top of the object for upward-propagating waves $\rho_{top} = 0$ (characteristic impedance of the object is equal to that of the channel, $Z_{ob} = Z_{ch}$), the ratio of the vertical electric fields at ground level for the strike-object and flat-ground cases (electric-field-attenuation factor) is $d/\sqrt{(d^2 + h^2)}$, where h is the height of the strike object and d is the horizontal distance from the object. The corresponding ratio for the azimuthal magnetic field is equal to unity. Baba and Rakov (2005b) showed that the ratio for either electric or magnetic field increases with decreasing ρ_{bot} ($\rho_{bot}<1$), decreasing ρ_{top} ($\rho_{top}<0$, except for the case of $\rho_{bot} = 0$), and decreasing v ($v<c$), and that at larger distances it becomes greater than unity.

It follows from the above that the presence of tall strike object reduces lightning electric fields relative to the case of strikes to flat ground at closer distances and enhances them at larger distances. Note that enhancement of remote lightning electric and magnetic fields by the presence of tall strike object was also discussed by Diendorfer and Schulz (1998), Rachidi *et al.* (2001), Rakov (2001), Kordi *et al.* (2003), and Bermudez *et al.* (2005).

Piantini and Janiszewski (1998) have shown that the magnitude of lightning-induced voltage at the midpoint of a 5-km-long horizontal wire (matched at either end) located 10 m above perfectly conducting ground and 50 m away from the strike object, increases with increasing the height of the object from 0 to 150 m if the risetime of lightning current is 0.5 µs, and decreases if the risetime is 1 µs or longer. Further, Piantini and Janiszewski (2003) have shown that the magnitude of lightning-induced voltage, at the midpoint of a 10-km-long horizontal wire located 10 m above perfectly conducting ground and 60 m away from the vertical lightning channel, decreases as the height of the junction point of the descending and upward connecting leaders gets larger when the risetime of lightning current is 3 µs. Note that an upward connecting leader launched from flat ground can be regarded as a grounded strike object. They assumed that the current reflection coefficient at the bottom of the strike object (Piantini and Janiszewski 1998) or at the bottom of the upward connecting leader (Piantini and Janiszewski 2003) was equal to zero. Induced voltages were computed using the Rusck field-to-wire electromagnetic coupling model (Rusck 1957) modified to take into account a tall strike object (modified Rusck model) and assuming that the current-propagation speeds along the vertical lightning channel and along the strike object were 0.3c and c, respectively. Note that Cooray (1994) showed that the Rusck model was incomplete (because of neglecting the portion of the horizontal electric field due to the vector potential) but yielded induced voltages that were identical to those calculated using the more accurate Agrawal model (Agrawal *et al.* 1980) for the case of an infinitely long horizontal wire and a vertical lightning strike to flat, perfectly conducting ground. Later, Michishita and Ishii (1997) showed that the Rusck model was equivalent to the Agrawal model even if the horizontal wire had a finite length. Piantini and Janiszewski (1998) demonstrated the validity of the modified Rusck model by comparing calculated voltages with those measured in experiments of Yokoyama *et al.* (1983, 1986).

Silveira and Visacro (2002) (see also Silveira *et al.* 2002) have shown that the magnitude of lightning-induced voltage on a 300-m-long horizontal wire (matched at either end) located 10 m above perfectly conducting ground and 100 m away from the vertical lightning channel increases with increasing the height of the junction point between the descending and upward connecting leaders. They employed a model based on the hybrid electromagnetic field/circuit-theory approach (Visacro *et al.* 2002), in which the current wave propagation speed along the leader channels both above and below the junction point was equal to c, and used a current waveform having a risetime of 1 μs. Results presented by Silveira and Visacro (2002) appear to be not consistent with those of Piantini and Janiszewski (1998, 2003). Further discussion is found in Section 5.2.4.

Voltages induced by lightning strikes to a tall object were also calculated by Michishita *et al.* (2003), who represented the strike object by an *R-L-C* transmission line and used the Agrawal model, and Pokharel *et al.* (2004), who represented the strike object by a vertical perfectly conducting wire and used the NEC-2 (Burke and Poggio 1980). Both groups employed Norton's approximation (Norton 1937) to take into account the lossy-ground effect and succeeded in reproducing the corresponding measured voltages (Michishita *et al.* 2003) induced on a test line by lightning strikes to a 200-m-high object (Fukui chimney).

In the rest of Section 5.2, we examine in detail the ratios of magnitudes of lightning-induced voltages for the cases of strikes to a tall object and to flat ground as a function of distance from the lightning channel, d, height of the strike object, h, the current reflection coefficients at the extremities of the strike object, ρ_{top} and ρ_{bot}, the current reflection coefficient at the channel base (ground) in the case of strikes to flat ground, ρ_{gr}, the risetime of lightning return-stroke current, and return-stroke speed, v. The current reflection coefficients, ρ_{top}, ρ_{bot}, and ρ_{gr} are given by

$$\rho_{top} = \frac{Z_{ob} - Z_{ch}}{Z_{ob} + Z_{ch}} \tag{5.1}$$

$$\rho_{bot} = \frac{Z_{ob} - Z_{gr}}{Z_{ob} + Z_{gr}} \tag{5.2}$$

$$\rho_{gr} = \frac{Z_{ch} - Z_{gr}}{Z_{ch} + Z_{gr}} \tag{5.3}$$

where Z_{ob} is the characteristic impedance of the strike object, Z_{ch} is the equivalent impedance of the lightning channel, and Z_{gr} is the grounding impedance. Table 5.4 summarizes relations between current reflection coefficients (ρ_{bot}, ρ_{top}, and ρ_{gr}) and pertinent impedances (Z_{ob}, Z_{ch}, and Z_{gr}) for four sets of ρ_{top}, and ρ_{bot} considered here. It is clear from Table 5.4 that ρ_{gr} is not an independent parameter; it is equal to ρ_{bot} (as long as $Z_{ch} \geq Z_{ob} \gg Z_{gr}$, which is expected in most practical situations).

In Section 5.2.2, we present the methodology for examining electromagnetic coupling between the lightning channel attached to a tall grounded object and a horizontal wire above ground. In Section 5.2.3, we compare induced voltages due to a lightning strike to a 100-m-high object with their counterparts due to the same strike to flat ground, calculated by Baba and Rakov (2006) for different values of d, ρ_{top}, ρ_{bot}, and ρ_{gr}. Further, we investigate the influences on the ratio of magnitudes of lightning-induced voltages for both the tall-object and flat-ground cases of the return-stroke speed, v, the height of strike object, h, and the risetime of lightning return-stroke current waveform. In Section 5.2.4, we compare the lightning-induced voltages calculated using the FDTD method (Yee 1966) with those calculated by Piantini and Janiszewski (1998; 2003) and Silveira and Visacro (2002). In Appendix 5.2.A, we show that, for the case of strikes to flat ground, the FDTD method used by Baba and Rakov (2006) yields reasonably accurate results by comparing lightning-induced voltages calculated using the FDTD method with those measured in a small-scale experiment by Ishii *et al.* (1999). In Appendix 5.2.B, the FDTD-calculated results are compared with those calculated using Rusck's formula (1958). In Appendix 5.2.C, we compare induced voltages due to lightning strikes to a 200-m-high object calculated by Baba and Rakov (2006) using the FDTD method with those measured by Michishita *et al.* (2003).

Table 5.4   Relations between current reflection coefficients (ρ_{top}, ρ_{bot}, and ρ_{gr}) and impedances (Z_{ob}, Z_{ch}, and Z_{gr}) for four different sets of ρ_{top} and ρ_{bot}

ρ_{top}	ρ_{bot}	**Impedances from (5.1) and (5.2)**	ρ_{gr}
−0.5	1	$Z_{gr} = 0$, $Z_{ob} = Z_{ch}/3$	1
0	1	$Z_{gr} = 0$, $Z_{ob} = Z_{ch}$	1
−0.5	0	$Z_{gr} = Z_{ob}$, $Z_{ob} = Z_{ch}/3$	0.5
0	0	$Z_{gr} = Z_{ob} = Z_{ch}$	0

Note that $\rho_{bot} = 1$ can be also achieved when $Z_{gr} \ll Z_{ob}$ (as opposed to $Z_{gr} = 0$). Also note that, in the case of $\rho_{bot} = 0$ ($Z_{gr} = Z_{ob}$), ρ_{gr} becomes equal to $-\rho_{top}$. Thus, the magnitude of current waves injected into both the lightning channel and the strike object in the case of strike to tall object, $(1-\rho_{top})I_{sc}/2$, becomes equal to that injected into the channel in the case of the same strike to flat ground, $(1+\rho_{gr})I_{sc}/2$. 2006 ©IEEE. Reprinted, with permission, from Baba and Rakov (2006, Table I).

5.2.2   Methodology

The model used by Baba and Rakov (2006) is presented in Figure 5.1, which shows a
horizontal perfectly conducting wire of length 1200 m and radius 5 mm, at distance
$d = 40$, 60, 100, or 200 m from a tall object of height $h = 100$ m struck by lightning.
The horizontal wire is located 10 m above ground. Each end of the wire is terminated in

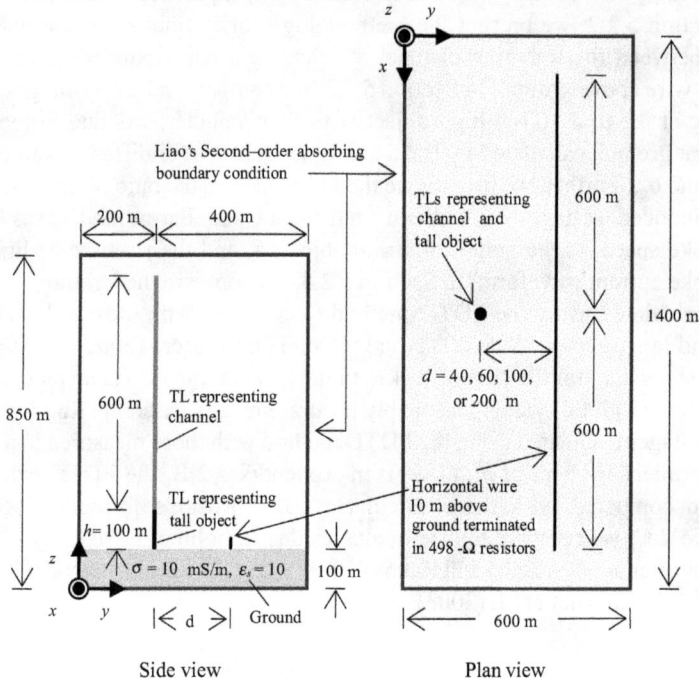

*Figure 5.1   Configuration used for the FDTD computations presented in
Section 5.2. A 1200-m-long horizontal perfectly conducting wire at
distances d = 40, 60, 100, or 200 m from a tall object of height
h = 100 m struck by lightning. The horizontal wire has a radius of
5 mm and is located 10 m above ground. Each end of the wire is
terminated in a 498-Ω matching resistor. The tall object and the
lightning channel are both represented by a vertical array of current
sources that are activated according to the TL model extended to
include a tall strike object (Baba and Rakov 2005). The working
volume of 1400 m × 600 m × 850 m, which is divided into
5 m × 5 m × 5 m cubic cells, is surrounded by six planes of Liao's
second-order absorbing boundary condition (Liao et al. 1984). 2006
©IEEE. Reprinted, with permission, from Baba and Rakov (2006,
Figure 1)*

a 498-Ω matching resistor. The conductivity, relative permittivity, and relative permeability of the ground are set to $\sigma = 10$ mS/m, $\varepsilon_r = 10$, and $\mu_r = 1$, respectively. A 600-m-long vertical lightning channel is connected to the top of the tall object. The influence of reflections from the upper end of the 600-m-long channel does not appear in calculated lightning-induced voltage waveforms (see Figures 5.2–5.4), shown on a 4-μs time scale. Lightning-induced voltage on the horizontal wire is evaluated by integrating the vertical electric field from the ground surface to the height of the wire. The electric field is calculated using the FDTD method of solving the discretized Maxwell's equations. Calculations are also carried out for the cases of lightning strike to flat lossy ground ($\sigma = 10$ mS/m) and to flat perfectly conducting ground ($\sigma = \infty$). The working volume of 1400 m × 600 m × 850 m (see Figure 5.1) is divided into

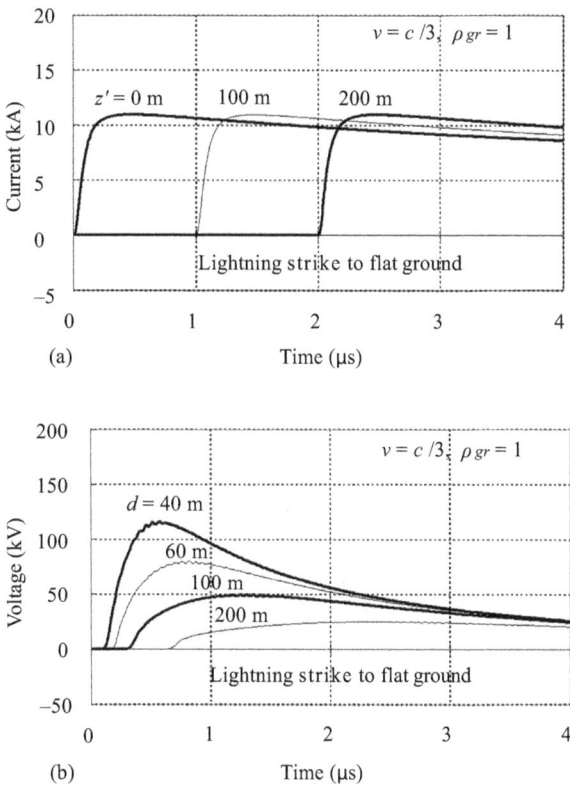

Figure 5.2 (a) Current waveforms for a strike to flat ground ($v = c/3$, $\rho_{gr} = 1$) at different heights, $z' = 0$, 100, and 200 m, along the lightning channel, calculated using (5.6). (b) Lightning-induced voltages at the midpoint of the horizontal wire at distances $d = 40$, 60, 100, and 200 m from the lightning channel, obtained by integrating the FDTD-calculated vertical electric field. 2006 ©IEEE. Reprinted, with permission, from Baba and Rakov (2006, Figure 2)

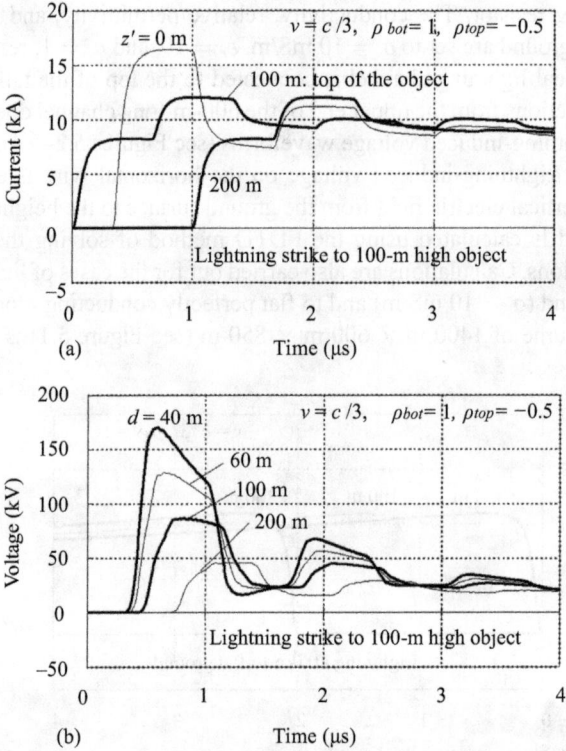

Figure 5.3 (a) Current waveforms for a strike to the 100-m-high object at
          different heights above ground, $z' = 0$ (bottom of the tall object), 100
          (top of the object and bottom of the channel), and 200 m (100 m above
          the top of the object), calculated using (5.4) and (5.5). (b) Lightning-
          induced voltages at the midpoint of the horizontal wire at distances
          $d = 40, 60, 100,$ and 200 m from the strike object, obtained by
          integrating the FDTD-calculated vertical electric field. Note that
          voltage magnitudes in (b) are higher than their counterparts for the
          flat-ground case shown in Figure 5.2(b) for all the distances
          considered. 2006 ©IEEE. Reprinted, with permission, from Baba and
          Rakov (2006, Figure 3)

5 m × 5 m × 5 m cubic cells and surrounded by six planes of Liao's second-order
absorbing boundary condition (Liao *et al.* 1984) to suppress unwanted reflections
there. The 5-mm-radius horizontal wire is represented in the FDTD procedure by a
zero-radius wire (simulated by forcing the tangential components of electric field
along the axis of the wire to zero) embedded in cells for which the relative permittivity
is set to an artificially lower value and the relative permeability to an artificially higher
value (Noda and Yokoyama 2002). Values of ε_r and μ_r are set to 0.213 and 1/0.213,
respectively (see Appendix 5.2.A).

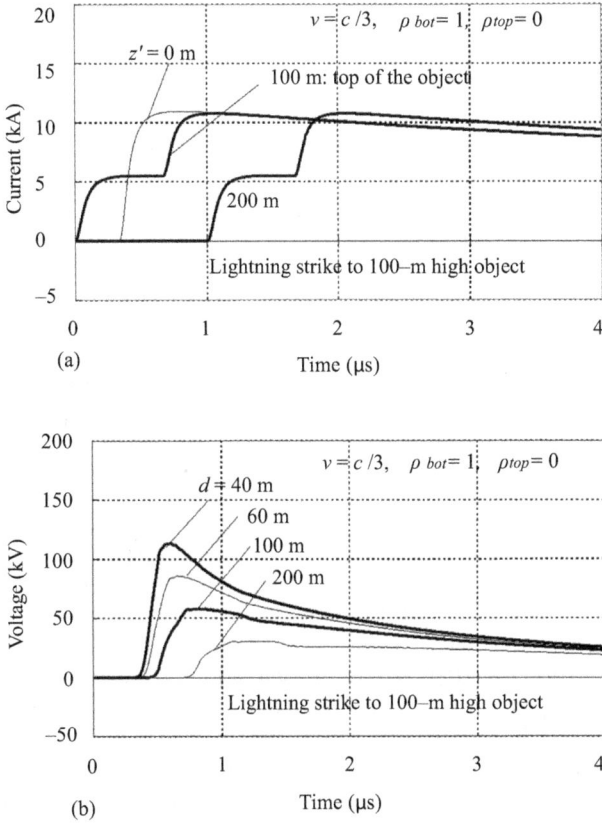

Figure 5.4 Same as Figure 5.3 but for $\rho_{top} = 0$. Note that voltage magnitudes in (b) are lower at d = 40 m and higher at d = 60, 100, and 200 m than their counterparts for the flat-ground case in Figure 5.2(b). 2006 ©IEEE. Reprinted, with permission, from Baba and Rakov (2006, Figure 4)

In order to find the distribution of current along both the lightning channel and the tall strike object, the "engineering" TL model extended to include a tall strike object (Baba and Rakov 2005a) was used. The reason for using the TL model instead of an electromagnetic return-stroke model (e.g., Rakov and Uman 1998) that would allow a self-consistent full-wave solution for both lightning-current distribution and fields needed to calculate voltages induced on the wire, is that the TL model allows one to set more directly the speeds of current waves along the strike object and the lightning channel, as well as reflection coefficients at the extremities of the strike object. Evaluation of the dependence of lightning-induced voltages on the assumed values of these speeds and reflection coefficients was one of the main objectives of Baba and Rakov (2006).

For the case of lightning strike to a tall object, equations for current, $I(z',t)$, along the tall object ($0 \leq z' \leq h$) and along the lightning channel ($z' \geq h$), proposed by Baba and Rakov (2005a), are given below.

$$I(z',t) = \frac{1 - \rho_{top}}{2} \sum_{n=0}^{\infty} \left[\begin{array}{l} \rho_{bot}{}^{n}\rho_{top}{}^{n}I_{sc}\left(h, t - \frac{h - z'}{c} - \frac{2nh}{c}\right) \\ + \rho_{bot}{}^{n+1}\rho_{top}{}^{n}I_{sc}\left(h, t - \frac{h + z'}{c} - \frac{2nh}{c}\right) \end{array} \right] \tag{5.4}$$

$$\text{for} \quad 0 \leq z' \leq h \quad \text{(along the strike object)}$$

$$I(z',t) = \frac{1 - \rho_{top}}{2} \left[\begin{array}{l} I_{sc}\left(h, t - \frac{z' - h}{c}\right) \\ + \sum_{n=1}^{\infty} \rho_{bot}{}^{n}\rho_{top}{}^{n-1}\left(1 + \rho_{top}\right)I_{sc}\left(h, t - \frac{z' - h}{v} - \frac{2nh}{c}\right) \end{array} \right]$$

$$\text{for} \quad z' \geq h \quad \text{(along the lightning channel)}$$

$$\tag{5.5}$$

where $I_{sc}(h,t)$ is the lightning short-circuit current (which is defined as the lightning current that would be measured at an ideally grounded strike object of negligible height), ρ_{bot} is the current reflection coefficient at the bottom of the tall object, ρ_{top} is the current reflection coefficient at the top of the object for upward-propagating waves, n is an index representing the successive multiple reflections occurring at the two ends of the tall object, c is the speed of light (current-propagation speed along the strike object), and v is the current-propagation speed along the lightning channel.

Equations (5.4) and (5.5) show that two current waves of the same magnitude, $(1-\rho_{top})I_{sc}(h,t)/2$, are initially injected downward, into the tall object, and upward, into the lightning channel.

The current distribution, $I(z',t)$, along the lightning channel for the case of strike to flat ground, is given by (Baba and Rakov 2005a)

$$I(z',t) = \frac{1 + \rho_{gr}}{2} I_{sc}\left(0, t - \frac{z'}{v}\right) \tag{5.6}$$

where $I_{sc}(0,t)$ is the lightning short-circuit current (same as $I_{sc}(h,t)$ in (5.4) and (5.5) but injected at $z' = 0$ instead of $z' = h$) and ρ_{gr} is the current reflection coefficient at the channel base (ground). Note that when h approaches zero (5.5) reduces to (5.6) and (5.4) reduces to (5.6) with $z' = 0$ (Baba and Rakov 2005a). When $h \rightarrow 0$, terms in (5.5) become $I_{sc}(h, t - (z' - h)/v) \approx I_{sc}(0, t - z'/v)$, $I_{sc}(h, t - (z' - h)/v - 2nh/c) \approx I_{sc}(0, t - z'/v)$, $\sum_{n=1}^{\infty} \rho_{bot}{}^{n-1}\rho_{top}{}^{n-1} \approx 1/(1 - \rho_{bot}\rho_{top})$, and when $h \rightarrow 0$ and $z' = 0$, terms in (5.4) become $I_{sc}(h, t - (h - z')/c - 2nh/c) \approx I_{sc}(0, t)$, $I_{sc}(h, t - (h + z')/c - 2nh/c) \approx I_{sc}(0, t)$, $\sum_{n=0}^{\infty} \rho_{bot}{}^{n}\rho_{top}{}^{n} \approx 1/(1 - \rho_{bot}\rho_{top})$. The total charge transfer to ground, calculated integrating current given by (5.4) at $z' = 0$, is the same as that calculated integrating current given by (5.6) at $z' = 0$ (Baba and Rakov 2005b). Therefore, current distributions for the case of strikes to a tall object ((5.4) and (5.5)) and for the case of strikes to flat ground ((5.6)) correspond to the same lightning discharge, as required for examining the influence of the strike object. On the other hand, currents injected into the lightning channel in these two cases are generally different: $I = (1 - \rho_{top})I_{sc}/2$ vs. $I = (1 + \rho_{gr})I_{sc}/2$, unless $\rho_{top} = 0$ and $\rho_{gr} = 0$ (matched conditions at the position of the source) or $\rho_{top} = -\rho_{gr}$ ($Z_{ob} = Z_{gr}$). Both these situations are physically unrealistic, since typically $\rho_{gr} = 1$ ($Z_{gr} \ll Z_{ob}$ and $Z_{gr} \ll Z_{ch}$).

In the FDTD calculations, the lightning channel and the tall strike object are each simulated by a vertical array of current sources (Baba and Rakov 2003). Each current source has a length of 5 m and is described by specifying the four magnetic-field vectors forming a square contour surrounding the cubic cell representing the current source (Baba and Rakov 2003).

Lightning-induced voltages are calculated at the midpoint of the horizontal wire with a time increment of 5 ns. Verification of the applicability of the FDTD approach to calculation of lightning-induced voltages is presented in Appendices 5.2.A, 5.2.B, and 5.2.C.

5.2.3 *Analysis and results*

In this section, induced voltages on the horizontal wire (see Figure 5.1) due to a lightning strike to the 100-m-high object are compared with their counterparts due to the same strike to flat ground. Presented first are results for perhaps the most realistic situation in which $v = c/3$ (e.g., Rakov 2007), the current reflection coefficient at the bottom of the object is $\rho_{bot} = 1$ (Z_{ob} is usually much larger than Z_{gr}), and the current reflection coefficient at the top of the tall object is $\rho_{top} = -0.5$. Note that Janischewskyj *et al.* (1996), from their analysis of five current waveforms measured 474 m above ground on the CN Tower, inferred ρ_{top} to vary from -0.27 to -0.49, and Fuchs (1998), from 13 simultaneous current measurements at the top and bottom of the Peissenberg tower, found ρ_{top} to vary from -0.39 to -0.68. In the case of lightning strike to flat ground, we assume that the current reflection coefficient at the channel base (ground) is $\rho_{gr} = 1$ (Z_{ch}, which is expected to be several hundred ohm or more, is much larger than Z_{gr} at the strike point; e.g., Rakov 2001). We describe $I_{sc}(h,t)$ or I_{sc} $(0,t)$ using a current waveform proposed by Nucci *et al.* (1990), which is thought to be typical for lightning subsequent return strokes. The zero-to-peak risetime, *RT*, of this current waveform is about 0.5 μs (the corresponding 10-to-90% *RT* is 0.15 μs).

Figure 5.2(a) shows current waveforms at different heights, $z' = 0$, 100, and 200 m, along the lightning channel for a lightning strike to flat ground, calculated using (5.6). Figure 5.2(b) shows corresponding lightning-induced voltages at the midpoint of the horizontal wire at distances $d = 40$, 60, 100, and 200 m from the lightning channel. As expected, the voltage magnitude decreases with increasing distance. Figure 5.3(a) and (b) is similar to Figure 5.2(a) and (b), respectively, but for the case of lightning strike to the 100-m-high object (see Figure 5.1). Figure 5.3(a) shows current waveforms at different heights, $z' = 0$ (bottom of the tall object), 100 (top of the object and bottom of the channel), and 200 m (100 m above the top of the object), calculated using (5.4) and (5.5), and Figure 5.3(b) shows corresponding lightning-induced voltages. The magnitude of lightning-induced voltage is always larger in the case of lightning strike to the 100-m-high object than in the case of the same strike to flat ground, regardless of the distance between the lightning channel/ strike object and the horizontal wire. The ratios of magnitudes of lightning-induced voltages for the strike-object case to that for the flat-ground case are 1.5, 1.6, 1.7, and 1.8 for $d = 40, 60, 100$, and 200 m, respectively. These ratio values for $\sigma = 10$ mS/m are not much different from their counterparts computed assuming perfectly conducting ground ($\sigma = \infty$): 1.5, 1.7, 2.0, and 2.2 for $d = 40, 60, 100$, and 200 m, respectively. The difference between magnitudes of lightning-induced voltages at the midpoint of the horizontal wire located 10 m above perfectly conducting ground for a lightning strike to flat ground based on the FDTD calculations and those calculated using Rusck's formula (1958) is within 5% at distances ranging from $d = 40$ to 200 m (see Appendix 5.2.B). In summary, it is clear that for $RT = 0.5 \,\mu s$, $v = c/3$, $\rho_{top} = -0.5$, and $\rho_{bot} = 1$ lightning-induced voltages at distances ranging from 40 to 200 m are enhanced by the presence of the 100-m-high strike object.

We consider next the case of $v = c/3$, $\rho_{top} = 0$, and $\rho_{bot} = 1$, which differs from the previously discussed (basic) case by the value of ρ_{top}. The assumption $\rho_{top} = 0$ implies that $Z_{ob} = Z_{ch}$ (matched conditions at the top of the object). Figure 5.4(a) shows current waveforms at different heights, $z' = 0, 100$, and 200 m, for a lightning strike to the 100-m-high object, and Figure 5.4(b) shows corresponding lightning-induced voltages. The magnitude of lightning-induced voltage at $d = 40$ m is a little smaller in the case of lightning strike to the tall object than in the case of the same lightning strike to flat ground (see Figure 5.2(b)), and larger at $d = 60, 100$, and 200 m. Thus, for $v = c/3$, $\rho_{top} = 0$, and $\rho_{bot} = 1$, the presence of the 100-m-high strike object leads to a decrease of lightning-induced voltages at $d = 40$ m (and at smaller distances) and to an increase for d ranging from 60 to 200 m.

We now consider the unrealistic but sometimes assumed case of $\rho_{bot} = 0$ and summarize all results of this section in Figure 5.5, which shows ratios of magnitudes of lightning-induced voltages for the strike-object and flat-ground cases for $v = c/3$ and different values of ρ_{top}, ρ_{bot}, and $\rho_{gr} = \rho_{bot}$ (except for $\rho_{bot} = 0$). In the case of $\rho_{bot} = 0$ ($Z_{gr} = Z_{ob}$), ρ_{gr} becomes equal to $-\rho_{top}$. Thus, for strikes to a tall object, the magnitudes of current waves injected into both the lightning channel and the strike object, $(1-\rho_{top}) I_{sc}/2$, become equal to that injected into the channel for strikes to flat ground, $(1+\rho_{gr}) I_{sc}/2$. As a result, the ratio of magnitudes of lightning-induced voltages for strike-object and flat-ground cases becomes independent of ρ_{top}. It is clear from Figure 5.5 that the ratio increases with decreasing ρ_{bot} ($\rho_{bot}<1$), decreasing ρ_{top} ($\rho_{top}<0$, except for the case of $\rho_{bot} = 0$), and with increasing distance, d. This tendency is similar to that observed by Baba and Rakov (2005b) for the vertical electric field or azimuthal magnetic field at ground level at distances $d = 40$ to 200 m. The ratio decreases with increasing v, as follows from a comparison of Figure 5.5 ($v = c/3$) with Figure 5.6, in which $v = c$ (the limiting value). As seen in Figure 5.6, the lightning-induced voltage is reduced at distances ranging from $d = 40$ m to $d = 200$ m due to the presence of the 100-m-high strike object when $v = c$, $\rho_{top} = 0$, and $\rho_{bot} = \rho_{gr} = 1$.

We now examine the magnitude of lightning-induced voltage as a function of strike-object (junction point) height, h, at $d = 100$ m. Figure 5.7(a) shows lightning-induced voltages based on FDTD calculations for $v = c/3$, $\rho_{top} = 0$, $\rho_{bot} = \rho_{gr} = 0$, and, $h = 0, 25, 50, 100, 200$, and 300 m. Figure 5.7(b) shows ratios

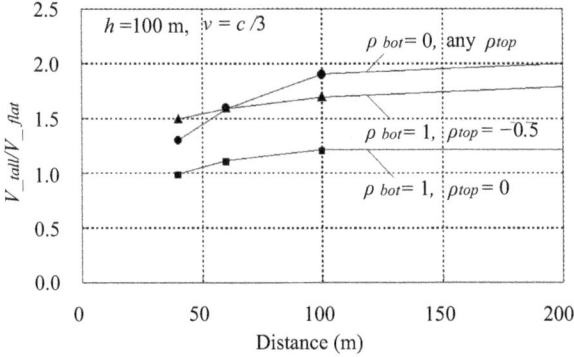

Figure 5.5   Ratios of magnitudes of lightning-induced voltages for strike-object (h = 100 m) and flat-ground cases for different values of ρ_{top} and ρ_{bot} Note that $\rho_{gr} = \rho_{bot}$, except for $\rho_{bot} = 0$. In the latter case ($Z_{gr} = Z_{ob}$), $\rho_{gr} = -\rho_{top}$ (see Table 5.4) and V_tall/V_flat is the same for any value of ρ_{top}. Current waves are assumed to propagate at speed c along the strike object and at speed v = c/3 along the lightning channel. 2006 ©IEEE. Reprinted, with permission, from Baba and Rakov (2006, Figure 5)

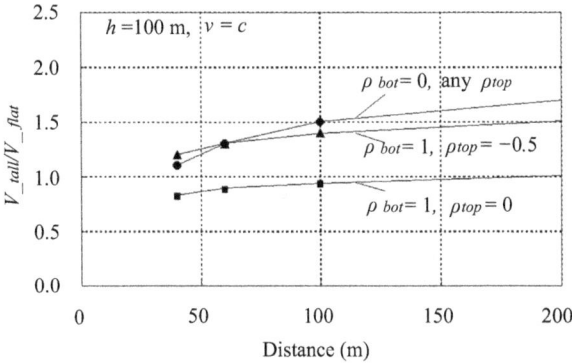

Figure 5.6   Same as Figure 5.5 but for the limiting case of v = c. 2006 ©IEEE. Reprinted, with permission, from Baba and Rakov (2006, Figure 6)

of magnitudes of lightning-induced voltages at $d = 100$ m for strike-object and flat-ground cases computed using different sets of ρ_{top} and ρ_{bot}. It is clear from Figure 5.7 that the ratio increases with increasing h up to 100 m and then decreases with increasing h. Figure 5.8, which is the same as Figure 5.7(b) but for $v = c$, suggests that, except for the case of $\rho_{top} = 0$ and $\rho_{bot} = 1$, the ratio at $d = 100$ m increases with increasing h up to 50–100 m and then decreases with increasing h.

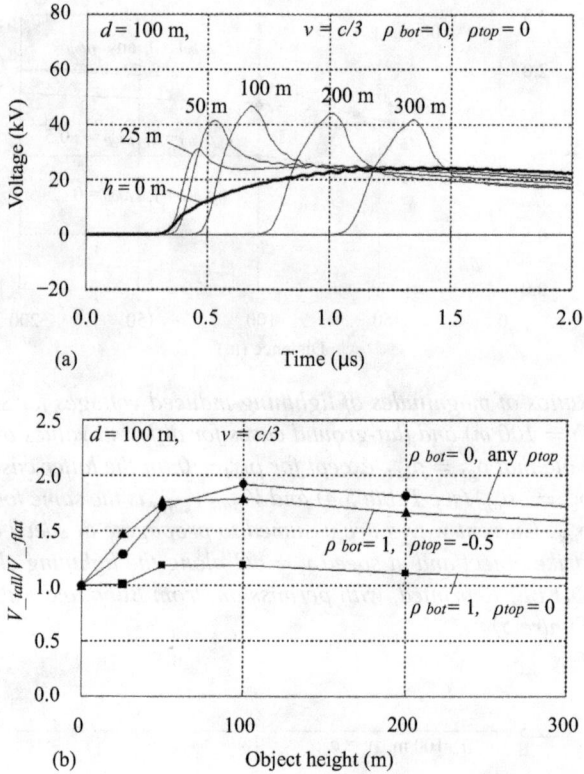

Figure 5.7 (a) *Lightning-induced voltages at the midpoint of the horizontal wire at a distance of d = 100 m from the strike object, based on the FDTD calculations for $\rho_{top} = 0$ and $\rho_{bot} = 0$, and different strike object heights, h = 0, 25, 50, 100, 200, and 300 m. (b) Ratios of magnitudes of lightning-induced voltages at d = 100 m for h ranging from 0 to 300 m to that for h = 0 (strike to flat ground) for different values of $\rho_{top}, \rho_{bot},$ and ρ_{gr}. Current waves are assumed to propagate at speed c along the strike object and at speed v = c/3 along the lightning channel. 2006 ©IEEE. Reprinted, with permission, from Baba and Rakov (2006, Figure 7)*

When $\rho_{top} = 0$ and $\rho_{bot} = 1$, the ratio decreases monotonically with increasing h. It follows from Figures 5.7 and 5.8 that the ratio decreases with increasing v.

Finally, we consider the lightning-induced voltage as a function of risetime (RT) of the lightning (short-circuit) current, I_{sc}. The waveform of I_{sc} is approximated by an expression containing the so-called Heidler function, and the zero-to-peak RTs are set to about 0.5 μs, as in the basic case, 1 μs, and 3 μs (the corresponding 10-to-90% RTs are 0.15, 0.39, and 1.42 μs, respectively). Figure 5.9(a) shows ratios of magnitudes of lightning-induced voltages at $d = 100$ m for the strike-object and flat-ground cases for

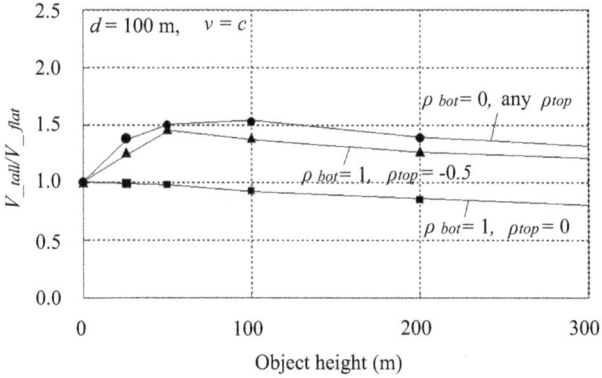

Figure 5.8   Same as Figure 5.7(b) but for v = c. 2006 ©IEEE. Reprinted, with permission, from Baba and Rakov (2006, Figure 8)

$\rho_{top} = -0.5$ and $\rho_{bot} = 1$ and different current risetimes. Figure 5.9(b) is the same as Figure 5.9(a) but for $\rho_{bot} = 0$. When $\rho_{bot} = 0$ and the risetime of I_{sc} is 3 µs, the ratio is less than unity and decreases monotonically with increasing h. It follows from Figure 5.9 that the ratio increases with decreasing the risetime of lightning current waveform.

5.2.4   Discussion

5.2.4.1   Comparison with calculations of Piantini and Janiszewski (1998, 2003)

Piantini and Janiszewski (1998), who considered a return stroke initiated at the top of a tall strike object without an upward connecting leader and used the modified Rusck model, have shown that the magnitude of lightning-induced voltage on a 5-km-long horizontal wire located 10 m above perfectly conducting ground and 50 m away from the strike object increases with increasing the height of the strike object for a lightning current waveform having a risetime of 0.5 µs (rising linearly to its maximum) and decreases for a current waveform having a risetime 1 µs or longer (rising linearly to its maximum). They used the TL model (Uman and McLain 1969) and assumed that the return-stroke speed $v = 0.3c$. Further, they assumed that no reflections occur at the top or at the bottom of the object. One can represent these conditions by setting $\rho_{top} = \rho_{bot} = \rho_{gr} = 0$, and $v = 0.3c$ in (5.4), (5.5), and (5.6). Figure 5.10 shows ratios of magnitudes of lightning-induced voltages on a 1200-m-long horizontal wire, matched at both ends, located 10 m above perfectly conducting ground at $d = 50$ m for the strike-object and flat-ground cases, calculated for the above conditions. Note that in these calculations the lightning current was assumed to rise linearly to its maximum in 0.5 or 3 µs. The ratios calculated by Piantini and Janiszewski (1998) at a distance of $d = 50$ m are also shown (see hollow triangles and circles in Figure 5.10). The trends predicted by the two models agree well except for the cases when $h \geq 50$ m and the lightning-current risetime is 0.5 µs. When the risetime of lightning current is 3 µs or longer and $\rho_{bot} = 0$, the ratios are less than unity, which indicates a decrease in induced voltage with increasing strike-object height. Piantini and Janiszewski (1998) attributed

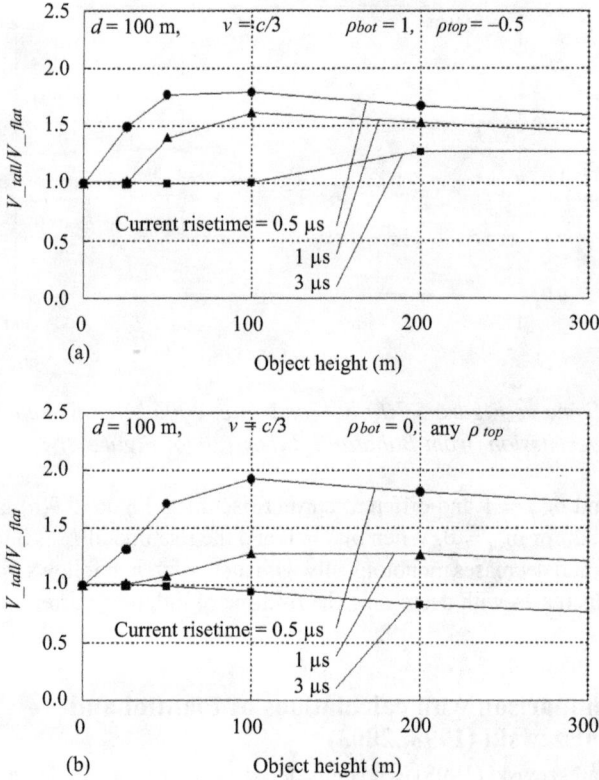

Figure 5.9   Ratios of magnitudes of lightning-induced voltages at d = 100 m for h
ranging from 0 to 300 m to that for h = 0 (strike to flat ground) for
different zero-to-peak risetimes of the lightning (short-circuit) current
I_{sc}, 0.5, 1, and 3 μs for (a) ρ_{top} = −0.5 and ρ_{bot} = 1 and (b) ρ_{bot} = 0
and any ρ_{top}. Current waves are assumed to propagate at speed c
along the strike object and at speed v = c/3 along the lightning
channel. 2006 ©IEEE. Reprinted, with permission, from Baba and
Rakov (2006, Figure 9)

the difference in trends for $RT = 0.5$ μs (hollow triangles in Figure 5.10) and $RT = 3$ μs (hollow circles in Figure 5.10) to different relative contributions from the "electro-static" (related to electric scalar potential) and "magnetic" (related to magnetic vector potential) components of lightning electric field to the induced voltage. Interestingly, the results of Piantini and Janiszewski (1998) for $RT = 0.5$ μs appear to be qualitatively consistent with those of Silveira and Visacro (2002) (discussed in Section 5.2.B; see Figure 5.12), although the latter are for $RT = 1$ μs, for which Piantini and Janiszewski found the opposite trend.

Piantini and Janiszewski (2003) have shown that the magnitude of lightning-induced voltage on a 5-km-long horizontal wire located 10 m above perfectly

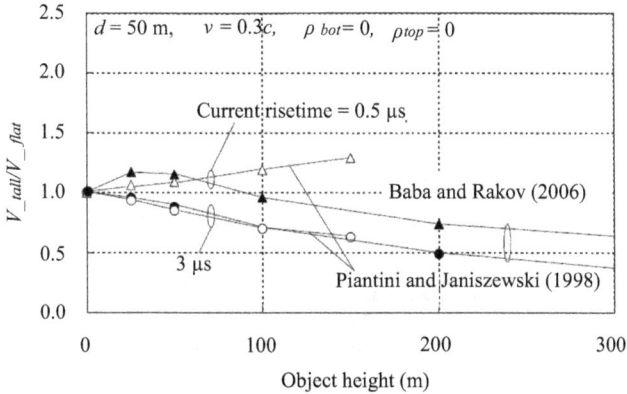

Figure 5.10 *Ratios of magnitudes of lightning-induced voltages at d = 50 m for h ranging from 0 to 300 m to that for h = 0 (strike to flat ground) based on the FDTD calculations (solid triangles and solid circles) for $\rho_{bot} = 0$ and $\rho_{top} = 0$. The lightning current is assumed to rise linearly to its maximum in 0.5 µs (triangles) or 3 µs (circles) and to propagate at speed c along the strike object and at speed v = 0.3c along the lightning channel. Ratios calculated for the same conditions by Piantini and Janiszewski (1998) are shown by hollow triangles and hollow circles. 2006 ©IEEE. Reprinted, with permission, from Baba and Rakov (2006, Figure 10)*

conducting ground and 60 m away from the lightning channel decreases as the height of the junction point between the descending and upward connecting leaders gets larger for a lightning current waveform having a risetime of 3 µs (rising linearly to its maximum). No strike object was considered, that is, they assumed that the upward connecting leader originated from the ground surface. Conceptually upward connecting leader originated from the ground can be viewed as a tall grounded strike object, which allows one to apply here the methodology described in Section 5.2.2. Piantini and Janiszewski (2003) assumed that both upward and downward current waves propagated from the junction point at the same speed, 0.3c. They seem to have assumed that no reflections occur at the top and bottom of the upward connecting leader. One can represent these conditions by setting $\rho_{top} = \rho_{bot} = \rho_{gr} = 0$ and replacing all the speeds (including c) in (5.4), (5.5), and (5.6) with 0.3c. Figure 5.11 shows ratios of magnitudes of lightning-induced voltages (based on FDTD calculations for $RT = 3$ µs) on a 1200-m-long horizontal wire, matched at both ends, located 10 m above perfectly conducting ground at d = 60 m for the strike-object and flat-ground cases, calculated for the above conditions. The ratios calculated by Piantini and Janiszewski (2003) are also shown (see hollow circles). The trends predicted by the two models agree well. The ratios are less than unity, which indicates a decrease in induced voltage with increasing the junction point height.

It is worth mentioning that a decrease in the induced voltage at a distance of about 50 m due to the presence of 30-m-long upward connecting leader was predicted more than 75 years ago by Wagner and McCann (1942, Figure 16).

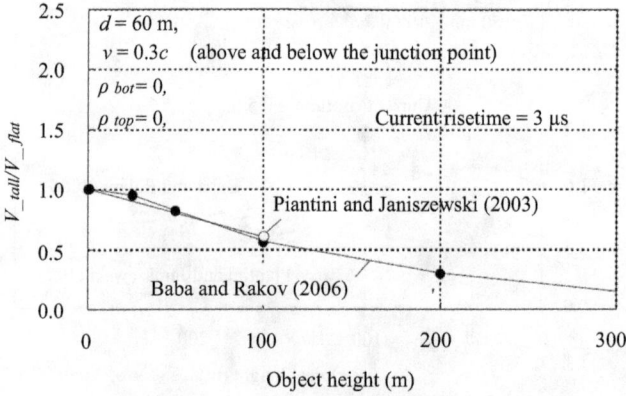

Figure 5.11   Ratios of magnitudes of lightning-induced voltages at d = 60 m for heights h of the junction point between the descending and upward connecting leaders ranging from 0 to 300 m to that for h = 0 (strike to flat ground without an upward connecting leader), based on the FDTD calculations (solid circles) for ρ_bot = 0 and ρ_top = 0. Lightning current is assumed to rise linearly to its maximum in 3 μs and to propagate at speed 0.3c along the leader channels both above and below the junction point. Ratios calculated for the same conditions by Piantini and Janiszewski (2003) are shown by two hollow circles, one of which (for h = 0) coincides with the solid circle for h = 0. 2006 ©IEEE. Reprinted, with permission, from Baba and Rakov (2006, Figure 11)

5.2.4.2   Comparison with calculations of Silveira and Visacro (2002)

As noted in Section 5.2.1, Silveira and Visacro (2002) (see also Silveira *et al.* (2002)), who considered a return stroke initiated at the junction point between the descending and upward connecting leaders, have found that the magnitude of lightning-induced voltage on a 300-m-long horizontal wire (matched at both ends) located 10 m above perfectly conducting ground and 100 m away from the vertical lightning channel increased with increasing the height of the junction point between the descending and upward connecting leaders. For example, according to Silveira and Visacro, the magnitude of lightning-induced voltage increases by a factor of 1.3 or 2.0 as the height of the junction point increases from $h = 0$ to 100 or 300 m, respectively. Silveira and Visacro (2002) used a model based on the hybrid electromagnetic field/circuit-theory approach (Visacro *et al.* 2002). They used a current waveform linearly rising to its maximum value in 1 μs and assumed that the current wave propagation speed along the leader channels both above and below the junction point was equal to c. Also, they apparently assumed that the current reflection coefficients at the top and bottom of the upward connecting leader channel were equal to zero. Thus, by setting $v = c$ and $\rho_{top} = \rho_{bot} = \rho_{gr} = 0$ in (5.4), (5.5), and (5.6), we can simulate the current distribution used by Silveira and Visacro (2002) and compute corresponding induced voltages on the overhead wire. Calculations were performed for $d = 100$ m and different values of h ranging from 0 to 300 m. Resultant ratios of magnitudes of lightning-induced voltages on a 300-m-long horizontal wire (matched at both ends) located 10 m above perfectly conducting ground for the strike-object and flat-ground cases are shown, along with Silveira and Visacro results, in Figure 5.12, both calculated for a current waveform linearly rising to its maximum in 1 μs. The increasing trend (voltage enhancement effect) for $h = 100$ m and 300 m reported by Silveira and Visacro (2002) (see hollow circles in Figure 5.12), who used a model based on the hybrid electromagnetic field/circuit-theory approach, is opposite to that based on FDTD calculations (see solid circles in Figure 5.12). Reasons for the discrepancy are presently unknown.

5.2.5 Summary

We have examined, using the FDTD method, the ratios of magnitudes of lightning-induced voltages for the cases of strikes to a 100-m-high object and to flat ground as a function of distance from the lightning channel, d, current reflection coefficients at the top and bottom of the strike object, ρ_{top} and ρ_{bot}, the current reflection coefficient at the channel base (in the case of strikes to flat ground), ρ_{gr}, and the return-stroke speed, v. The ratio of magnitudes of lightning-induced voltages for tall-object and flat-ground cases increases with increasing d, decreasing ρ_{bot} (<1), decreasing ρ_{top} (<0, except for the case of $\rho_{bot} = 0$), and decreasing v (<c). The ratio is larger than unity (strike object serves to enhance the induced voltage) for $d = 40$ to 200 m and

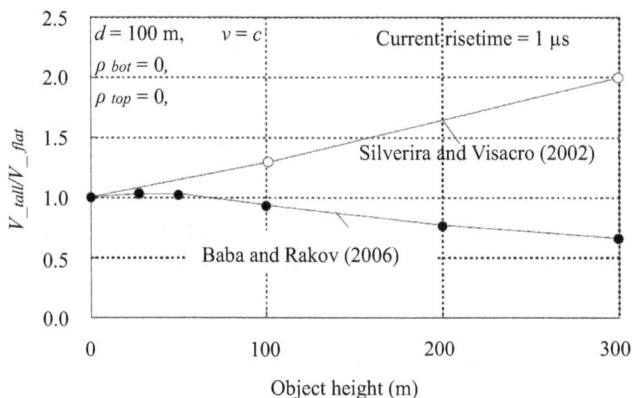

Figure 5.12 *Ratios of magnitudes of lightning-induced voltages at $d = 100$ m for junction point heights h ranging from 0 to 300 m to that for $h = 0$ (strike to flat ground without an upward connecting leader), based on the FDTD calculations (solid circles), and those calculated for the same conditions by Silveira and Visacro (2002), who used a model based on the hybrid electromagnetic field/circuit-theory approach (hollow circles). 2006 ©IEEE. Reprinted, with permission, from Baba and Rakov (2006, Figure 12)*

realistic conditions such as $\rho_{bot} = (\rho_{gr}) = 1$, $\rho_{top} = -0.5$, and $v = c/3$, but becomes smaller than unity (lightning-induced voltage for the tall-object case is smaller than for the flat-ground case) under some special conditions such as $\rho_{bot} = (\rho_{gr}) = 1$, $\rho_{top} = 0$, and $v = c$.

Further, the influence of the strike-object height, h, at a distance of $d = 100$ m was investigated. It was found that, in perhaps the most realistic case, $\rho_{bot} = (\rho_{gr}) = 1$, $\rho_{top} = -0.5$, and $v = c/3$, the ratio of magnitudes of lightning-induced voltages increased with increasing h from 0 to 100 m, and then decreased with increasing h. In a less realistic case, $\rho_{bot} = (\rho_{gr}) = 1$, $\rho_{top} = 0$, and $v = c$, the ratio was less than unity and decreased monotonically with increasing h. Also, the ratio was found to increase with decreasing the risetime of lightning return-stroke current waveform. The results presented here for relatively long current risetimes (3 μs) are in good agreement with those of Piantini and Janiszewski (1998), who used the modified Rusck model, but for relatively short risetimes (0.5 μs) different trends are observed. Both the results based on FDTD calculations and those of Piantini and Janiszewski (1998) for the current risetime equal to 1 μs disagree with the corresponding results of Silveira and Visacro (2002), who used a model based on the hybrid electromagnetic field/circuit-theory approach.

The above findings regarding the lightning-induced voltages in the presence of a tall strike object have important implications for optimizing lightning protection means for telecommunication and power distribution lines.

Appendix 5.2.A   Testing the validity of the FDTD calculations against experimental data (strikes to flat ground)

The FDTD method solves the discretized Maxwell's equations to find lightning electromagnetic fields and the reaction (scattered fields) of the overhead wire to these fields. The induced voltage on a horizontal wire is calculated by integrating the vertical electric field from the ground surface to the wire height. It is shown in this appendix that the FDTD method yields reasonably accurate lightning-induced voltages for the case of strikes to flat ground. In order to do this, lightning-induced voltages calculated using the FDTD method are compared with those measured by Ishii *et al.* (1999) in a small-scale experiment. In their experiment, a lightning return-stroke channel was represented by a coiled wire of length 28 m. One end of this coiled wire was connected to a pulse generator and the other end was kept open. The current waveform injected into the wire was measured using a current transformer. The apparent propagation speed of current wave along this wire was 125 m/μs. Another wire, 0.25 mm in radius and 25 m in length, was horizontally stretched, away from the simulated lightning channel, at a height of 0.5 m above ground. The close (to the simulated channel) end of this horizontal wire was either terminated in a 430-Ω resistor or left open and the remote end was terminated in a 430-Ω resistor. The lightning-induced voltages at both ends of the wire were measured using voltage probes having 20-pF input capacitance.

Figure 5.13 shows the configuration of Ishii *et al.*'s (1999) small-scale experiment that was simulated using the FDTD method. The conductivity and relative permittivity of ground were set to $\sigma = 0.06$ S/m and $\varepsilon_r = 10$, respectively. Note that Ishii *et al.* (1999) successfully reproduced lightning-induced voltages measured in their experiment with Agrawal *et al.*'s field-to-wire electromagnetic coupling model (Agrawal *et al.* 1980),

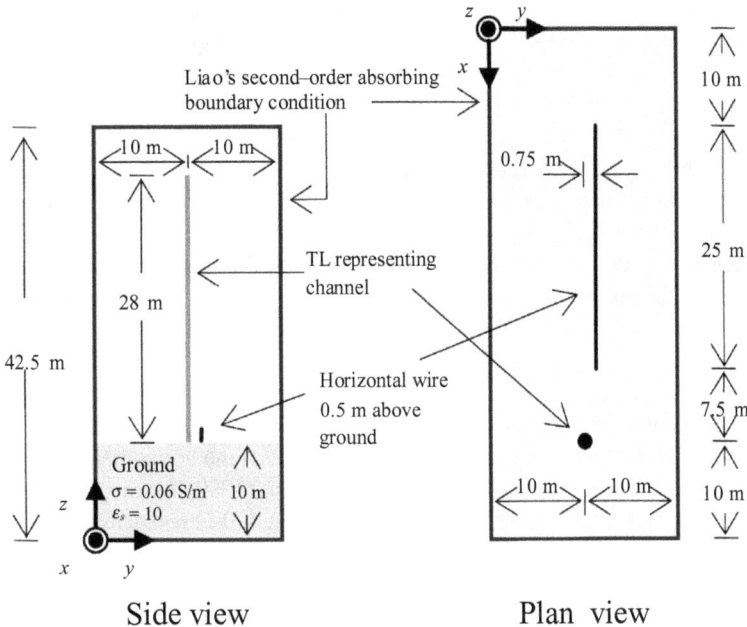

Figure 5.13 *A 25-m-long horizontal wire, one end of which is at distances x = 7.5 m and y = 0.75 m from a simulated lightning channel, as in Ishii* et al.'s *(1999) small-scale experiment, simulated here using the FDTD method. The close (to the simulated channel) end of the horizontal wire is either terminated in a 430-Ω resistance in parallel with a 20-pF capacitance (representing the input capacitance of voltage probe) or in a 20-pF capacitance, and the remote end is terminated in a 430-Ω resistance in parallel with a 20-pF capacitance. The lightning channel is represented by a vertical array of current sources that are specified using the TL model (Uman and McLain 1969), and the return-stroke speed is set to 125 m/µs. The working volume of 52.5 m × 20 m × 42.5 m, which is divided into 0.25 m × 0.25 m × 0.25 m cubic cells, is surrounded by six planes of Liao's second-order absorbing boundary condition (Liao* et al. *1984). 2006 © IEEE. Reprinted, with permission, from Baba and Rakov (2006, Figure 13)*

and Pokharel *et al.* (2003) reproduced them with NEC-2 (Burke and Poggio 1980), both assuming $\sigma = 0.06$ S/m and $\varepsilon_r = 10$. In the FDTD calculations, the lightning return-stroke channel was represented by a vertical array of current sources (Baba and Rakov 2003) that were specified using the TL model (Uman and McLain 1969) and the return-stroke speed was set to 125 m/μs. The horizontal wire of radius 0.25 mm in the rectangular-geometry FDTD procedure was represented by employing a method proposed by Noda and Yokoyama (2002). They found that a thin wire in air had an equivalent radius of $0.23\Delta s$ (Δs is the side length of cubic cells used in FDTD simulations) in the case that the electric field along the axis of the thin wire was set to zero in an orthogonal and uniform Cartesian grid for FDTD simulations. They further showed that a thin wire having an arbitrary radius r_0^* could be equivalently represented by placing a zero-radius wire in an artificial rectangular prism, coaxial with the thin wire, having a cross-sectional area of $2\Delta s \times 2\Delta s$ and the modified relative permittivity and permeability given by $\varepsilon_r^* = \ln(1/0.23)/\ln(\Delta s/r_0^*)$ and $\mu_r^* = \ln(\Delta s/r_0^*)/\ln(1/0.23)$, respectively. For calculations presented in this appendix, $\Delta s = 0.25$ m and $r_0^* = 0.25$ mm and, hence, $\varepsilon_r^* = 0.213$ and $\mu_r^* = 1/0.213 = 4.69$ $\left(\varepsilon_r^*\varepsilon_0\mu_r^*\mu_0 = \varepsilon_0\mu_0 = 1/c^2\right)$.

In order to test the validity of the FDTD method, lightning-induced voltages were calculated at both ends of the horizontal wire (Ishii *et al.* 1999) measured induced voltages only at the ends of the horizontal wire up to 300 ns with a time increment of 0.25 ns.

Figure 5.14 shows the injected current waveform measured by Ishii *et al.*, which was used as the channel-base current waveform in the TL model. Figure 5.15(a) shows induced voltage waveforms at the close and remote ends of the horizontal wire based on FDTD calculations and those measured by Ishii *et al.* (1999) for the case of both ends being terminated in a lumped circuit composed of a 430-Ω resistance in parallel

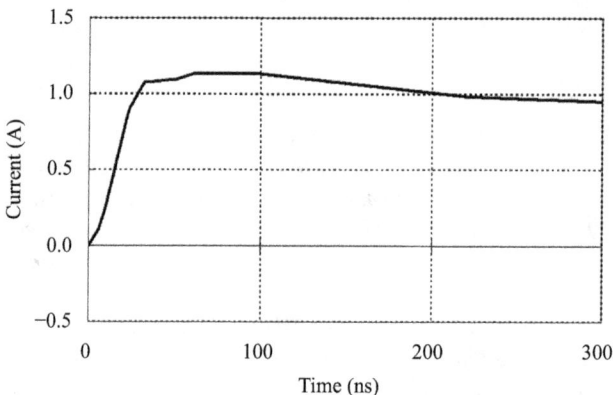

Figure 5.14   Injected current waveform measured by Ishii et al. (1999), which was used as the channel-base current waveform in the TL model. The TL-model-predicted distribution of current along the lightning channel was used in FDTD-based calculations of lightning-induced voltages on the horizontal wire shown in Figure 5.15. 2006 ©IEEE. Reprinted, with permission, from Baba and Rakov (2006, Figure 14)

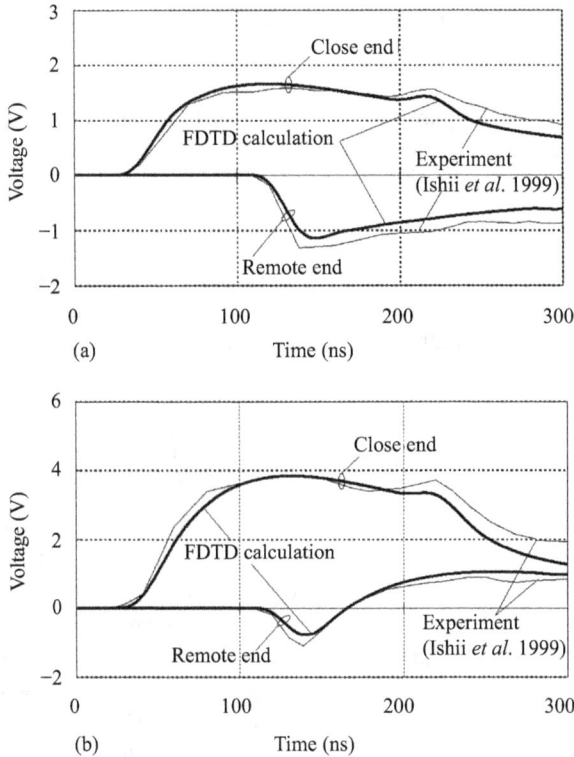

Figure 5.15 *Induced-voltage waveforms at the close and remote ends of the horizontal wire based on FDTD calculations vs. those measured by Ishii et al. (1999), (a) for the case of both ends of the horizontal wire being terminated in a 430-Ω resistance in parallel with a 20-pF capacitance, and (b) for the case of the close end being terminated in a 20-pF capacitance and the remote end in a 430-Ω resistance in parallel with a 20-pF capacitance. 2006 ©IEEE. Reprinted, with permission, from Baba and Rakov (2006, Figure 15)*

with a 20-pF capacitance. Figure 5.15(b) is the same as Figure 5.15(a), but for the case of the close end being terminated in a 20-pF capacitance and the remote end being terminated in a 430-Ω resistance in parallel with a 20-pF capacitance. It is clear from Figure 5.15(a) and (b) that induced voltages calculated using the FDTD method agree reasonably well with those measured. Note that voltage waveforms based on FDTD calculations are also in good agreement with those calculated for the same config-uration using Agrawal's field-to-wire coupling model (Ishii *et al.* 1999) and those calculated using NEC-2 (Pokharel *et al.* 2003). Thus, one can conclude that the FDTD method yields reasonably accurate lightning-induced voltages on a horizontal wire above ground, at least for the case of strikes to flat ground.

Appendix 5.2.B   Comparison with Rusck's formula (strikes to flat ground)

In this appendix, we compare the magnitudes of lightning-induced voltages at the midpoint of a 1200-m-long horizontal wire (matched at both ends) above perfectly conducting ground, calculated using the FDTD method, with those calculated using Rusck's formula (Rusck 1958). Rusck derived the following expression for the magnitude of lightning-induced voltage, V_{R_flat}, at the midpoint of an infinitely long horizontal wire at height, h_l, above perfectly conducting ground for a return-stroke current represented by a step function propagating at speed v along the vertical lightning channel attached to flat ground.

$$V_{R\_flat} = \frac{30 I_{max} h_l}{d} \left(1 + \frac{1}{\sqrt{2}} \frac{v}{c} \frac{1}{\sqrt{1 - (v/c)^2/2}} \right) \tag{5.7}$$

where I_{max} is the magnitude of the return-stroke current and d is the horizontal distance from the lightning channel to the wire. Since a short-front current waveform rising from zero to its maximum in about 0.5 µs (the corresponding 10-to-90% risetime is 0.15 µs) was used in calculating lightning-induced voltages shown in Figure 5.2(b), one can expect the magnitude of lightning-induced voltage based on FDTD calculations (Figure 5.2(b)) to be similar to that calculated using (5.7). Table 5.5 shows the magnitudes of lightning-induced voltages calculated using the FDTD method and (5.7) for the case of perfectly conducting ground ($\sigma = \infty$). It is clear from Table 5.5 that the magnitudes of induced voltages calculated using these two methods are in good agreement. Note that the magnitudes of lightning-induced voltages for a ground having $\sigma = 10$ mS/m and $\varepsilon_r = 10$ are 116, 80.4, 49.7, and 25.2 kV at distances $d = 40$, 60, 100, and 200, respectively, which are only 8 to 24% higher than those for $\sigma = \infty$.

Table 5.5   Magnitudes of lightning-induced voltages at the midpoint of a 1200-m-long horizontal wire 10 m above perfectly conducting ground, calculated using the FDTD method for a lightning strike to flat ground, and those calculated using Rusck's formula (1958). The magnitude of return-stroke current is assumed to be 11 kA and the return-stroke speed is set to c/3

	Induced voltage (kV)		
d (m)	FDTD	Rusck	Difference (%)
40	107	102	4.9
60	71.6	68.0	5.3
100	42.4	40.8	3.9
200	20.4	20.4	0.0

2006 ©IEEE. Reprinted, with permission, from Baba and Rakov (2006, Table II).

Appendix 5.2.C   Testing the validity of the FDTD calculations against experimental data (strikes to a tall object)

Michishita *et al.* (2003) measured lightning-induced voltages on an overhead test distribution line simultaneously with lightning currents at the top of 200-m-high strike object (Fukui chimney). Figure 5.16 shows the configuration of their experiment that is simulated here using the FDTD method. A horizontal wire 2.5 mm in radius and about 300 m in length was stretched 11 m above ground. Both ends of this horizontal wire were terminated in 400-Ω resistors. Lightning-induced voltages at each end of the horizontal wire were measured. Michishita *et al.* (2003) reasonably well reproduced their measured lightning-induced voltages using the Agrawal model (Agrawal 1980). They represented the strike object by a lossless uniform transmission line with characteristic impedance $Z_{ob} = 250$ Ω, terminated in the frequency-dependent grounding impedance Z_{gr}, a 100-Ω resistance in parallel with a 10-Ω resistance and a 0.3 mH inductance (connected in series) resulting in $\rho_{bot} = 0.42, 0.45,$ and 0.52 for frequencies equal to ∞, 1, and 0.2 MHz, respectively. Lightning channel was represented by a lossless uniform transmission line whose characteristic impedance $Z_{ch} = 1000$ Ω (corresponding to $\rho_{top} = -0.6$) and the current-propagation speed along the channel was set to $v = c/3$. The conductivity and relative permittivity of ground were set to $\sigma = 10$ mS/m (or ∞) and $\varepsilon_r = 10$, respectively.

For the FDTD simulations presented in Figure 5.17, the TL model extended to include a tall strike object was used to represent the Fukui chimney and the lightning channel. Following, Michishita *et al.* (2003), $\rho_{top} = -0.6, \rho_{bot} = 0.42,$ and $v = c/3$ were used in the calculations. The current distribution along the object and lightning channel is given by (5.4) and (5.5), respectively. The conductivity and relative permittivity of ground were set to $\sigma = 5$ or 10 mS/m and $\varepsilon_r = 10$, respectively. One of the values of conductivity (10 mS/m) was used by Michishita *et al.* (2003) and the other (5 mS/m) was additionally selected because it provided a better agreement between model-predicted and measured voltages at the remote end of the wire. The horizontal wire, a portion of which near the remote end was neither parallel nor perpendicular to *x*- or *y*-axis (see Figure 5.16), was simulated using a staircase approximation in the FDTD calculations. Lightning-induced voltages at each end of the horizontal wire were calculated up to 6 μs with a time increment of 5 ns.

Figure 5.17(a) shows the current waveform measured by Michishita *et al.* (2003) at the top of the Fukui chimney, which was employed as $I(h,t)$ in (5.4) and (5.5). Current waveforms at $z' = 0$ (bottom of the Fukui chimney) and $z' = 400$ m (200 m above the top of the Fukui chimney) calculated for $\rho_{top} = -0.6, \rho_{bot} = 0.42,$ and $v = c/3$ are also shown in Figure 5.17(a). Figure 5.17(b) shows induced voltage waveforms at the close and remote ends of the horizontal wire calculated using the FDTD method for two different values of ground conductivity, along with those measured by Michishita *et al.* (2003). It is clear from Figure 5.17(b) that induced voltages calculated using the FDTD method are in good agreement with measured ones. Thus, one can conclude that the FDTD method yields reasonably accurate

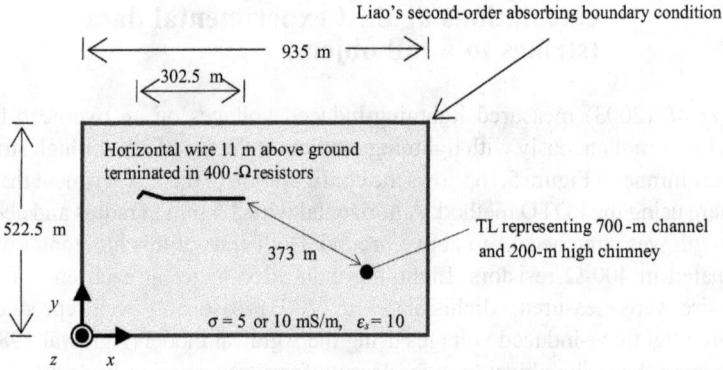

Liao's second-order absorbing boundary condition

|← ——————— 935 m ——————→|

|←— 302.5 m —→|

Horizontal wire 11 m above ground
terminated in 400-Ω resistors

522.5 m

TL representing 700-m channel
and 200-m high chimney

373 m

σ = 5 or 10 mS/m, ε_s = 10

y

z   x

Plan view

*Figure 5.16   About 300-m-long horizontal wire, each end of which is terminated in
a 400-Ω resistor, as in Michishita et al.'s (2003) field experiment,
whose interaction with lightning striking the 200-m-high Fukui
chimney is simulated using the FDTD method. Both the lightning
channel and the 200-m-high strike object are represented by the TL
model extended to include a tall object (Baba and Rakov 2005a). The
working volume of 935 m × 522.5 m × 1045 m, which is divided into
5.5 m × 5.5 m × 5.5 m cubic cells, is surrounded by six planes of
Liao's second-order absorbing boundary condition (Liao et al. 1984)
in order to avoid reflections there. The conductivity and relative
permittivity of ground are set to σ = 5 or 10 mS/m and ε_r = 10,
respectively. 2006 ©IEEE. Reprinted, with permission,
from Baba and Rakov (2006, Figure 16)*

lightning-induced voltages on a horizontal wire above ground for the case of strikes
to a tall object.

5.3   Lightning-induced voltages on an overhead
two-conductor distribution line

5.3.1   Introduction

In order to optimize lightning protection means of telecommunication and power
distribution lines, one needs to know voltages that can be induced on overhead
wires by nearby lightning strikes. Lightning-induced voltages on overhead multiconductor lines have been computed reasonably accurately using field-to-wire
electromagnetic coupling models (e.g., Paolone *et al.* 2009) and approximate
expressions for electric fields over lossy ground, such as the Norton approximate

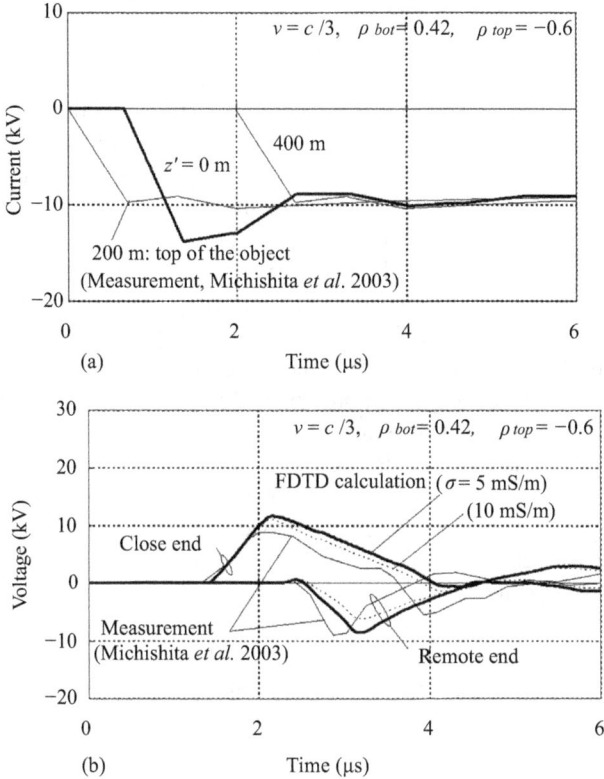

Figure 5.17 (a) Current waveform measured at the top of the 200-m-high object (Michishita et al. 2003), which was used as I(h,t) in (5.4) and (5.5), and current waveforms at z' = 0 (bottom of the object) and 400 m (200 m above the top of the object), calculated using (5.4) and (5.5) for v = c/3, $\rho_{top} = -0.6$ and $\rho_{bot} = 0.42$. (b) Induced voltage waveforms measured by Michishita et al. (2003) at the close (to the simulated channel) and remote ends of the horizontal wire and those calculated using the FDTD method. 2006 ©IEEE. Reprinted, with permission, from Baba and Rakov (2006, Figure 17)

expressions (Norton 1937) and the Cooray–Rubinstein formula (Cooray 1992; Rubinstein 1996). Also, lightning-induced voltages have recently been computed with a similar accuracy using the MoM (e.g., Pokharel *et al.* 2003) and the hybrid electromagnetic/circuit model (e.g., Silveira *et al.* 2009; Yutthagowith *et al.* 2009). In order to consider the effects of lossy ground on electromagnetic fields, Pokharel *et al.* (2003) and Silveira *et al.* (2009) employed the Norton approximate expressions, while Yutthagowith *et al.* (2009) used the Cooray–Rubinstein formula.

The FDTD method (Yee 1966) has been applied to analyzing lightning-induced voltages by Baba and Rakov (2006) and Ren *et al.* (2008). Baba and Rakov (2006) used the 3D-FDTD method for comparing voltages induced on a single overhead wire above lossy ground due to nearby lightning strikes to flat ground and to a vertically extended object (see Section 5.2). Ren *et al.* (2008) used the 2D-FDTD method for evaluating the electric fields over lossy ground, with the 2D-FDTD-computed fields being used for evaluating the lightning-induced voltages with Agrawal *et al.*'s field-to-wire coupling model (Agrawal *et al.* 1980). One of the reasons for using the latter, hybrid approach is related to a difficulty of representing closely spaced overhead thin wires in the 3D-uniform-grid FDTD method. One of the advantages of the use of the FDTD method in analyzing lightning-induced voltages is that it yields electromagnetic fields in the presence of lossy ground directly in the time domain without using any approximate formula in the frequency domain. Another advantage is that it does not necessarily require a field-to-wire coupling model. One of the disadvantages is that it is computationally expensive.

In this section, we show calculations, based on the 3D-FDTD method, of lightning-induced voltages on a 738-m-long overhead two-wire line for different return-stroke speeds, 60, 130, and 200 m/μs, ground conductivity values, 0.35, 3.5, and 35 mS/m, and vertical ground rod lengths, 5.4, 11.7, and 23.4 m. The employed 3D-FDTD method uses a subgrid model, in which spatial discretization is fine (cell side length is 0.9 m) in the vicinity of overhead wires and downconductors (90 m × 828 m × 108 m rectangular space that accommodates overhead wires, downconductors, and ground rods) and coarse (cell side length is 4.5 m) in the rest of the computational domain. Overhead wires and downconductors are simulated using the thin-wire representation (Noda and Yokoyama 2002). We compare the 3D-FDTD-computed waveforms of lightning-induced voltages (voltages were obtained by integrating the FDTD-computed vertical electric fields) with the corresponding waveforms measured in a rocket-triggered-lightning experiment by Barker *et al.* (1996).

5.3.2   Methodology

Figure 5.18 shows the experimental configuration, including about 740-m-long two-wire test distribution line and a rocket-triggered lightning channel at a distance of 145 m from the line. This configuration was tested (used for measuring lightning-induced voltages) in 1993 at the lightning-triggering facility at Camp Blanding, Florida (see Barker *et al.* 1996). One of the conductors was grounded at 3 out of 15 poles of the line. The radius of both upper and lower wires was about 5 mm and the radius of downconductors was about 2 mm. Note that the line length (682 m) given in Barker *et al.* (1996) is incorrect. The correct line length (given, for example, in Mata *et al.* (2000)) is about 740 m.

Figure 5.19(a) and (b) shows side and plan views, respectively, of the representation of Barker *et al.*'s experimental configuration for 3D-FDTD computations. The two horizontal wires of length 738 m at heights 7.2 and 5.4 m are parallel to the *y*-axis (wire sag is neglected). Note that this horizontal wire length (738 m = 164 × 4.5 m) is chosen to match the size of coarse cubic cells of side length 4.5 m. The difference in horizontal

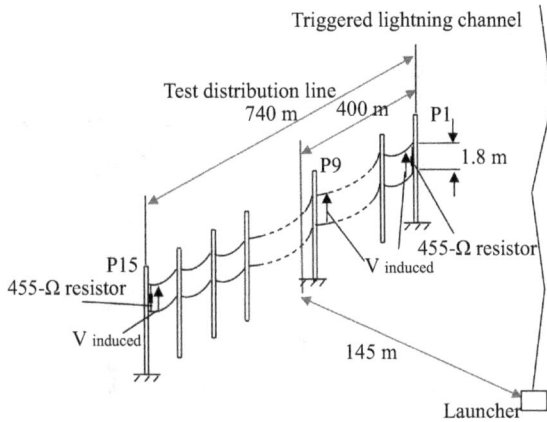

Figure 5.18   *Experimental configuration employed by Barker et al. (1996),*
including 740-m-long two-wire test distribution line at a distance of
145 m from rocket-triggered-lightning channel. The lower conductor
is grounded at poles P1, P9, and P15. 2012 ©IEEE. Reprinted, with
permission, from Sumitani et al. (2012, Figure 1)

wire lengths, 740 vs. 738 m, does not cause a significant difference in computed induced voltages. The working volume of 504 m × 1098 m × 2367 m for 3D-FDTD compu-tations is divided into cubic cells of two different sizes and is surrounded by six planes of Liao's second-order absorbing boundary condition (Liao *et al.* 1984) to minimize unwanted reflections there. The rectangular space of 90 m × 828 m × 108 m, which accommodates overhead wires, downconductors, and ground rods, is divided into cubic cells whose side length is 0.9 m (see Figure 5.19(c) and (d)). The rest of the working volume is divided into cubic cells whose side length is 4.5 m. The total number of cells in the working volume is about 2.5×10^7 ($\approx$504/4.5 × 1098/4.5 × 2367/4.5 + 90/ 0.9 × 828/0.9 × 108/0.9).

The upper and lower horizontal wires of radius $a = 5$ mm and three down-conductors of radius $a = 2$ mm are represented, as per Noda and Yokoyama (2002), by placing a wire having an equivalent radius of $a_0 = 0.207$ m ($\approx 0.23\Delta x$ $= 0.23\Delta y = 0.23\Delta z = 0.23 \times 0.9$ m) in the center of an artificial rectangular prism having a cross-sectional area of $(2 \times 0.9$ m$) \times (2 \times 0.9$ m$)$ and the modified (relative to air) constitutive parameters: lower electric permittivity $\varepsilon_0' = m\varepsilon_0$ and higher magnetic permeability $\mu_0' = \mu_0/m$ (ε_0 and μ_0 are the permittivity and permeability of air, respectively, as shown in Figure 5.19(e)). The modification coefficient m is given by $\ln(\Delta s/ a_0)/\ln(\Delta s / a)$, where $\Delta s(= \Delta x = \Delta y = \Delta z = 0.9$ m in this case) is the lat-eral side length of cells, $a_0 (\approx 0.23\Delta s)$ is the equivalent radius, and a is the radius to be specified (Noda and Yokoyama 2002). For $a = 5$ and 2 mm, $m = 0.283$ and 0.241, respectively.

At each of the termination poles (P1 and P15), the upper wire is connected to the lower wire via a 455-Ω resistor (see Figure 5.19(c)). At P1, P9 (pole located

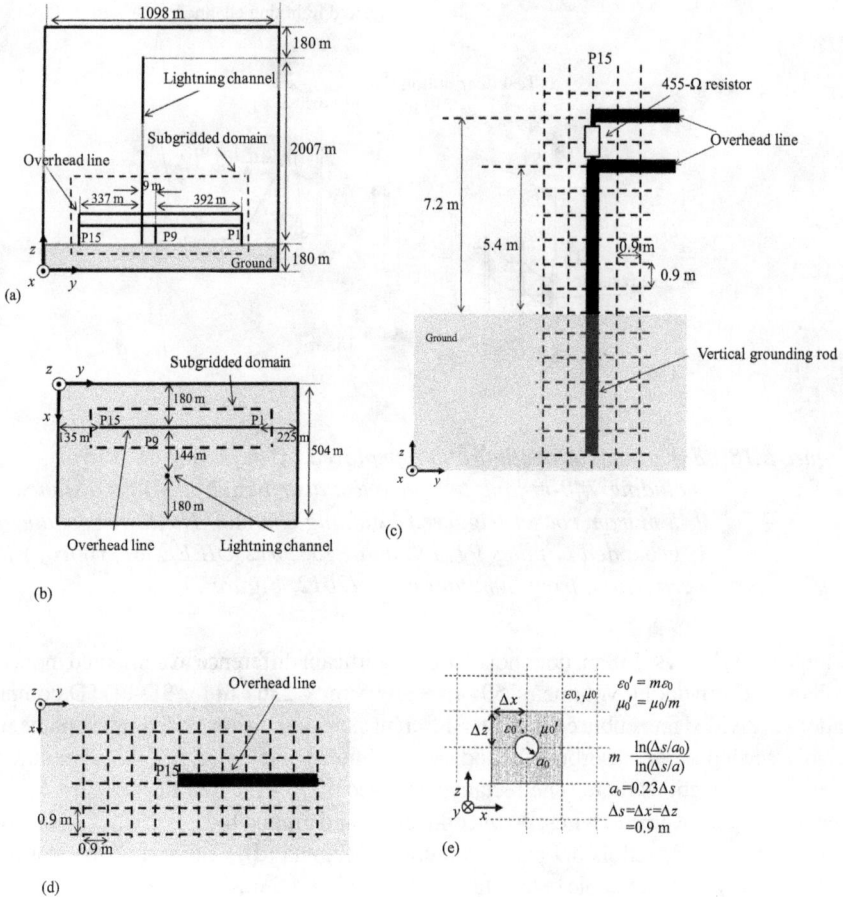

Figure 5.19   (a) Side and (b) plan views of Barker et al.'s experimental
configuration for FDTD computations. The 738-m-long and 5-mm-
radius wires at heights 7.2 and 5.4 m are parallel to the y-axis. P1,
P9, and P15 denote instrumented poles. At P1 and P15, the upper
wire is connected to the lower wire via a 455-Ω resistor. At P1, P9,
and P15, the lower wire is connected to the ground rod. The
lightning channel is represented by a 2000-m-long vertical phased
array of current sources. Also shown are magnified (c) side and
(d) plan views of the configuration around P15, and (e) cross-
sectional view of a horizontal wire of radius a, represented by the
wire having an equivalent radius of $a_0 = 0.23\Delta s$ in the center of an
artificial rectangular prism having a cross-sectional area of
$2\Delta z \times 2\Delta x$ and the modified (relative to air) constitutive parameters:
$\varepsilon_0' = m\varepsilon_0$ and $\mu_0' = \mu_0/m$. 2012 ©IEEE. Reprinted, with permission,
from Sumitani et al. (2012, Figure 2)

approximately at the midpoint of the line), and P15, the lower wire is connected via a downconductor to a vertical ground rod of radius $r = 7.9$ mm and assumed length $l = 5.4$, 11.7, or 23.4 m (the actual length of ground rods used in the experiment is unknown). The distance from the ground surface to the bottom absorbing boundary is set to 180 m. The ground relative permittivity is set to $\varepsilon_r = 10$ and the ground conductivity is set to $\sigma = 0.35$, 3.5, or 35 mS/m. When $\sigma = 3.5$ mS/m and $r = 7.9$ mm, the grounding resistance value evaluated with Sunde's formula (1968) ($R = [\ln(4l/r) -1]/(2\pi\sigma l)$) for $l = 5.4$, 11.7, and 23.4 m is $R = 62$, 31 or 17 Ω. Note that, although the ground conductivity was not given by Barker *et al.* (1996), the grounding impedance values were measured at P1, P9, and P15 at a frequency of 4 kHz and ranged from 30 to 75 Ω. Therefore, one can infer that the ground had conductivity of about a few millisiemens per meter.

Current variation along the lightning channel is specified by the modified TL model with linear current decay (MTLL) with height (Rakov and Dulzon 1987), assuming that the channel height $H = 7000$ m. The channel is represented by a 2000-m-long vertical phased array of current sources (Baba and Rakov 2003) located 145 m away from the line. Each current source is activated by the arrival of an upward-propagating return-stroke front whose speed is $v = 60$, 130, or 200 m/μs. The absence of current above 2000-m altitude does not influence the induced voltages until about 22 μs for $v = 130$ m/μs. Note that the lightning return-stroke speed was not measured in the experiment of Barker *et al.* (1996).

5.3.3 Analysis and results

Figure 5.20 shows waveform of one of the lightning channel-base currents measured by Barker *et al.* (1996) and its approximation used for all the FDTD computations presented in Section 5.3. Figures 5.21, 5.22, and 5.23 show the influence on the induced voltages at P9, P1, and P15, respectively, of ground conductivity σ, return-stroke speed v, and ground rod length l. Voltage waveforms measured at P9 and P1 by Barker *et al.* are also shown in Figures 5.21 and 5.22, respectively.

Figure 5.21(a) shows FDTD-computed waveforms of lightning-induced voltages between the two wires at P9 (located approximately at the midpoint of the line) for $v = 130$ m/μs, $l = 11.7$ m, and different values of σ. Peak values of the computed lightning-induced voltage are 56, 51, and 44 kV for $\sigma = 0.35$, 3.5, and 35 mS/m, respectively. As expected, higher voltages correspond to lower σ.

Figure 5.21(b) shows FDTD-computed waveforms for $l = 11.7$ m, $\sigma = 3.5$ mS/m, and different values of v. Peak values of the computed lightning-induced voltage are 42, 51, and 52 kV for $v = 60$, 130, and 200 m/μs, respectively. The higher the speed, the higher the induced voltage, although the difference for more common speeds (130 and 200 m/μs) is small.

Figure 5.21(c) shows FDTD-computed waveforms for $v = 130$ m/μs, $\sigma = 3.5$ mS/m, and different values of l. Peak values of the computed lightning-induced voltage are 49, 51, and 53 kV for $l = 5.4$, 11.7, and 23.4 m, respectively. The reason why the peak voltage between the upper and lower wires increases with increasing ground rod length (decreasing grounding impedance) is that the voltage

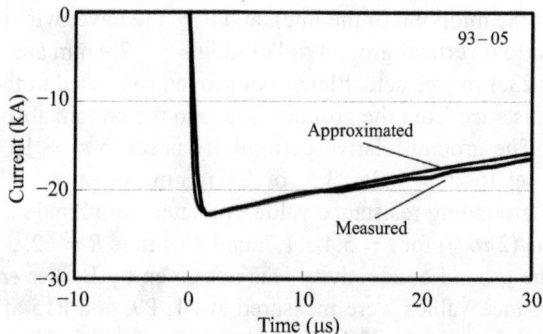

Figure 5.20   Waveform of channel-base current measured for one of lightning return strokes (93–05) by Barker et al. (1996) and its approximation used for FDTD computations. 2012 ©IEEE. Reprinted, with permission, from Sumitani et al. (2012, Figure 3)

on the lower wire, directly connected to the ground rod, decreases, while the upper-wire voltage remains essentially the same.

In each part of Figure 5.21, the voltage waveform (corresponding to current shown in Figure 5.20) measured at P9 by Barker *et al.* (1996) is also shown. The measured voltage peak value is 51 kV.

It is clear from Figure 5.21 that the magnitude of lightning-induced voltage at P9 increases with decreasing the ground conductivity, increasing the return-stroke speed, and increasing the ground rod length. When $\sigma = 3.5$ mS/m, $v = 130$ or 200 m/μs, and $l = 11.7$ m (corresponding grounding resistance $R = 31$ Ω), the FDTD-computed initial part of waveform and its peak value agree best with those of the measured waveform. This conclusion is consistent with that of Ren *et al.* (2008), who used 2D-FDTD computations with Agrawal *et al.*'s field-to-wire coupling model, and that of Yutthagowith *et al.* (2009), who used hybrid electromagnetic/circuit model computations with the Cooray–Rubinstein formula.

Figure 5.22(a) is the same as Figure 5.21(a), but for P1. The measured peak voltage is 18 kV. Peak values of the computed lightning-induced voltage are 23, 21, and 22 kV for $\sigma = 0.35$, 3.5, and 35 mS/m, respectively. The peak value of lightning-induced voltage at P1 is not much influenced by the ground conductivity.

Figure 5.22(b) is the same as Figure 5.21(b), but for P1. Peak values of the lightning-induced voltage computed for $v = 60$, 130, and 200 m/μs are 23, 21, and 17 kV, respectively. The peak voltage at P1 decreases with increasing the return-stroke speed, which is opposite to the trend shown by the peak voltage at P9. The disparity is probably related to current attenuation with height, which is more important at more distant P1.

Figure 5.22(c) is the same as Figure 5.21(c), but for P1. Peak values of the lightning-induced voltage computed for $l = 5.4$, 11.7, and 23.4 m are 22, 21, and 21 kV, respectively. The peak voltage at P1 is not much influenced by the ground rod length.

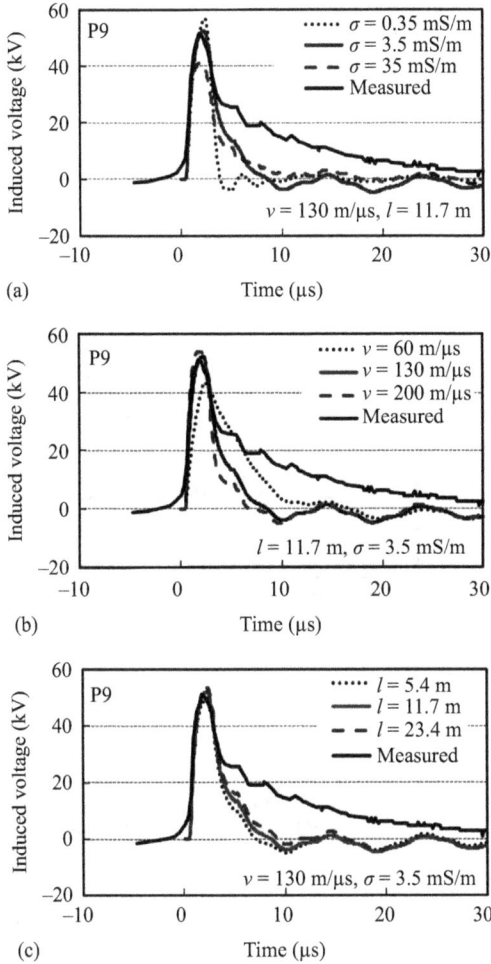

Figure 5.21 *Waveforms of lightning-induced voltage at P9 based on FDTD computations for (a) v = 130 m/μs, l = 11.7, m, σ = 0.35, 3.5, or 35 mS/m, (b) l = 11.7 m, σ = 3.5 mS/m, v = 60, 130, or 200 m/μs, and (c) v = 130 m/μs, σ = 3.5 mS/m, l = 5.4, 11.7, or 23.4 m (which correspond to grounding resistance values R = 62, 31, or 17 Ω, respectively). Also shown is the measured voltage waveform (Barker et al. 1996). 2012 ©IEEE. Reprinted, with permission, from Sumitani et al. (2012, Figure 4)*

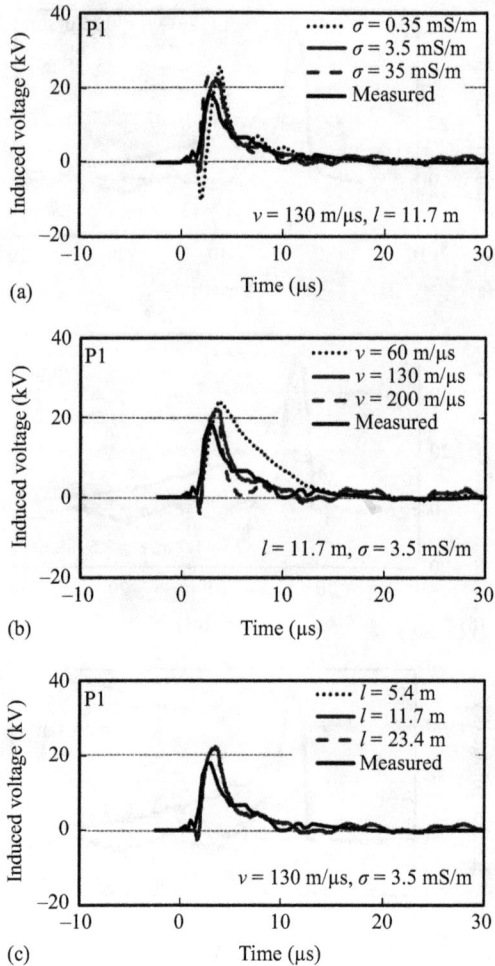

Figure 5.22   Same as Figure 5.21, but for pole P1. 2012 ©IEEE. Reprinted, with permission, from Sumitani et al. (2012, Figure 5)

For both P9 and P1, when $\sigma = 3.5$ or 35 mS/m and $v = 130$ or 200 m/μs, the FDTD-computed waveform (including its peak value) agrees reasonably well with the corresponding measured one.

Figure 5.23(a) is the same as Figure 5.21(a), but for P15. The corresponding measured waveform is not available. Peak values of the lightning-induced voltage computed for $\sigma = 0.35$, 3.5, and 35 mS/m are 31, 23, and 23 kV, respectively, increasing with decreasing ground conductivity.

Figure 5.23(b) is the same as Figure 5.21(b), but for P15. Peak values of the lightning-induced voltage computed for $v = 60$, 130, and 200 m/μs are 26, 23, and

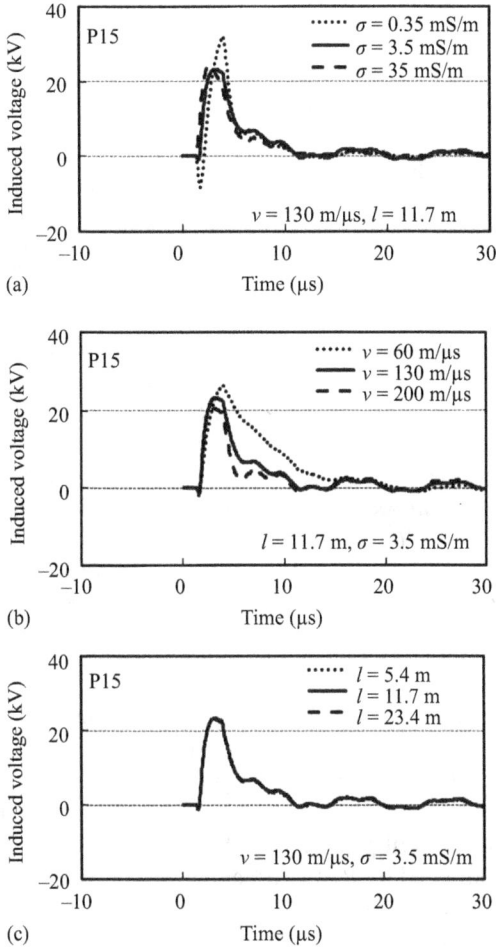

Figure 5.23   Same as Figure 5.21, but for pole P15. 2012 ©IEEE. Reprinted, with permission, from Sumitani et al. 2012, (Figure 6)

19 kV, respectively, increasing with decreasing the return-stroke speed. This trend is similar to that at pole P1.

Figure 5.23(c) is the same as Figure 5.21(c), but for P15. Peak values of the lightning-induced voltage computed for $l = 5.4$, 11.7, and 23.4 m are all 23 kV.

Figure 5.24 shows FDTD-computed waveforms of lightning-induced voltages between the two wires at P9 for $\sigma = 3.5$ mS/m, $v = 130$ m/μs, $l = 11.7$ m, and two different models of the lightning return stroke: the MTLL model (with $H = 7000$ m) and the TL model (Uman and McLain 1969). It is clear from Figure 5.24 that the lightning-induced voltage at P9 is not materially influenced

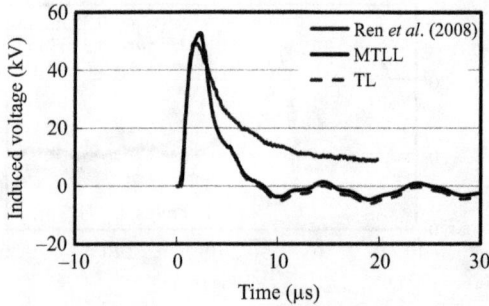

Figure 5.24   *FDTD-computed waveforms of lightning-induced voltages between the two wires at P9 for σ = 3.5 mS/m, v = 130 m/μs, l = 11.7 m, and two different models of the lightning return stroke: the MTLL model (with H = 7000 m) and the TL model. In both cases, the lightning channel is represented by a 2000-m-long vertical phased array of current sources. Also shown is the waveform computed by Ren et al. (2008). All three waveforms have similar peaks, but Ren et al.'s waveform shows larger values at later times. 2012 ©IEEE. Reprinted, with permission, from Sumitani et al. (2012, Figure 7)*

by the lightning current attenuation with height. The corresponding voltage waveform computed by Ren *et al.* (2008), who used the 2D-FDTD method with the MTLL model ($H = 7000$ m) and Agrawal *et al.*'s field-to-wire coupling model, is also shown in Figure 5.24. Initial parts of all three waveforms including peaks agree well with each other. Note that the falling part of the waveform at P9 computed by Ren *et al.* agrees better with that of the measured waveform, although they could not reproduce with their model the voltage waveform measured at P1, while it is well reproduced in FDTD-based calculations shown in Figure 5.22.

5.3.4   Summary

Lightning-induced voltages on an overhead two-wire line were computed using the 3D-FDTD method, in which spatial discretization is fine in the vicinity of small-radius wires and coarse in the rest of the computational domain. Variation of current along the lightning channel is specified by the MTLL model, assuming that $H = 7000$ m. The overhead wires having radii of some millimeters are simulated using thin-wire representations. For one of the triggered-lightning strokes studied by Barker *et al.* (1996), FDTD-computed waveforms (at least their initial parts) of lightning-induced voltages computed at the midpoint of the line and at its one end (at which measured voltage is available) for the return-stroke speed of 130 or 200 m/μs, the ground conductivity of 3.5 mS/m, and the vertical ground rod length ranging from 5.4 to 23.4 m agree reasonably well with the corresponding measurements.

5.4   Lightning-induced voltages on an overhead single conductor in the presence of corona discharge from the conductor

5.4.1   Introduction

Nucci *et al.* (2000) and Dragan *et al.* (2010) have computed lightning-induced voltages on a single overhead wire in the presence of corona discharge, using a distributed-circuit model with electromagnetic coupling between the lightning channel and the wire being represented by sources distributed along the line. In their simulations, a 5-mm-radius, 1-km-long horizontal wire, located 7.5 m above ground was employed. Corona was taken into account by means of dynamic capacitance, based on an assumed charge–voltage (*q–V*) diagram. Two ground strike points (with different lightning parameters) were considered. It has been found that corona serves to increase the magnitude of lightning-induced voltages up to a factor of 2.

In this section, we consider application of a simplified (engineering) model of corona discharge, developed by Thang *et al.* (2012a, 2012b) for FDTD computations (Yee 1966), to the analysis of lightning-induced voltages on a single wire above perfectly conducting and lossy ground, which simulates the configurations employed by Nucci *et al.* (2000) and Dragan *et al.* (2010). In the corona model, the progression of corona streamers from the wire is represented as the radial expansion of cylindrical weakly conducting (40 µS/m) region around the wire. Note that the assumed corona conductivity is low enough for the longitudinal current to remain essentially on the wire. As a result, only the capacitance of the line is materially affected by corona, while its inductance remains the same. In this regard, the FDTD model considered here and those of Nucci *et al.* (2000) and Dragan *et al.* (2010) are similar. The validity of the FDTD model (including the assumed value of corona conductivity) has been tested by Thang *et al.* (2012a, 2012b) against experimental data found in works of Noda *et al.* (2003), Inoue (1983), and Wagner *et al.* (1954). Specifically, it has been shown by Thang *et al.* (2012a) that the waveform of radial current and the relation between the total charge (the sum of charge on the wire and corona space charge in the surrounding air) and applied voltage (*q–V* curve) computed using the FDTD method including the corona model for 22- and 44-m-long horizontal wires agree reasonably well with the corresponding measured ones. Further, it has been shown by Thang *et al.* (2012a) that the computed increase of coupling between the energized wire and another wire nearby due to corona discharge agrees well with the corresponding one measured by Noda *et al.* (2003). Finally, it has been shown by Thang *et al.* (2012b) that computed waveforms (including wavefront distortion and attenuation at later times) of fast-front surge voltages at different distances from the energized end of 1.4- and 2.2-km-long overhead wires agree reasonably well with the corresponding waveforms measured by Inoue (1983) and Wagner *et al.* (1954).

5.4.2   Methodology

Figure 5.25(a) and (b) shows the plan (*xy*-plane) and side (*yz*-plane) views of a 5-mm-radius, 1-km-long overhead horizontal perfectly conducting wire located

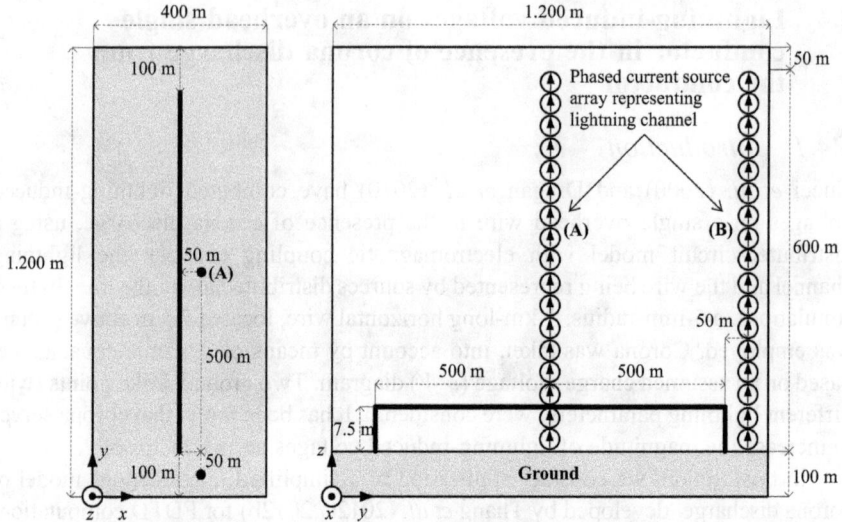

*Figure 5.25   (a) Plan (xy-plane) and (b) side (yz-plane) views of a 5-mm-radius,
1-km-long overhead horizontal wire located 7.5 m above ground.
Corona discharge is assumed to occur only on the horizontal wire.
Ground strike points are shown in (a) and simulated lighting
channels are shown in (b). 2014 ©IEEE. Reprinted, with permission,
from Thang et al. (2014, Figure 1)*

7.5 m above ground that was assumed to be either perfectly conducting or lossy
with conductivity values of 0.01 or 0.001 mS/m. These values are the same as those
used by Nucci *et al.* (2000) and Dragan *et al.* (2010), respectively. Lightning
channel is represented by a 600-m-long, vertical phased ideal current source array
(Baba and Rakov 2003). The array simulates a current pulse that propagates upward
at a speed of 130 m/µs. Lightning is assumed to terminate on ground and two
ground strike point locations are considered: A (against the midpoint of the wire)
and B (close to one of the line terminations). For stroke location A, both ends of the
wire are connected to the ground via 480-Ω matching resistors. For stroke location
B, the close end of the wire is open-circuited and the far end is connected to the
ground via a 480-Ω matching resistor. Note that *x, y,* and *z* coordinates are defined
here so that the wire is parallel to the *y*-axis and the ground surface is parallel to
both *x*- and *y*-axes (and, therefore, perpendicular to the *z*-axis).

   For FDTD computations, this conductor system is accommodated in a working
volume of 400 m × 1200 m × 750 m, which is divided nonuniformly into rectangular
cells and is surrounded by six planes of Liao's second-order absorbing boundary
condition (Liao *et al.* 1984) to minimize unwanted reflections there. Cell sides along *x*-
, *y*-, and *z*-axes are not constant: 2.2 cm in the vicinity (1.0 m × 1.0 m) of the horizontal
and vertical conductors, and increasing gradually to 10 and 200 cm beyond that region,
as shown in Figure 5.26. The equivalent radius (Noda and Yokoyama 2003) of the

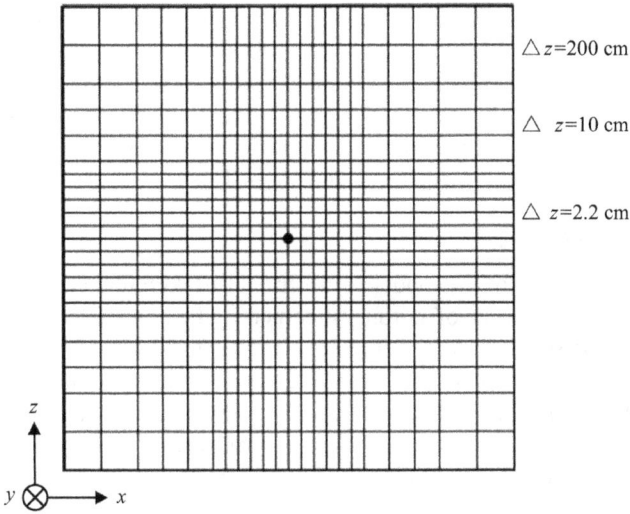

Figure 5.26 *Cross-sectional (zx-plane) view of the discretized space around horizontal conductor used in the FDTD computations of lightning-induced voltages in the presence of corona. 2014 ©IEEE. Reprinted, with permission, from Thang et al. (2014, Figure 2)*

horizontal wire is $r_0 \approx 5$ mm ($\approx 0.23\Delta x = 0.23\Delta z = 0.23 \times 2.2$ cm). Corona discharge is assumed to occur only on the horizontal wire.

The critical electric field E_0 on the surface of a cylindrical wire of radius r_0 for initiation of corona discharge is given by equation of Hartmann (1984), which is reproduced below:

$$E_0 = m \cdot 2.594 \times 10^6 \left(1 + \frac{0.1269}{r_0^{0.4346}}\right) \qquad [\text{V/m}] \qquad (5.8)$$

where m is a coefficient depending on the wire surface conditions. We assumed that $m = 0.5$. When $r_0 = 5$ mm, E_0 is 2.9 MV/m for $m = 0.5$, which is the same as the corona threshold field that was used in the model of Nucci *et al.* (2000).

We set the critical background electric field necessary for streamer propagation (e.g., Cooray 2003) to $E_{cn} = 1.5$ MV/m and $E_{cp} = 0.5$ MV/m (Waters *et al.* 1987) for negative and positive polarities, respectively. The corona ionization process is simulated by expanding the weakly conducting region of constant conductivity ($\sigma_{cor} = 40$ μS/m) to the corona radius r_c. The corona radius r_c is obtained, using analytical expression (5.9) based on E_{cp} (0.5 MV/m) or E_{cn} (1.5 MV/m) and the FDTD-computed charge per unit length (q). Then, the conductivity of the cells located within r_c is set to $\sigma_{cor} = 40$ μS/m.

$$E_c = \frac{q}{2\pi\varepsilon_0 r_c} + \frac{q}{2\pi\varepsilon_0(2h - r_c)} \qquad [\text{V/m}] \qquad (5.9)$$

The corona radius for each meter along the wire is calculated at each time step. As a result, the corona radius has a nonuniform distribution along the wire. Note that the critical background electric field for positive or negative polarity is selected at each time step, so that the model works for both unipolar and bipolar voltage waveforms. As stated in Section 5.4.1, the model and its parameters described earlier have been validated against experimental data in Thang *et al.* (2012a, 2012b).

5.4.3   *Analysis and results*

One of the parameters of the model is the critical background electric field that is necessary for propagation of corona streamers, which is different for different polarities. In this section, we consider both negative (the most common type) and positive (relatively rare, but more energetic type) cloud-to-ground strokes. Nucci *et al.* (2000) and Dragan *et al.* (2010) do not specifically state which stroke polarity they considered (their distributed-circuit model does not contain any explicit polarity-sensitive input parameters), although their assumed return-stroke current waveforms with maximum rates-of-rise of 42 and 66 kA/μs are characteristic of negative strokes. In the present analysis, one needs to determine the polarity of corona streamers, during the return-stroke stage, corresponding to nearby strokes transporting either negative or positive charge to ground.

For direct strikes to overhead conductors, the polarity of corona on the conductor during the return-stroke stage is clear: if negative charge is injected into the conductor (negative stroke), corona streamers are also negative and they are positive for a positive stroke. For nearby strikes, we can use the following considerations. In the case of negative stroke, the descending leader moves negative charge closer to the grounded conductor. At some point, the conductor will go to corona, with the corona streamers being positive. Once the negative leader attaches to ground, the electric field causing the positive corona collapses and, as a result, the positive corona space charge will tend to move back into the conductor. The collapse of positive corona (formed during the leader stage) probably occurs via the so-called reverse, negative corona (during the return-stroke stage). So, for a negative nearby stroke, corona streamers during the return-stroke stage are negative (same as for the negative direct-strike case), and for a positive nearby stroke they are positive.

We use here the same two lightning return-stroke current waveforms at the channel base as in Nucci *et al.* (2000) and Dragan *et al.* (2010) and apply them to both negative and positive stroke cases (although typical positive return-stroke current waveforms may have different parameters). These two current waveforms are shown in Figure 5.27. Lightning current has a peak of 35 kA and a maximum time derivative of 42 kA/μs for stroke location A, and these parameters are 55 kA and 66 kA/μs, respectively, for stroke location B.

5.4.3.1   Negative lightning return stroke

Figures 5.28 and 5.29 illustrate induced voltages at different points along the overhead wire with corona above perfectly conducting ground and lossy ground whose conductivity is $\sigma_{gr} = 0.01$ and 0.001 S/m, computed using the FDTD method for a negative

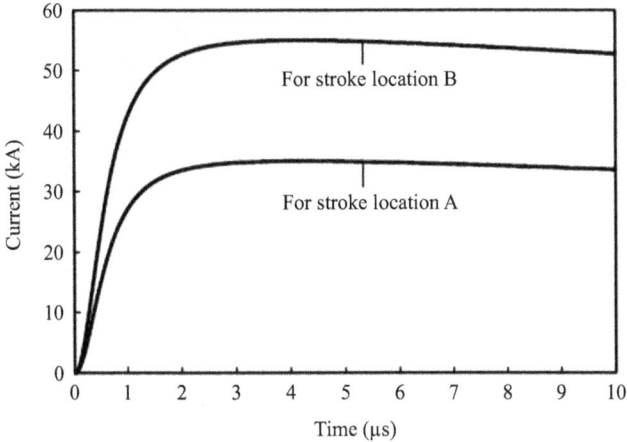

Figure 5.27 *Waveforms of injected negative lightning return-stroke current (positive charge moving up). The peak of the injected current is 35 kA and a maximum time derivative is 42 kA/μs for stroke location A, and they are 55 kA and 66 kA/μs, respectively, for stroke location B. 2014 ©IEEE. Reprinted, with permission, from Thang et al. (2014, Figure 3)*

lightning return stroke. Figure 5.28 is for stroke location A (35-kA current) and $d = 0$, 250, and 500 m from either end (due to symmetry) of the wire, and Figure 5.29 is for stroke location B (55-kA current) and $d = 0$, 250, and 500 m from the closer (to the lightning channel) end of the wire. The FDTD-computed waveforms of induced voltages without considering corona are also shown in these figures.

For stroke location A, peak values of lightning-induced voltages at $d = 500$ m computed without considering corona are about 140, 185, and 300 kV for ground conductivity equal to infinity, 0.01, and 0.001 S/m, respectively. For stroke location B, peak values of lightning-induced voltages at $d = 0$ m without considering corona, are about 135, 195, and 335 kV for ground conductivity equal to infinity, 0.01, and 0.001 S/m, respectively. These results agree fairly well with the corresponding results based on the distributed-circuit-theory approach presented by Nucci *et al.* (2000) and Dragan *et al.* (2010): about 130, 160, 250 kV for stroke location A, and about 130, 200, 400 kV for stroke location B. One possible reason for the relatively small (7 to 25%) discrepancies is the difference between the models employed.

It follows from Figures 5.28 and 5.29 that the induced voltage magnitudes are larger and the risetimes are longer in the presence of corona discharge on the horizontal wire. This trend agrees with what was first reported by Nucci *et al.* (2000) and subsequently confirmed by Dragan *et al.* (2010) using a different corona model, although the increase predicted by the full-wave model (up to 5%) is less significant than in their studies based on the circuit-theory approach (up to a factor of 2).

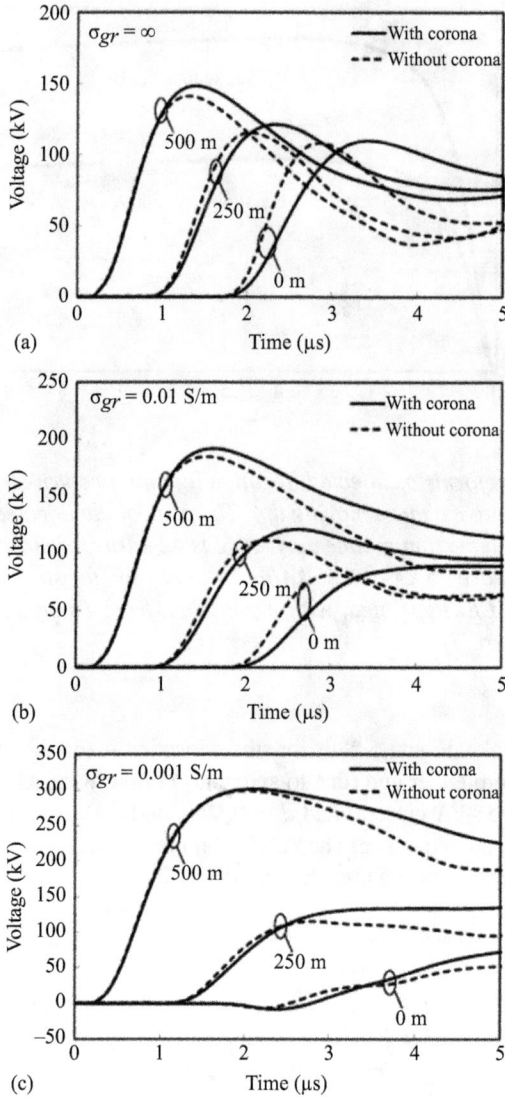

Figure 5.28   Negative stroke at location A: FDTD-computed (for $\sigma_{cor} = 40\,\mu S/m$, $E_0 = 2.9\,MV/m$, and $E_{cn} = 1.5\,MV/m$) waveforms of induced voltages at $d = 0$, 250, and 500 m from either end of the 5-mm-radius, 1.0-km-long horizontal wire. The computations were performed for (a) perfectly conducting ground ($\sigma_{gr} = \infty$), (b) $\sigma_{gr} = 0.01\,S/m$, and (c) $\sigma_{gr} = 0.001\,S/m$. Relative ground permittivity was set to 10. 2014 ©IEEE. Reprinted, with permission, from Thang et al. (2014, Figure 4)

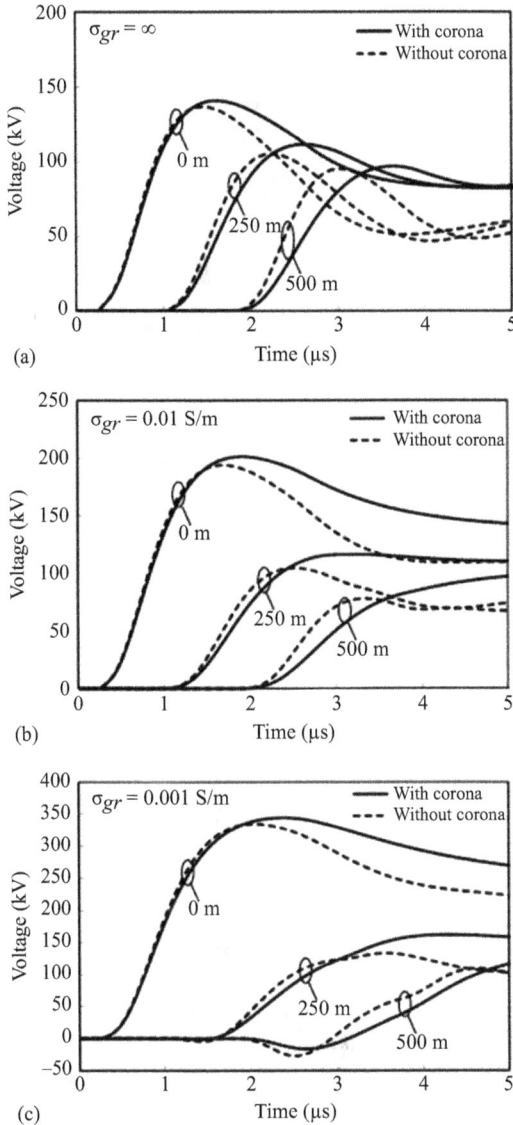

Figure 5.29 *Negative stroke at location B: FDTD-computed (for $\sigma_{cor} = 40\ \mu S/m$, $E_0 = 2.9\ MV/m$, and $E_{cn} = 1.5\ MV/m$) waveforms of induced voltages at $d = 0$, 250, and 500 m from the closer (to the lightning channel) end of the 5-mm-radius, 1.0-km-long horizontal wire. The computations were performed for (a) perfectly conducting ground ($\sigma_{gr} = \infty$), (b) $\sigma_{gr} = 0.01\ S/m$, and (c) $\sigma_{gr} = 0.001\ S/m$. Relative ground permittivity was set to 10. 2014 ©IEEE. Reprinted, with permission, from Thang et al. (2014, Figure 5)*

Note that voltage risetimes appreciably increase in the presence of corona, particularly at larger distances from the lightning channel. This corona effect is similar to that known to occur in the case of direct lightning strikes to overhead conductors (e.g., Nucci *et al.* 2000; Thang *et al.* 2012b).

Figures 5.30 and 5.31 show the variation with time of corona radius at different points along the wire and the variation of corona radius along the wire at time 5 μs, computed using the FDTD method for the case of perfectly conducting ground. Figure 5.30 is for stroke location A and $d = 500, 450, 400, 350, 300$, and 250 m from either end of the wire, and Figure 5.31 is for stroke location B and $d = 0, 50, 100, 150$, and 200 m from the near end of the wire. The maximum radius of corona region around the wire for stroke location A and 35-kA current peak is 19.8 cm; for stroke location B and 55-kA current peak, it is 13.2 cm.

It follows from Figures 5.30 and 5.31 that the presence of lightning-induced corona on the wire makes the transmission line (formed by the wire and its image) nonuniform. Note that corona radius variation is step-like due to the size of square cells employed in the FDTD computations.

5.4.3.2   Positive lightning return stroke

In this section, we consider only the perfectly conducting ground case. Figures 5.32 and 5.33 are the same as Figure 5.28(a) and 5.29(a), respectively, but for the case of positive lightning return stroke (negative charge moving up). Figure 5.32 is for stroke location A (35-kA current) and Figure 5.33 is for stroke location B (55-kA current). The FDTD-computed waveforms of induced voltages without considering corona are also shown in these figures.

It follows from Figures 5.32 and 5.33 that the induced voltage peaks are larger and the risetimes are longer in the presence of corona on the horizontal wire. This trend agrees with that reported by Nucci *et al.* (2000) and Dragan *et al.* (2010) based on the circuit-theory approach, although the voltage peaks increase predicted by the full-wave model (up to 9%) is less significant than that reported by Nucci *et al.* (2000) and Dragan *et al.* (2010).

Figures 5.34 and 5.35 show the variation with time of corona radius at different points along the wire and the variation of corona radius along the wire at time 5 μs, computed using the FDTD method for the case of perfectly conducting ground. Figure 5.34 is for stroke location A and $d = 500, 450, 400, 350, 300$, and 250 m from either end of the wire, and Figure 5.35 is for stroke location B and $d = 0, 50, 100, 150, 200$, and 250 m from the near end of the wire. The maximum radius of corona region around the wire for stroke location A and 35-kA current peak is 61.6 cm; for stroke location B and 55-kA current peak, it is 44 cm. It follows from the comparison of Figures 5.30, 5.31, 5.34, and 5.35 that the positive corona around the 5-mm-radius wire is appreciably larger than the negative corona, as expected, but significant differences in corona radius translate into relatively small differences in voltage peaks (compare Figures 5.28(a) and 5.29(a) with Figures 5.32 and 5.33, respectively).

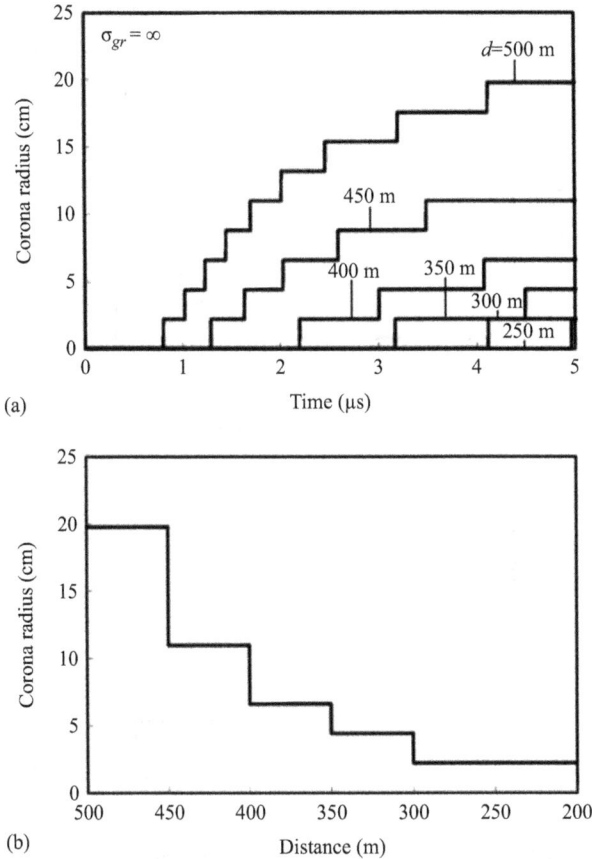

Figure 5.30 Negative stroke at location A: (a) Time variation of corona radius at d = 500, 450, 400, 350, 300, and 250 m from either end of the 5-mm-radius, 1.0-km-long horizontal wire located above perfectly conducting ground, and (b) corona radius as a function of distance from either end of the wire at time 5 μs. 2014 ©IEEE. Reprinted, with permission, from Thang et al. (2014, Figure 6)

5.4.4 Discussion

We now check whether just a thicker wire of constant radius would similarly experience higher lightning-induced voltages; that is, if corona just increases the effective radius of the wire. For this test, we compute (for the case of negative stroke and $\sigma_{gr} = \infty$) the lightning-induced voltage for a 20-cm-radius, horizontal perfectly conducting wire (without corona) for the same configuration as shown in Figure 5.25. Note that the wire radius is increased by a factor of 40 relative to that used in the calculations presented above, to a value similar to or larger than the maximum corona radius (19.8 cm for stroke location A and 13.2 cm for stroke location B). Figures 5.36 and 5.37 show the

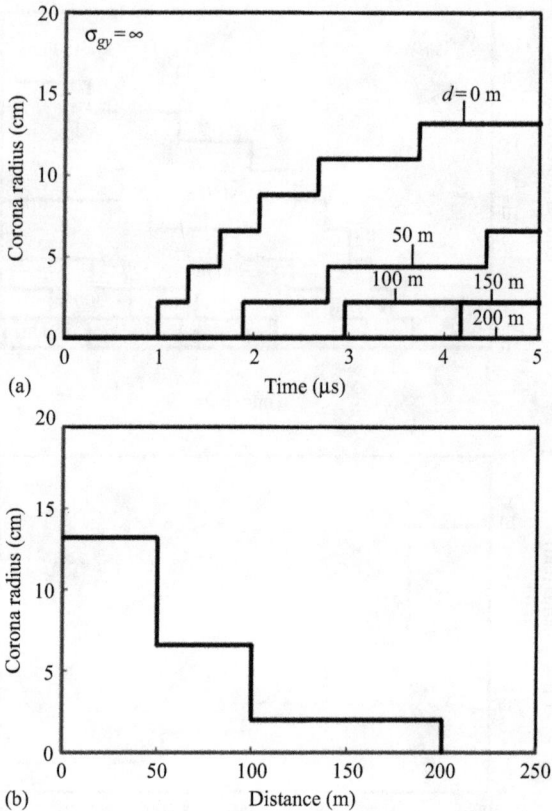

Figure 5.31 Negative stroke at location B: (a) Time variation of corona radius at
d = 0, 50, 100, 150, and 200 m from the near end of the 5-mm-radius,
1.0-km-long horizontal wire located above perfectly conducting
ground, and (b) corona radius as a function of distance from the
closer end of the wire at time 5 μs. 2014 ©IEEE. Reprinted, with
permission, from Thang et al. (2014, Figure 7)

resultant waveforms of induced voltage at $d = 500$ m from either end of the wire (for
stroke location A) and at $d = 0$ m from the near end of the wire (for stroke location B),
respectively, computed using the FDTD method. Also shown in Figures 5.36 and 5.37
are induced voltage waveforms computed for a 5-mm-radius, perfectly conducting wire
with and without corona. It follows from these figures that the induced voltages are
larger for the thicker wire, although the increase of the peak voltages due to the increase
of wire radius from 5 to 200 mm is not as large as that due to relatively low-conductivity
corona on the thinner wire developing to a maximum radius similar to or larger than the
radius of the thicker wire. Note that, in each simulation, both ends of the wire are
connected to the ground via 480-Ω matching resistors for stroke location A, and one end
of the wire is open-circuited and the other end is connected to the ground via a 480-Ω

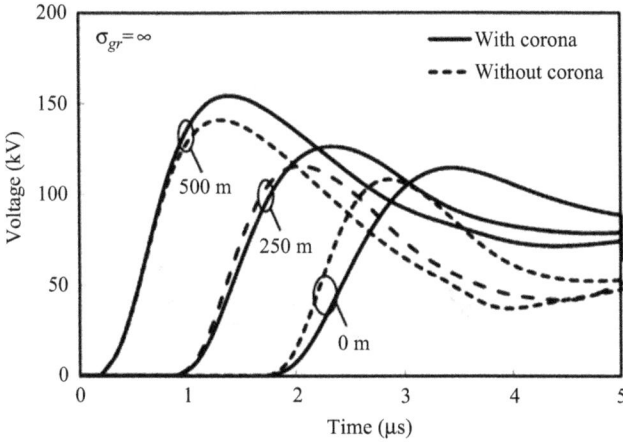

Figure 5.32 *Positive stroke at location A: FDTD-computed (for $\sigma_{cor} = 40\ \mu S/m$, $E_0 = 2.9\ MV/m$, and $E_{cp} = 0.5\ MV/m$) waveforms of induced voltages at d = 0, 250, and 500 m from either end of the 5-mm-radius, 1.0-km-long horizontal wire located above perfectly conducting ground. 2014 ©IEEE. Reprinted, with permission, from Thang et al. (2014, Figure 8)*

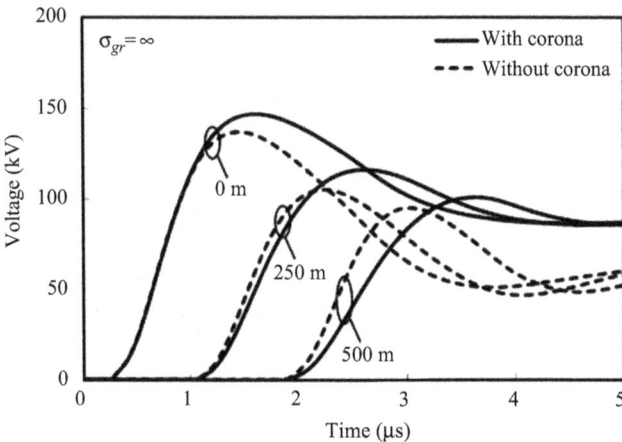

Figure 5.33 *Positive stroke at location B: FDTD-computed (for $\sigma_{cor} = 40\ \mu S/m$, $E_0 = 2.9\ MV/m$, and $E_{cp} = 0.5\ MV/m$) waveforms of induced voltages at d = 0, 250, and 500 m from the near end of the 5-mm-radius, 1.0-km-long horizontal wire located above perfectly conducting ground. 2014 ©IEEE. Reprinted, with permission, from Thang et al. (2014, Figure 9)*

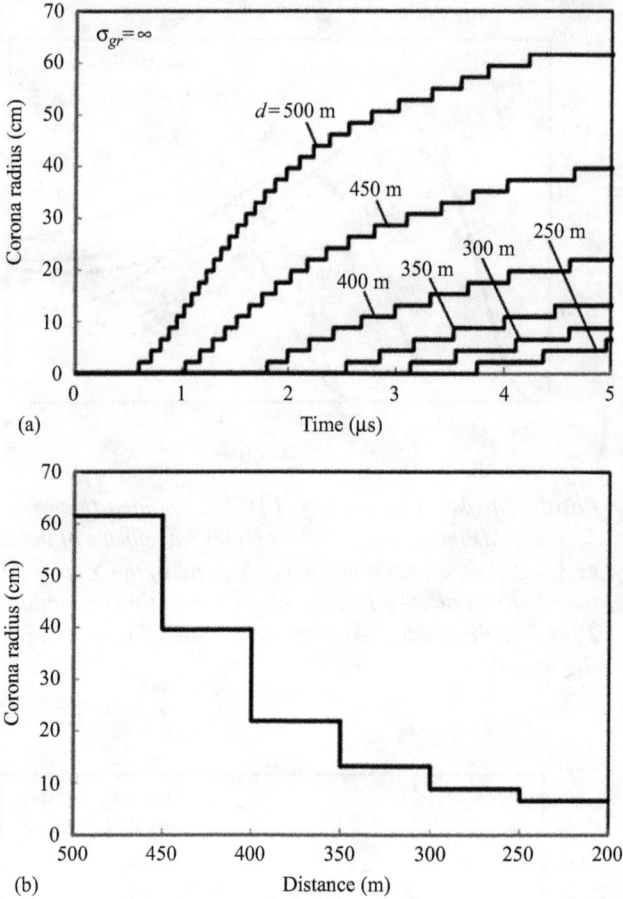

Figure 5.34   *Positive stroke at location A: (a) Time variation of corona radius at*
d = 500, 450, 400, 350, 300, and 250 m from either end of the 5-mm-
radius, 1.0-km-long horizontal wire located above perfectly
conducting ground, and (b) corona radius as a function of distance
from either end of the wire at time 5 μs. 2014 ©IEEE. Reprinted, with
permission, from Thang et al. (2014, Figure 10)

matching resistor for stroke location B. In summary, the corona effect cannot be fully explained by a larger effective radius of the wire when the line remains uniform.

We now consider a nonuniform wire without corona. Figure 5.38 shows the waveforms of induced voltages, computed for stroke location B, at $d = 0$, 250, and 500 m from the near end of a horizontal single wire with nonuniform-radius (varying from 13 cm to 5 mm) above perfectly conducting ground. The radius of this nonuniform-radius wire at each point along the wire is equal to the maximum radius of corona at that point shown in Figure 5.31. Also shown in this figure are the waveforms of induced voltage computed for a 5-mm-radius, perfectly

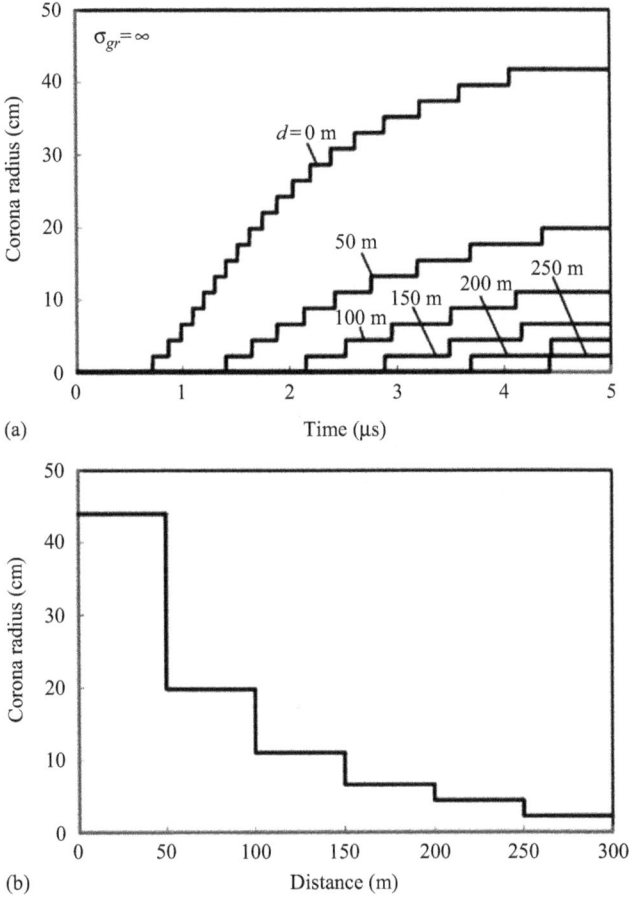

Figure 5.35 *Positive stroke at location B: (a) Time variation of corona radius at d = 0, 50, 100, 150, 200, and 250 m from the near end of the 5-mm-radius, 1.0-km-long horizontal wire located above perfectly conducting ground, and (b) corona radius as a function of distance from the closer end of the wire at time 5 μs. 2014 ©IEEE. Reprinted, with permission, from Thang et al. (2014, Figure 11)*

conducting wire with and without corona. It follows from Figure 5.38 that for the variable-radius wire, the induced voltage peak is somewhat higher than for the case of 5-mm-radius wire with corona. This implies that the distributed characteristic impedance discontinuity (causing distributed reflections) along the wire also serves to increase lightning-induced voltage peaks. Note that the nonuniform transmission line representation with variable-radius conductor reproduces fairly well the increase in risetime (associated with corona) with increasing distance d from the line end that is close to the strike location. Thus, the distributed impedance

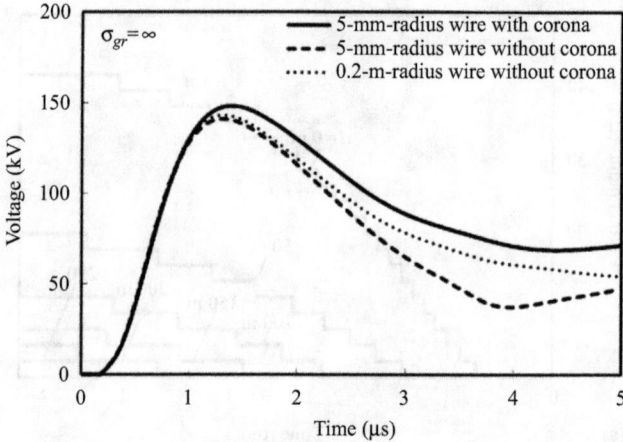

Figure 5.36   Negative stroke at location A: FDTD-computed (for $\sigma_{cor} = 40 \, \mu S/m$, $E_0 = 2.9 \, MV/m$, and $E_{cn} = 1.5 \, MV/m$) waveforms of induced voltage at $d = 500 \, m$ from either end of the 1.0-km-long horizontal wire located above perfectly conducting ground. The computations were performed for a thin wire (5-mm radius) with and without corona, and for a thick (20-cm radius) wire without corona. 2014 ©IEEE. Reprinted, with permission, from Thang et al. (2014, Figure 12)

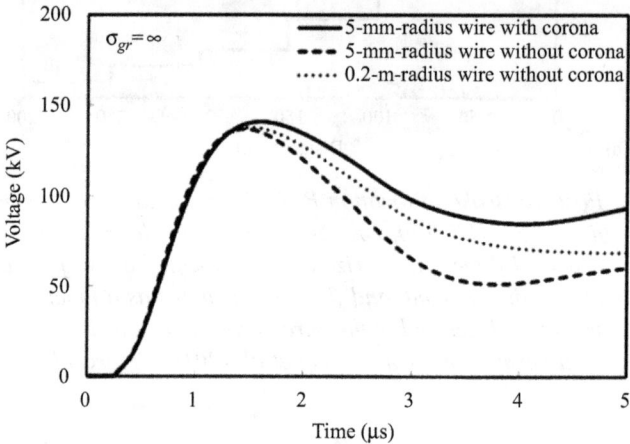

Figure 5.37   Negative stroke at location B: FDTD-computed (for $\sigma_{cor} = 40 \, \mu S/m$, $E_0 = 2.9 \, MV/m$, and $E_{cn} = 1.5 \, MV/m$) waveforms of induced voltage at $d = 0 \, m$ from the near end of the 1.0-km-long horizontal wire located above perfectly conducting ground. The computations were performed for a thin (5-mm radius) wire with and without corona, and for a thick (20-cm radius) wire without corona. 2014 ©IEEE. Reprinted, with permission, from Thang et al. (2014, Figure 13)

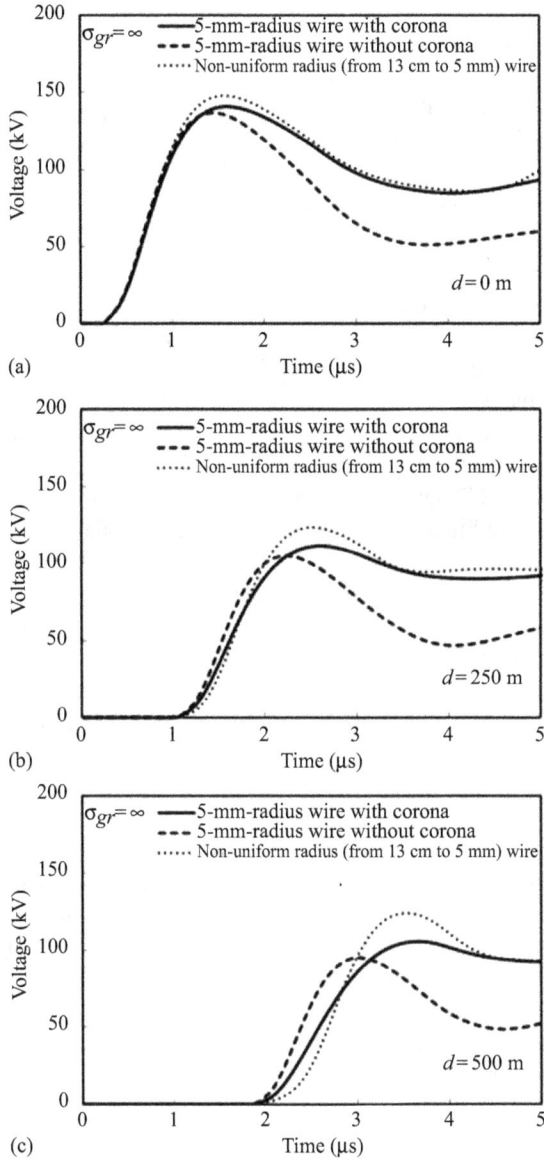

Figure 5.38 *Negative stroke at location B: FDTD-computed (for $E_0 = 2.9$ MV/m,*
and $E_{cn} = 1.5$ MV/m) waveforms of induced voltage at (a) $d = 0$ m, (b)
$d = 250$ m, and (c) $d = 500$ m from the end of a 1.0-km-long horizontal
wire above perfectly conducting ground. The computations were
performed for a thin (5-mm radius) wire with and without corona and
for a wire with a radius varying from 13 cm at the point closest to the
lightning channel to 5 mm at the far point. 2014 ©IEEE. Reprinted,
with permission, from Thang et al. (2014, Figure 14)

discontinuity is likely to be the primary mechanism of the enhancement of voltage peak and lengthening voltage risetime in the presence of corona.

An alternative explanation of the effect of corona on induced voltages was given by Nucci *et al.* (2000). According to that explanation, corona causes a decrease of the propagation speed of certain induced-voltage components, which, in turn, makes it possible for the total induced voltage to reach larger magnitudes.

Both the full-wave and distributed-circuit models predict that corona serves to increase voltages induced by nearby lightning strokes on overhead conductors, relative to the case of no corona. However, the increase predicted by the full-wave model (up to 5% for negative strokes and up to 9% for positive strokes) is small compared to that reported by Nucci *et al.* (2000) and Dragan *et al.* (2010), who used the distributed-circuit model with sources specified using the electromagnetic field theory. In Nucci *et al.* (2000), the increase was up to a factor of 2; in Dragan *et al.* (2010), it was up to 18%, with the primary difference between these two studies being the charge-voltage diagram. It is likely that the disparity between the full-wave-model results and those of Nucci *et al.* (2000) and Dragan *et al.* (2010) is also related to the differences in charge-voltage diagrams (see Figure 5.39). Unfortunately, as of today, no experimental data are available to confirm the enhancement effect of corona on voltages induced on overhead lines, but its prediction by very different models gives us confidence that the effect is real. Based on the predictions of the full-wave model, we feel that the effect should be relatively small.

The model-predicted increase of induced voltages in the presence of corona is in contrast with corona effect on voltages resulting from direct strikes. As an example, Figure 5.40 shows FDTD-computed voltages at $d = 0$, 250, and 500 m from the open-circuited end of the line that is close to point B (see Figure 5.25), with the other end being matched. Lightning current (negative charge) was injected into the open-circuited end. The current waveform parameters (peak value of 0.8 kA and maximum rate-of-rise of 0.96 kA/us) were adjusted to achieve corona radius increasing up to 11 cm, similar to that observed in induced-voltage calculations for a negative stroke at point B (see Figure 5.31). The wire radius was 5 mm and ground was assumed to be perfectly conducting. It is clear from Figure 5.40 that direct-strike voltages in the presence of corona are considerably (about a factor of 2) lower than in the absence of corona. The voltage waveform at the strike point ($d = 0$) has essentially the same shape as the injected current waveform, as expected. Since the voltage is the product of the injected current and characteristic impedance of the line, the decrease of voltage in the presence of corona implies that the characteristic impedance of the line with corona is significantly (about a factor of 2) lower than without corona, and this impedance reduction (from 490 to 230 Ω) is the primary cause of voltage reduction. In contrast, for induced voltages the coupling mechanism does not involve the characteristic impedance of the line and corona effect is dominated by distributed reflections from distributed impedance discontinuity associated with corona development, as discussed earlier. The distributed reflections should also occur in the case of direct strikes, but the dominant effect of lower characteristic impedance on voltage magnitude should make those reflections relatively insignificant.

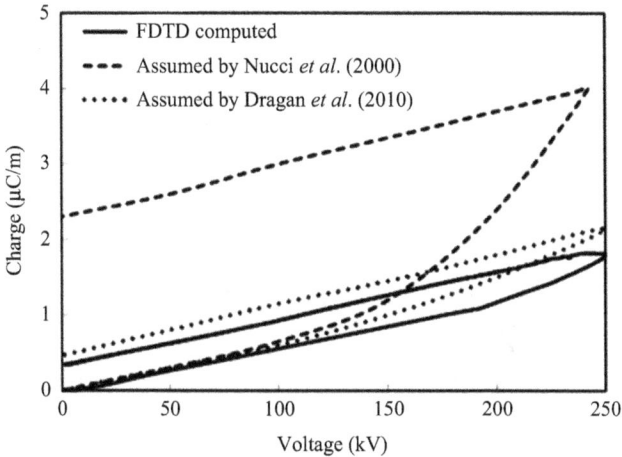

Figure 5.39 *Charge per unit length vs. voltage diagram for the 5-mm-radius, 1-km-long horizontal wire located 7.5 m above perfectly conducting ground computed using the full-wave FDTD model (solid line). The computations were performed for $\sigma_{cor} = 40$ $\mu S/m$, $E_0 = 2.2$ MV/m, $E_{cp} = 0.5$ MV/m (positive polarity). Also shown are the corresponding diagrams assumed by Nucci et al. (2000) and Dragan et al. (2010) in their distributed-circuit model with sources specified using the electromagnetic field theory (dashed and dotted lines, respectively). 2014 ©IEEE. Reprinted, with permission, from Thang et al. (2014, Figure 15)*

5.4.5 Summary

A simplified model of corona discharge for the FDTD computations has been applied to analysis of lightning-induced voltages at different points along the 5-mm-radius, 1-km-long single overhead wire in the presence of corona. Both perfectly conducting and lossy ground cases are considered. FDTD calculations were performed using a 3D-nonuniform grid. The progression of corona streamers from the wire is represented as the radial expansion of cylindrical weakly conducting (40 $\mu S/m$) region around the wire. The critical electric field on the surface of the wire for corona initiation is set to $E_0 = 2.9$ MV/m. The critical background electric field for streamer propagation is set to $E_{cn} = 1.5$ MV/m and $E_{cp} = 0.5$ MV/m for negative and positive polarities, respectively. The magnitudes of FDTD-computed lightning-induced voltages in the presence of corona discharge are larger than those computed without considering corona, which is in contrast with the corona effect in the case of direct lightning strikes. The observed trend is in agreement with that reported by Nucci *et al.* (2000) and Dragan *et al.* (2010), although the increase (up to 5 and 9% for negative and positive polarities, respectively) predicted by the full-wave model is less significant than that (up to a factor of 2) in their studies based on the distributed-circuit model with sources

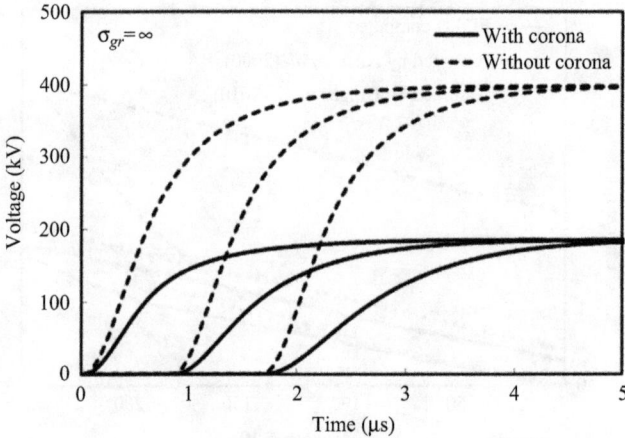

Figure 5.40   Direct strike (negative polarity) to the end of the line near point B:
FDTD-computed (for $E_0 = 2.9$ MV/m, and $E_{cn} = 1.5$ MV/m) voltage
waveform at $d = 0$, 250, and 500 m from the strike point. The
computations were performed for a 5-mm-radius wire with and
without corona. Peak current and maximum current rate-of-rise were
0.8 kA and 0.96 kA/μs, respectively. 2014 ©IEEE. Reprinted, with
permission, from Thang et al. (2014, Figure 16)

specified using electromagnetic field theory. The disparity is likely to be related to the use of different charge-voltage diagrams in different models. In the presence of corona, induced-voltage risetimes tend to be longer. It appears that the distributed impedance discontinuity, associated with corona development on the wire, is the primary reason for the enhancement of voltage peak and lengthening voltage risetime, compared to the case without corona.

5.5   Lightning-induced voltages on overhead multiconductor lines with surge arresters and pole transformers

5.5.1   Introduction

Lightning-induced voltages on overhead multiconductor lines have been computed reasonably accurately using field-to-wire electromagnetic coupling models (e.g., Paolone *et al.* 2009) and approximate expressions for electric fields over lossy ground, such as the Norton approximate expressions (Norton 1937) and the Cooray–Rubinstein formula (Cooray 1992; Rubinstein 1996). Further, lightning-induced voltages have recently been computed with a similar accuracy using the MoM (Harrington 1968) by Pokharel *et al.* (2003), the hybrid electromagnetic/ circuit model (HEM) (Visacro *et al.* 2005) by Silveira *et al.* (2009), the partial-element equivalent-circuit (PEEC) model (Ruehli 1974) by Yutthagowith *et al.*

(2009), and the finite element method (FEM) (Sadiku 1989) by Akbari *et al.* (2013), Sheshyekani and Akbari (2014), and Sheshyekani and Paknahad (2015). Finally, the FDTD method (Yee 1996) has been applied to analyze lightning-induced voltages by Baba and Rakov (2006), Ren *et al.* (2008), Yang *et al.* (2011), Sumitani *et al.* (2012), Tatematsu and Noda (2014), Thang *et al.* (2014), and Zhang, Q. *et al.* (2014a).

Sumitani *et al.* (2012) have computed lightning-induced voltages on an over-head two-wire line using the 3D-FDTD method, in which spatial discretization is fine in the vicinity of overhead wires and coarse in the rest of the computational domain (see Section 5.3). They used the so-called subgrid model, which is different from a nonuniform grid model and more efficient. Tatematsu and Noda (2014) have analyzed lightning-induced voltages on an overhead three-phase distribution line with an overhead ground wire and lightning arresters above perfectly conducting and lossy ground using the 3D-FDTD method. The nonlinear $V-I$ relation of the arrester was represented by piecewise linear approximation, based on the measured $V-I$ curve. The working volume was divided nonuniformly into rectangular cells (non-uniform grid model).

Ren *et al.* (2008), using the 2D-cylindrical FDTD method, have evaluated the electric fields over lossy ground, and calculated the lightning-induced voltages on an overhead two-wire distribution line using the FDTD-computed fields and Agrawal *et al.*'s field-to-wire coupling model (Agrawal *et al.* 1980). One of the reasons for using the two-step or hybrid approach is apparently related to a difficulty of representing closely spaced overhead thin wires in the 3D-uniform-grid FDTD method.

Piantini *et al.* (2007) have measured lightning-induced voltages on a 1/50 small-scale model of complex power distribution networks. In their experiments, a system composed of 1-cm-radius, 1.4-km-long four-conductor (three phase conductors and neutral) lines with distribution networks was considered. The heights above ground of the phase and neutral conductors were 10 and 8 m, respectively. The lightning-induced voltages were measured for different simulated lightning currents and different place-ments of surge arresters and transformers.

In this section, we present lightning-induced voltages on multiconductor lines with surge arresters and pole transformers computed using the 3D-FDTD method. The FDTD method employs a subgrid model, the same as the one used by Sumitani *et al.* (2012), in which spatial discretization is fine (cell side length is 0.5 m) in the vicinity of wires (1455 m × 320 m × 30 m) and coarse (cell side length is 5 m) in the rest of the computational domain. The wires are simulated using thin-wire representation (Noda and Yokoyama 2002), in which one places a wire having an equivalent radius of about 0.12 m ($\approx 0.23 \times 0.5$ m) in the center of an artificial rectangular prism having a cross-sectional area of 1 m × 1 m (= 2 cells × 2 cells) and the modified (relative to air) constitutive parameters: lower electric permittivity $\varepsilon_0' = m\varepsilon_0$ and higher magnetic per-meability $\mu_0' = \mu_0/m$ (ε_0 and μ_0 are the permittivity and permeability of air), as illu-strated (for a different study) in Figure 5.19(e). The modification coefficient m is given by $\ln(\Delta s/a_0)/\ln(\Delta s/a)$, where Δs ($= \Delta x = \Delta y = \Delta z = 0.5$ m) is the lateral side length of cells, a_0 ($\approx 0.23\Delta s$) is the equivalent radius, and a is the radius to be reproduced. When a wire of radius $a < a_0 \approx 0.23\Delta s$ is represented, m needs to be smaller than 1. For

$a = 1$ cm for $\Delta s = 0.5$ m, $m = 0.376$. When a wire of radius $a > a_0$ is represented, m needs to be larger than 1. In the latter case, however, the wire radius a needs to be smaller than the cell side length Δs since $m = \ln(\Delta s/a_0)/\ln(\Delta s/a)$ would be infinity for $a = \Delta s$.

The goal in this section is to compare the FDTD-computed waveforms of lightning-induced voltages with the corresponding ones measured by Piantini *et al.* (2007). Note that, using a nonuniform grid model, Tatematsu and Noda (2014) analyzed lightning-induced voltages on a multiconductor distribution line with surge arresters, but without pole transformers.

5.5.2   Methodology and configurations studied

We describe here two configurations (distribution networks A and B), which represent the experiments carried out by Piantini *et al.* (2007), who used a 1/50 small-scale model in order to simulate lightning-induced voltages on multiconductor lines with surge arresters and pole transformers. In this section, unless otherwise indicated, the values of all parameters in those experiments are referred to the full-scale system.

5.5.2.1   Distribution Network A

Figure 5.41 shows the experimental configuration of 1-cm-radius and 1.4-km-long three-phase conductors and neutral conductor located above perfectly conducting ground, which represents a 15-kV non-energized distribution line. Figure 5.41(a) shows the cross-sectional (*yz*-plane) view of the overhead conductors and Figure 5.41(b) shows the plan (*xy*-plane) view of the entire distribution network A. Three-phase conductors and one neutral conductor are located 10 and 8 m above ground, respectively. The distance between adjacent phase conductors is 0.75 m and each end of each phase conductor is connected to ground via a 455-Ω matching resistor. The distance between the lightning strike point and the voltage measuring point, M1, is about 82 m (20 m along the *x*-axis and 80 m along the *y*-axis).

Figure 5.42 shows the circuit representation of each component of the distribution network presented in Figure 5.41(b). Lightning-induced voltages were measured at node M1. The *V–I* characteristic of the surge arrester model (see Figure 5.42(a)) is fairly similar to that of actual ZnO distribution arrester with rated voltage and current of 12 kV and 5 kA, respectively (Piantini *et al.* 2007). Each element such as nonlinear or linear resistor, inductor, or capacitor is represented by one side of the cell in the FDTD simulation. Although Piantini *et al.* (2007) represented the grounding downconductors by 10-μH inductors, in the FDTD simulation they are represented by vertical perfectly conducting wires of radius 1 cm.

Figure 5.43 shows a 3D view of the computational domain showing the lightning channel, represented by a 900-m-long vertical phased-current-source-array model (Baba and Rakov 2003), simulating the TL model (Uman and McLain 1969), and distribution network A. The upward propagation speed of current along the simulated lightning channel is set to $0.11c$, where c is the speed of light, following the experimental condition. Note that the length of the simulated lightning channel in the experiment was 600 m (12 m in the 1/50 small-scale model), which will not affect the lightning-induced voltages within the first 18 μs or so.

0.75 m

2 cm

10 m

8 m

z

x   y

(a)

90 m 210 m 210 m 148 m 42 m 132 m 210 m 210 m 170 m

150 m 148 m   346 m   284 m   152 m 150 m

300 m

70 m

Lightning strike

20 m

M1

75 m

y

z   x

○ Surge arrester   ⊣⊢ Grounding point   (M1) Measuring point
▲ Transformer

(b)

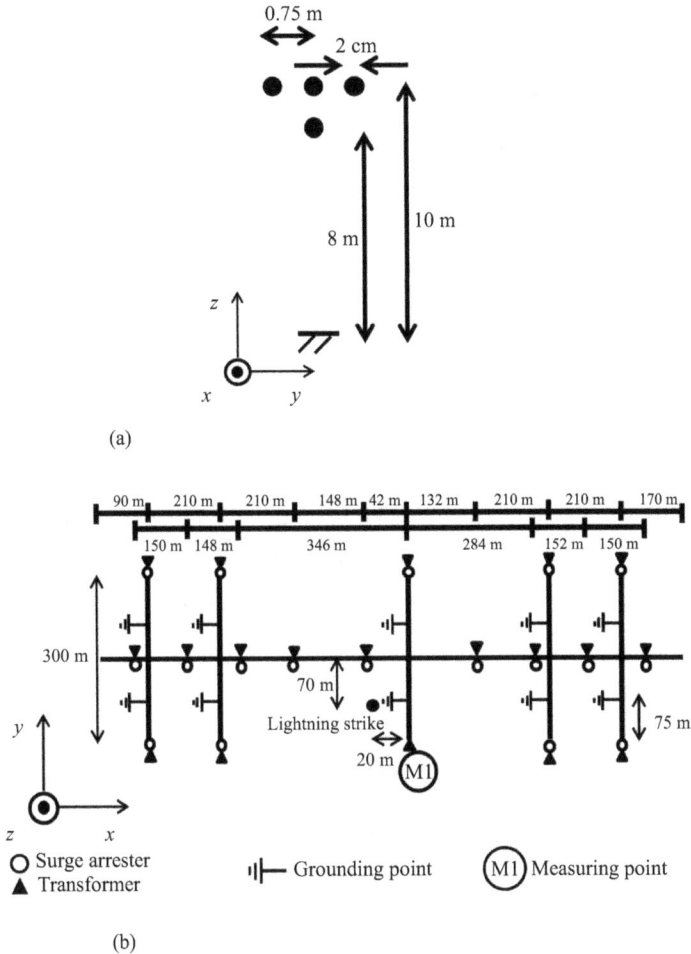

Figure 5.41   *Configuration of distribution network A: (a) cross-sectional (yz-plane) view of four overhead conductors, and (b) plan (xy-plane) view of the entire network. Induced-voltage waveforms at point M1 are shown in Figure 5.46. 2015 ©IEEE. Reprinted, with permission, from Thang et al. (2015a, Figure 2)*

For FDTD computations, the working volume of 1480 m × 500 m × 1000 m (see Figure 5.43) is divided into cubic cells of 5 m × 5 m × 5 m, except for the space in the vicinity of the distribution network (1455 m × 320 m × 30 m), where cubic cells of 0.5 m × 0.5 m × 0.5 m are employed. The total number of cells in the working volume is about 11.8×10^7 ($\approx 1480/5 \times 500/5 \times 1000/5 + 1455/0.5 \times 320/0.5 \times 30/0.5$). Liao's second-order absorbing boundary condition (Liao *et al.* 1984) is applied to five planes (the top plane and four side planes) to minimize unwanted reflections there. The bottom plane is set to be a perfect conductor. Note that the 100-m thickness of ground,

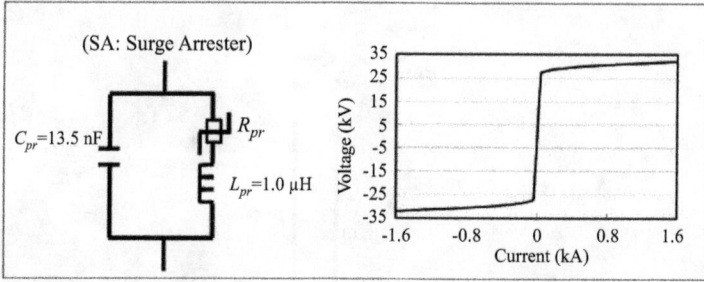

(a)

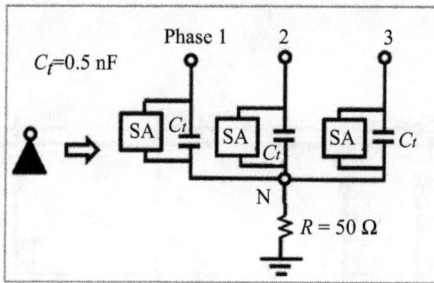

(b)

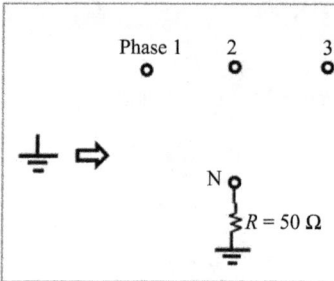

(c)

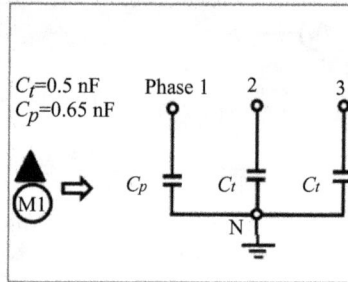

(d)

Figure 5.42   *Circuit representation of different system elements in the model: (a) surge arrester equivalent circuit and voltage-current characteristic of the nonlinear resistor R_{pr}, (b) surge arrester and transformer point, (c) grounding point, (d) transformer and measuring point. 2015 ©IEEE. Reprinted, with permission, from Thang et al. (2015a, Figure 3)*

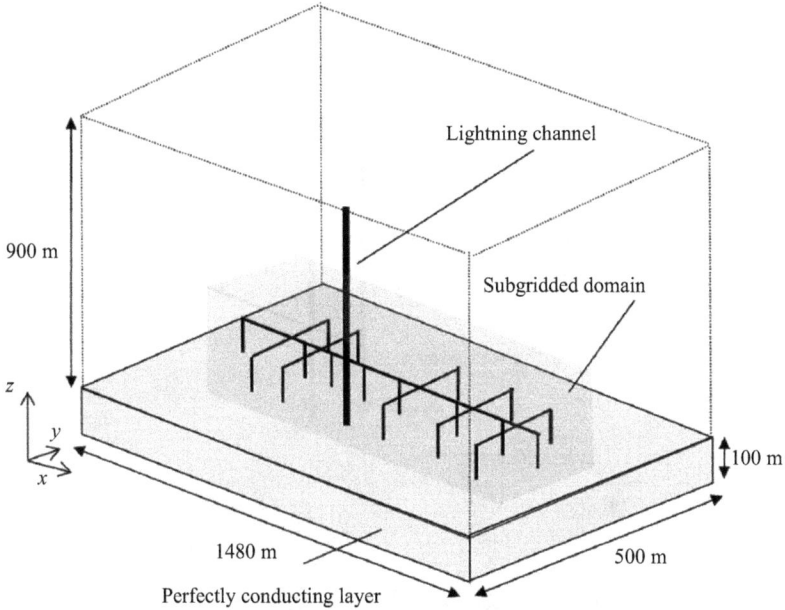

Figure 5.43   3D view of computational domain showing the lightning channel and distribution network A, analyzed using the 3D-FDTD method. 2015 ©IEEE. Reprinted, with permission, from Thang et al. (2015a, Figure 4)

which is not needed for representing flat perfectly conducting ground, is employed here because of the plans to study the influence of lossy ground in the future. Although the distance between the phase conductors was 0.75 m in the experiment, it is set to 1 m in the FDTD simulations. This difference has little influence on the lightning-induced voltages on phase conductors.

5.5.2.2   Distribution Network B

Figure 5.44(a) and (b) shows the line geometry (cross-sectional view of overhead conductors) and an overview of distribution network B, respectively. The three phase conductors are located 10 m above ground and the neutral conductor is located 8 m above ground. Each end of each conductor was connected to ground via a 455-Ω matching resistor. The lightning-induced voltage was measured at node M2. Similar to distribution network A, for FDTD computations, this system is accommodated in the working volume of 1480 m × 500 m × 1000 m, which is divided into cubic cells of 5 m × 5 m × 5 m, except for the space (1455 m × 320 m × 30 m) in the vicinity of the distribution network, where cubic cells of 0.5 m × 0.5 m × 0.5 m are employed. Again, the distance between the phase conductors is set to 1 m (2 cells), although it was 0.75 m in the measurement. The distance between the lightning strike point and the voltage measuring point, M2, is about 191 m (80 m along the *y*-axis (same as for distribution

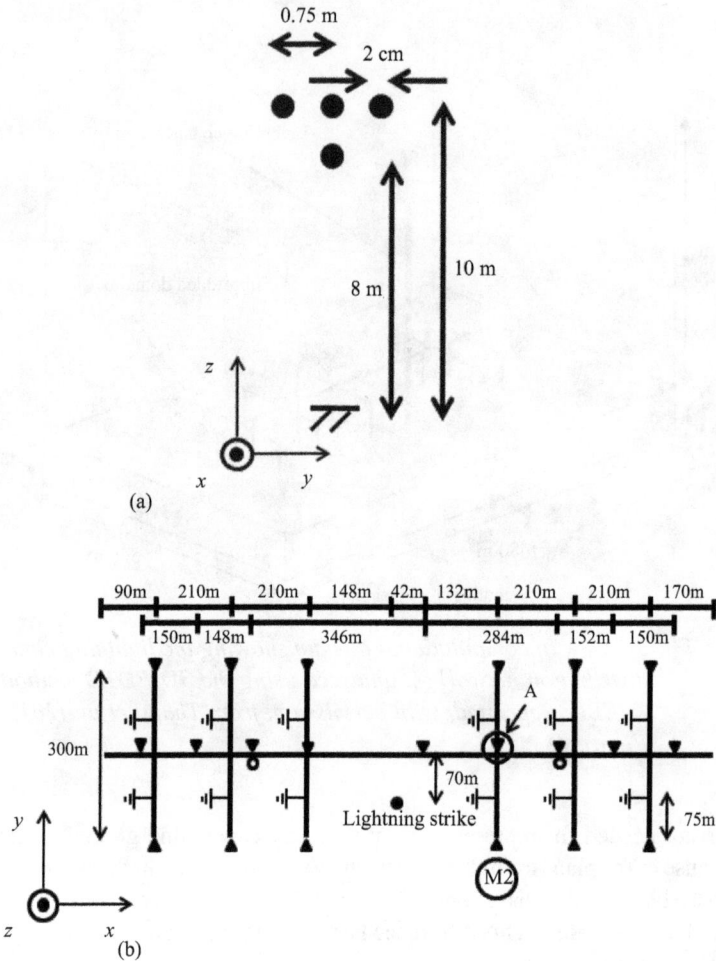

Figure 5.44   Configuration of distribution network B: (a) side (yz-plane) view of four overhead conductors, (b) plan (xy-plane) view of the entire network. Induced-voltage waveforms at point M2 are shown in Figures 5.47 to 5.49. 2015 ©IEEE. Reprinted, with permission, from Thang et al. (2015a, Figure 5)

network A) and 174 m along the *x*-axis), which is greater than the distance (82 m) between the strike point and M1.

5.5.3   Analysis and results

The current injected into the simulated lightning channel was represented by a triangular waveform with peak value of 34 kA, front time (risetime) of 2 μs, and half-peak width of 85 μs, which was used by Piantini *et al.* (2007). The initial part of the current is shown in

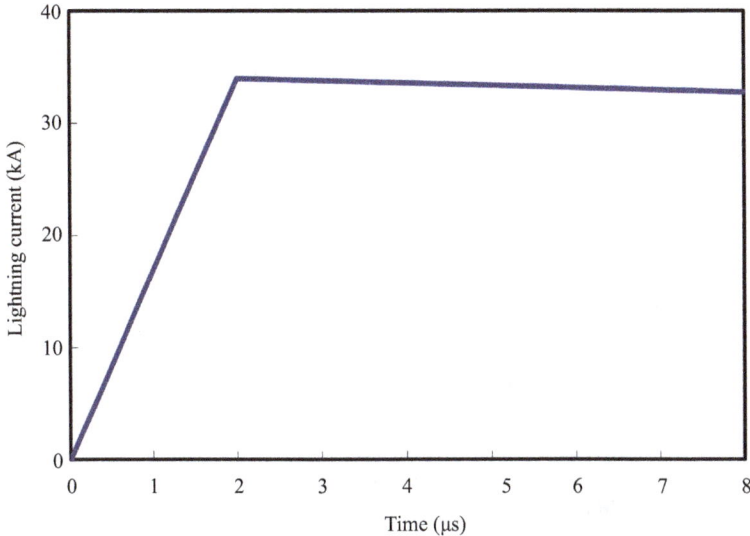

*Figure 5.45   Initial (8-µs) part of the simulated lightning current waveform with a
peak of 34 kA, front time of 2 µs, and half-peak width of 85 µs, which
was used in the experiments of Piantini et al. (2007) and in the FDTD
computations presented in Section 5.5. 2015 ©IEEE. Reprinted, with
permission, from Thang et al. (2015a, Figure 6)*

Figure 5.45. Additionally, currents with the same waveshape, but with peaks of 50 and
70 kA (not shown here) were used in the simulations.

Figure 5.46 shows the lightning-induced voltage waveform at node M1 of network A presented in Figure 5.41(b), computed using the FDTD method for the
triangular lightning current with 34-kA peak. Also shown in this figure is the corresponding voltage waveform measured by Piantini *et al.* 2007. Note that the starting
time ($t = 0$ in the plot) of the measured waveform was adjusted because sometimes,
due to the noise, the exact time of $t = 0$ was unknown in the measured results (other
measured waveforms in Figures 5.47 to 5.49 were also similarly adjusted). It follows
from Figure 5.46 that the FDTD-computed lightning-induced voltage waveform
agrees reasonably well with the corresponding measured one. Magnitudes of current
flowing through the surge arresters located closest and second closest to the lightning strike point (see Figure 5.41(b)) are about 800 and 100 A, respectively.
Magnitudes of currents through other arresters are smaller than 50 A. It follows
that arresters located closer to the lightning strike point are more involved in
suppressing the lightning-induced voltages.

Figures 5.47 to 5.49 show FDTD-computed lightning-induced voltage waveforms and the corresponding measured ones at node M2 of distribution network B
(see Figure 5.44(b)) for different lightning current peak values and different surge
arrester and transformer placements.

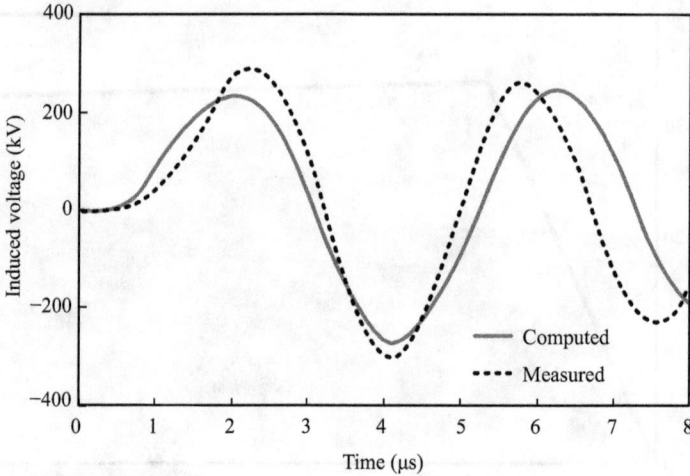

Figure 5.46   Distribution network A: FDTD-computed waveform of lightning-induced voltage (solid line) and the corresponding measured one (broken line) at point M1 (see Figure 5.41(b)). The simulated lightning current waveform is shown in Figure 5.45. 2015 ©IEEE. Reprinted, with permission, from Thang et al. (2015a, Figure 7)

Figure 5.47 shows the lightning-induced voltage waveforms at node M2 for the lightning current peak of 70 kA. The reason why the magnitude of lightning-induced voltage at node M2 for 70-kA current (see Figure 5.47) is smaller than that at node M1 for 34-kA current (see Figure 5.46) is that the distance between the lightning strike point and node M2 is greater than that between the strike point and node M1. Magnitudes of current through two surge arresters located at roughly the same distance from the lightning strike point (see Figure 5.44(b)) are both about 150 A. It follows that these two arresters work equally to suppress the lightning-induced voltages. Note that in some cases high-frequency oscillations associated mainly with operation of the switching device in the experimental setup were present in the wavefront of the simulated lightning current, causing small oscillations superimposed on the measured induced-voltage waveform, as seen in Figure 5.47.

Figure 5.48 shows the lightning-induced voltage waveforms for a current peak of 34 kA and surge arresters additionally placed at the measuring point M2 (in all phases) in parallel with the transformer, the latter being simulated as shown in Figure 5.42(d). Figure 5.49 shows the lightning-induced voltage waveforms for a current peak of 50 kA and all surge arresters removed, except for those installed in parallel with the transformer at point A (encircled and labeled in Figure 5.44). It follows from these figures that FDTD-computed lightning-induced voltage waveforms agree reasonably well with the corresponding measured ones.

Piantini *et al.* (2007) have compared the measured waveforms shown in Figures 5.46 to 5.49 with those computed using a field-to-wire coupling model. The

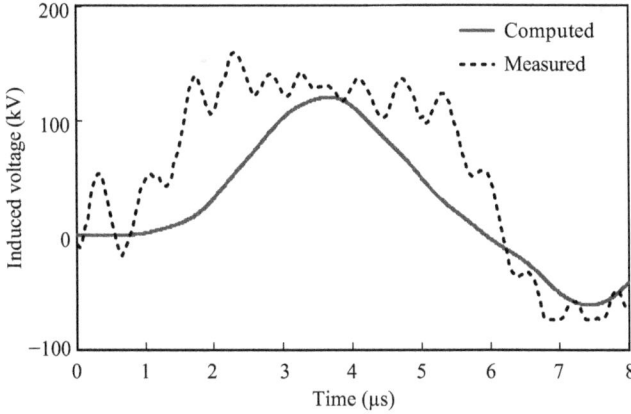

Figure 5.47 *Distribution network B: FDTD-computed waveform of lightning-induced voltage (solid line) and the corresponding measured one (broken line) at point M2 (see Figure 5.44(b)) for a triangular lightning current waveform with peak of 70 kA, front time of 2 μs, and half-peak width of 85 μs. 2015 ©IEEE. Reprinted, with permission, from Thang et al. (2015a, Figure 8)*

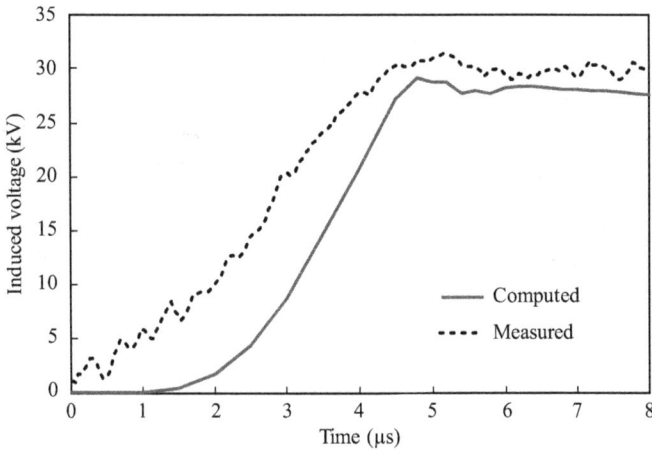

Figure 5.48 *Distribution network B: FDTD-computed and the corresponding measured lightning-induced voltage waveforms at point M2 (see Figure 5.44(b)) for a triangular lightning current waveform with peak of 34 kA, front time of 2 μs, and half-peak width of 85 μs. The network additionally had surge arresters placed at the measurement point in parallel with the transformer. 2015 ©IEEE. Reprinted, with permission, from Thang et al. (2015a, Figure 9)*

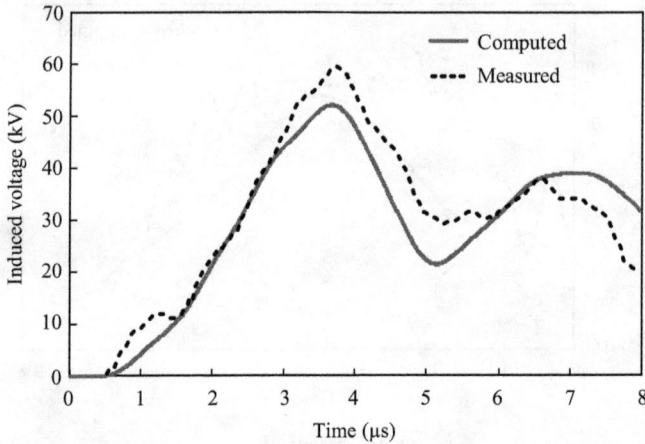

Figure 5.49   FDTD-computed and the corresponding measured lightning-induced voltage waveforms at point M2 (see Figure 5.44(b)) for a triangular lightning current waveform with peak of 50 kA, front time of 2 μs, and half-peak width of 85 μs. All surge arresters were removed except for those installed in parallel with the transformer located at point A (encircled in Figure 5.44). 2015 ©IEEE. Reprinted, with permission, from Thang et al. (2015a, Figure 10)

waveform computed using the field-to-wire coupling model for the case shown in Figure 5.47 agrees appreciably better with the corresponding measured one (except for the superimposed high-frequency oscillations, which are an artefact caused by operation of the switch in the experimental setup) than the FDTD-computed waveform. On the other hand, the FDTD-computed waveforms shown in Figures 5.46, 5.48, and 5.49 agree well with the corresponding ones computed using the field-to-wire coupling model.

Overall results indicate that lightning-induced voltages in multiconductor line networks with surge arresters and pole transformers can be studied using the FDTD method. In the simulations presented in this section, a personal computer with an OS of 64-bit Windows 7 and a CPU of 3.46-GHz Intel Core i7 was used. The time increment was set to $\triangle t = 0.5$ ns, and the maximum observation time was set to 8 μs. The computation time needed for one run was about 24 h and the memory required was 2.1 GB.

5.5.4   Summary

In this section, we have presented lightning-induced voltages on multiconductor lines with surge arresters and pole transformers computed using the 3D-FDTD method. This method uses a subgrid model in which spatial discretization is fine in the vicinity of the wires and coarse in the rest of the computational domain. The wires are simulated using the thin-wire representation. FDTD-computed waveforms of lightning-induced voltages agree reasonably well with the corresponding waveforms measured in the small-scale experiment performed by Piantini *et al.*

(2007). This indicates that lightning-induced voltages on multiconductor networks with surge arresters and pole transformers can be studied using the FDTD method.

5.6 Lightning-induced voltages on overhead multiconductor lines in the presence of nearby buildings

5.6.1 Introduction

Piantini *et al.* (2000, 2007) used a 1/50 small-scale experimental setup to model complex power distribution networks and investigate the effect of nearby buildings on the lightning-induced voltages on power lines. The distribution networks were composed of the main feeder and several laterals. Buildings with heights of 5 or 15 m were represented by grounded aluminum boxes. The lightning-induced voltages were measured at different points of the line relative to the locations of surge arresters and transformers.

In this section, using the 3D-FDTD method, we examine lightning-induced voltages on multiconductor lines with surge arresters and pole transformers in the presence of nearby buildings. The FDTD method employs a subgrid model, the same one as used by Sumitani *et al.* (2012), in which spatial discretization is fine in the vicinity of wires and coarse in the rest of the computational domain. The overhead wires are simulated using thin-wire representation (Noda and Yokoyama 2002). We compare the FDTD-computed waveforms of lightning-induced voltages with the corresponding waveforms measured by Piantini *et al.* (2000, 2007).

5.6.2 Methodology and configurations studied

The methodology used here is similar to that described in Section 5.5 (see the computational domain shown in Figure 5.43 and simulated lightning current waveform shown in Figure 5.45).

Figures 5.50 and 5.51 show two experimental configurations, studied by Piantini *et al.* (2000, 2007), where the dimensions are referred to the full-scale system. It includes the 1.4-km-long main feeder and several laterals. Each line has three 1-cm-radius phase conductors and a neutral conductor located above perfectly conducting ground. Figures 5.50(a) and 5.51(a) represent the cases of no buildings (building height $h_e = 0$ m), Figures 5.50(b) and 5.51(b) are for $h_e = 5$ m, and Figures 5.50(c) and 5.51(c) are for $h_e = 15$ m. The three horizontally arranged phase conductors and one neutral conductor are located 10 and 8 m above ground, respectively (see Figure 5.41(a)). The distance between adjacent phase conductors is 0.75 m and either end of each phase conductor is connected to ground via a 455-Ω matching resistor.

In Figure 5.50, the distance between the lightning strike point and the voltage measuring point, M, is 20 m. The distances between the measuring point and the closest set of surge arresters located on its left and right sides are labeled s_e and s_d, respectively. Two cases with different values of s_e and s_d were considered: case 1 with $s_e = 75$ m and $s_d = 75$ m, and case 2 with $s_e = 148$ m and $s_d = 174$ m. In

(a)

(b)

(c)

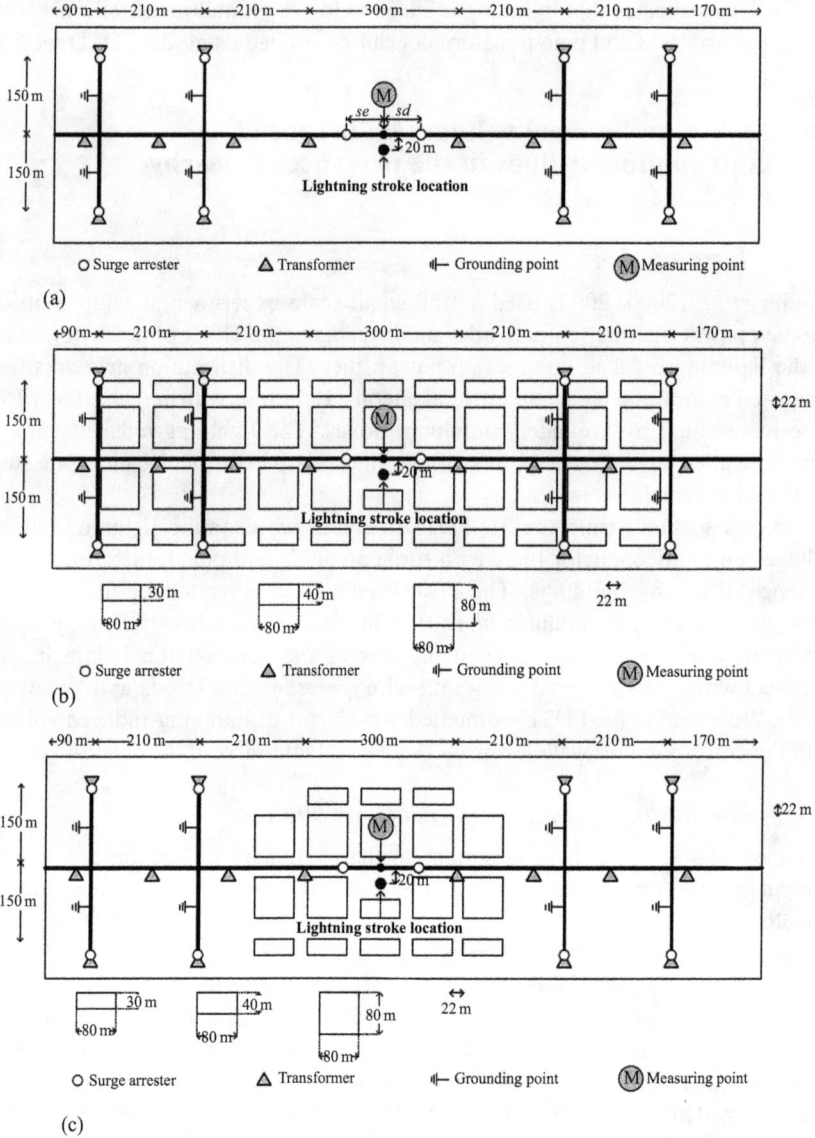

Figure 5.50   *Experimental configuration 1: Plan views of the main feeder with eight laterals in the absence and in the presence of buildings: (a) no buildings, (b) 44 buildings of height $h_e = 5$ m, and (c) 18 buildings of height $h_e = 15$ m. Distance between the stroke location and measuring point (M) is 20 m. The distances between the measuring point and the closest set of surge arresters located on its left and right sides are labeled s_e and s_d in (a). Indicated dimensions refer to the full-scale system. 2015 ©IEEE. Reprinted, with permission, from Thang et al. (2015b, Figure 1)*

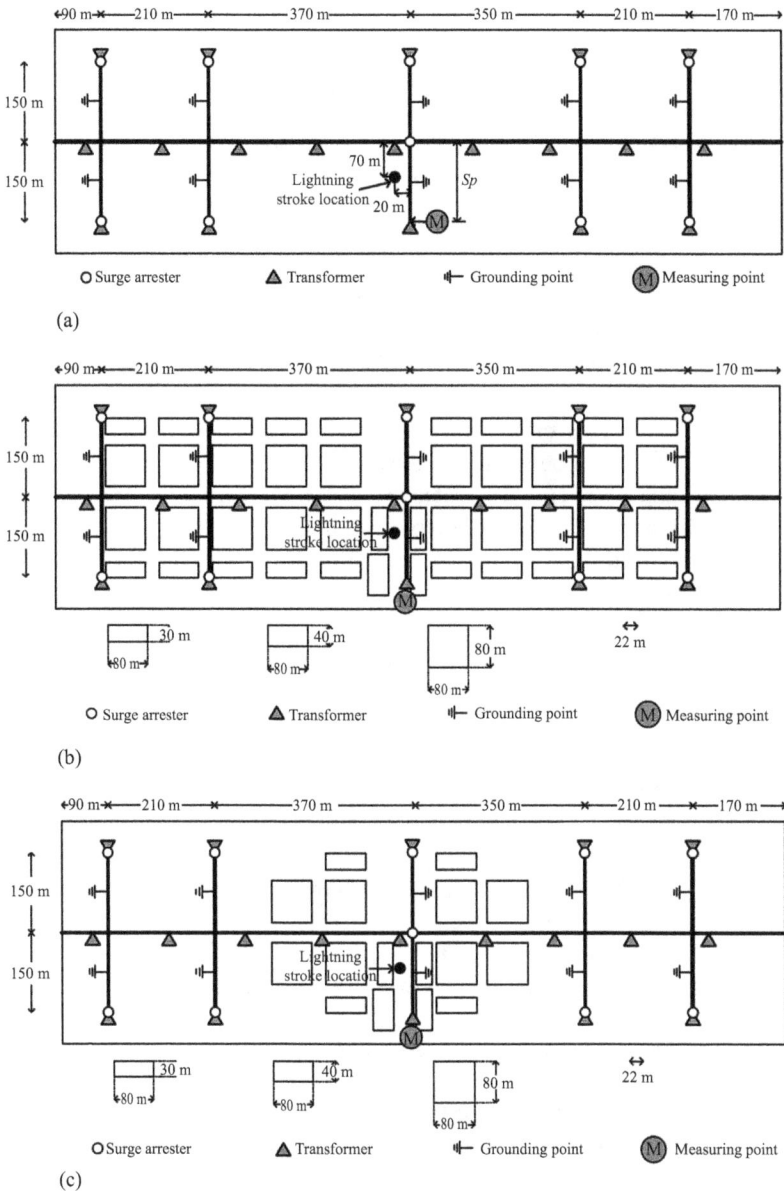

(a)

(b)

(c)

Figure 5.51   *Experimental configuration 2: Plan views of the main feeder with ten laterals in the absence and in the presence of buildings: (a) no buildings, (b) 44 buildings of height $h_e = 5$ m, and (c) 16 buildings of height $h_e = 15$ m. Distance between the stroke location and the main feeder is 70 m. The distance between the stroke location and the closest lateral is 20 m. Arresters are placed at the ends of the laterals and at distance s_r (labeled in (a)) from the measuring point. Indicated dimensions refer to the full-scale system. 2015 ©IEEE. Reprinted, with permission, from Thang et al. (2015b, Figure 2)*

Figure 5.51, the distance between the lightning strike point and the main feeder is 70 m, and between the lightning strike point and the closest lateral it is 20 m. The distance between the measuring point, M and the closest set of surge arresters is labeled s_r, which was set to either 75 or 150 m.

For FDTD computations, this system is accommodated in a working volume of 1480 m $\times$ 500 m $\times$ 1000 m (see Figure 5.43), which is divided into cubic cells of 5 m $\times$ 5 m $\times$ 5 m, except for the space in the vicinity of the distribution network (1455 m $\times$ 320 m $\times$ 30 m), where cubic cells of 0.5 m $\times$ 0.5 m $\times$ 0.5 m are employed. The total number of cells in the working volume is about 11.8 $\times$ 10^7 ($\approx$1480/5 $\times$ 500 /5 $\times$ 1000/5 + 1455/0.5 $\times$ 320/0.5 $\times$ 30/0.5). Liao's second-order absorbing boundary condition (Liao *et al.* 1984) is applied to five planes (the top plane and four side planes) to minimize unwanted reflections there. Each element such as nonlinear or linear resistor, inductor, or capacitor is represented by one side of the cell. More details on the circuit representation of each component of the distribution network can be found in the work of Thang *et al.* (2015a).

The lightning channel is simulated by a 900-m-long vertical phased-current-source-array (Baba and Rakov 2003) representing the TL model (Uman and McLain 1969). The simulated lightning current waveform used in both the experiment and FDTD calculations is shown in Figure 5.45. Additionally, another current with the same waveshape, but with a peak of 50 kA (not shown here) was used in the simulations. The upward propagation speed of current along the simulated lightning channel is set to 0.11c, where c is the speed of light, to match the speed in the experiment of Piantini *et al.* (2000, 2007).

5.6.3   Analysis and results

Figure 5.52(a), (b), and (c) shows lightning-induced voltage waveforms at node M of the network shown in Figure 5.50 (Experimental configuration 1), computed using the FDTD method for buildings having heights of 0 (no buildings), 5, and 15 m. In the simulations, the distances between the measuring point (M) and the closest set of surge arresters were set to $s_e = 75$ m and $s_d = 75$ m. Also shown in these figures are the corresponding measured voltage waveforms (Piantini *et al.* 2000, 2007). Figure 5.53(a), (b), and (c) is the same as Figure 5.52(a), (b), and (c), but for $s_e = 148$ m and $s_d = 174$ m. From these figures, the ratios of the calculated voltage peaks for buildings heights $h_e = 5$ m and 15 m to that for $h_e = 0$ m are 0.85 and 0.53 for Figure 5.52 and 0.84 and 0.50 for Figure 5.53, whereas for the measured voltages the corresponding ratios are 0.75 and 0.59 for Figure 5.52 and 0.73 and 0.34 for Figure 5.53.

Similarly, Figure 5.54(a), (b), and (c) shows lightning-induced voltage wave-forms at node M of the network shown in Figure 5.51 (Experimental configuration 2), computed using the FDTD method for buildings having heights of 0, 5, and 15 m. The distance between the measuring point (M) and the closest set of surge arrester was set to $s_r = 150$ m. Figure 5.55(a), (b), and (c) is the same as Figure 5.54 (a), (b), and (c), but for $s_r = 75$ m. The corresponding measured voltage waveforms are also shown in these figures. From these figures, the ratios of the calculated

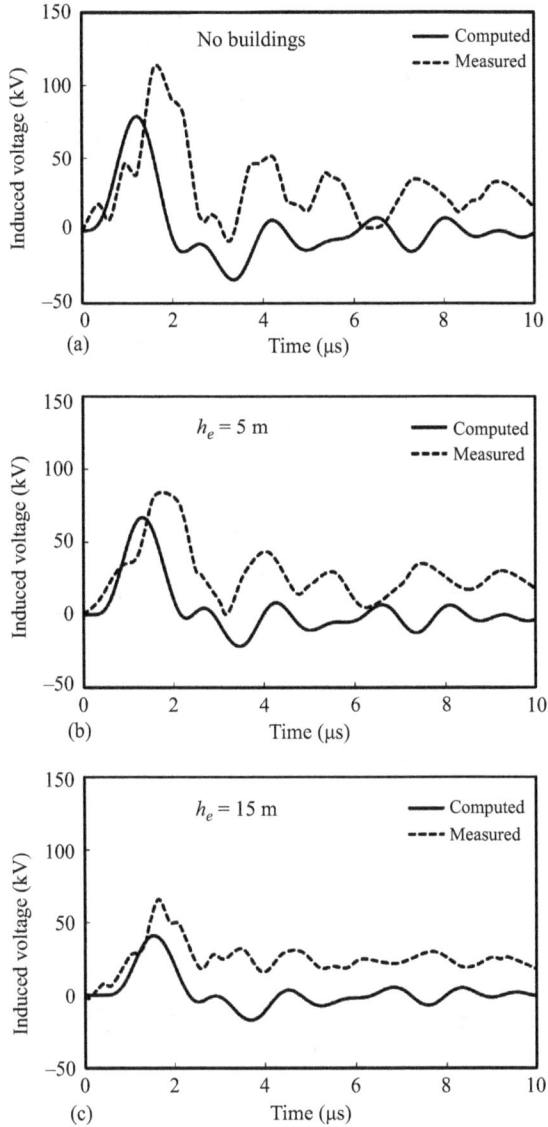

Figure 5.52 Experimental configuration 1: FDTD-computed and corresponding measured (Piantini et al. 2000, 2007) lightning-induced voltage waveforms for a triangular lightning current pulse with peak of 34 kA, risetime of 2 μs, and time to half value of 85 μs and buildings having heights of (a) 0, (b) 5, and (c) 15 m. The distances between the measuring point (M) and the closest set of surge arresters were $s_e = s_d = 75$ m (see Figures 5.50(a), (b), and (c)). 2015 ©IEEE. Reprinted, with permission, from Thang et al. (2015b, Figure 5)

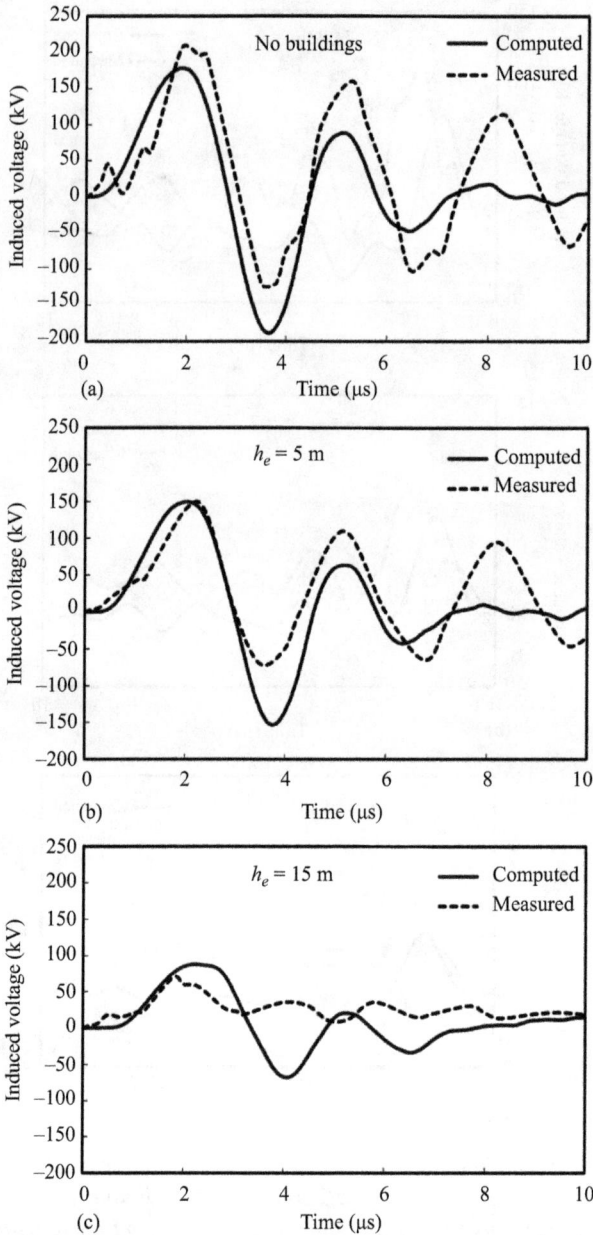

*Figure 5.53   Experimental configuration 1: FDTD-computed and corresponding
measured (Piantini et al. 2000, 2007) lightning-induced voltage
waveforms for a triangular lightning current pulse with peak of 34 kA,
risetime of 2 μs, and time to half value of 85 μs and buildings having
heights of (a) 0, (b) 5, and (c) 15 m. The distances between the measuring
point (M) and the closest set of surge arresters were $s_e = 148$ m and
$s_d = 174$ m (see Figures 5.50(a), (b), and (c)). 2015 ©IEEE. Reprinted,
with permission, from Thang et al. (2015b, Figure 6)*

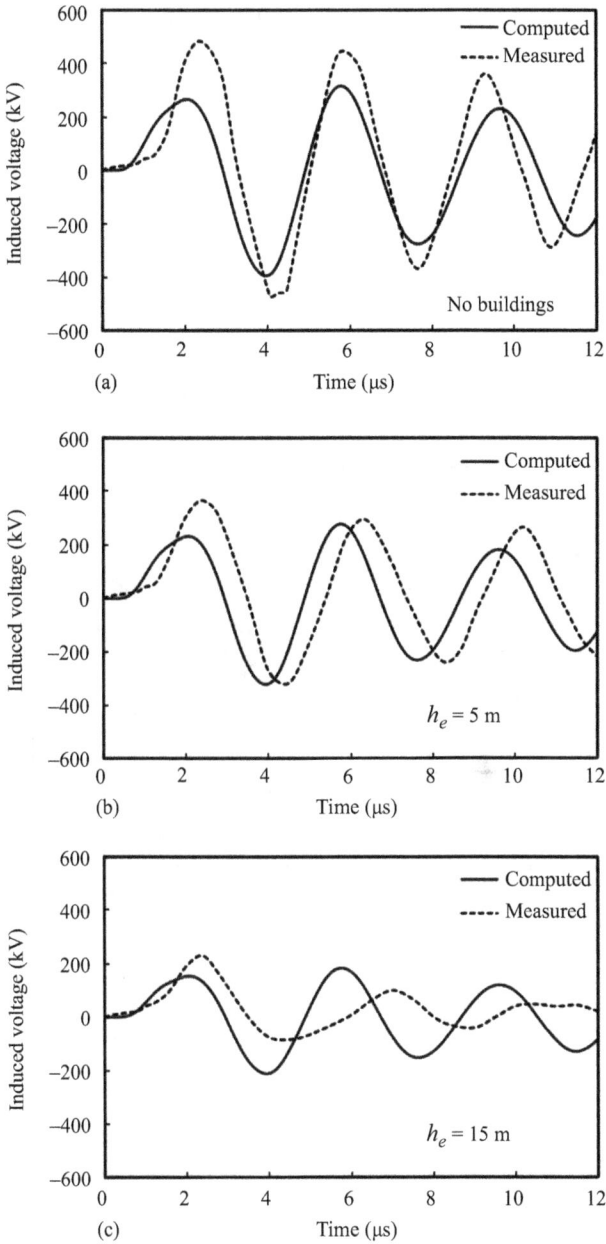

Figure 5.54   Experimental configuration 2: FDTD-computed and corresponding measured (Piantini et al. 2000, 2007) lightning-induced voltage waveforms for a triangular lightning current pulse with peak of 50 kA, risetime of 2 μs, and time to half value of 85 μs and buildings having heights of (a) 0, (b) 5, and (c) 15 m. The distance between the measuring point (M) and the closest set of surge arresters was $s_r = 150$ m (see Figures 5.51(a), (b), and (c)). 2015 ©IEEE. Reprinted, with permission, from Thang et al. (2015b, Figure 7)

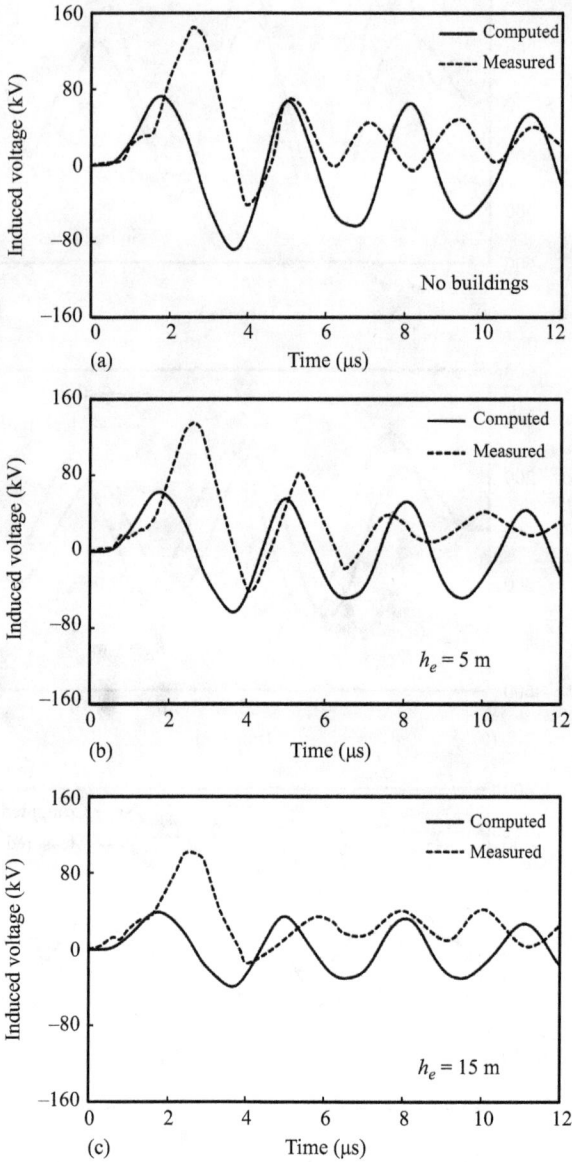

*Figure 5.55   Experimental configuration 2: FDTD-computed and corresponding
measured (Piantini et al. 2000, 2007) lightning-induced voltage
waveforms for a triangular lightning current pulse with peak of 34
kA, risetime of 2 μs, and time to half value of 85 μs and buildings
having heights of (a) 0, (b) 5, and (c) 15 m. The distance between the
measuring point (M) and the closest set of surge arrester was
$s_r = 75$ m (see Figures 5.51(a), (b), and (c)). 2015 ©IEEE.
Reprinted, with permission, from Thang et al. (2015b, Figure 8)*

voltages peaks for building height $h_e = 5$ m and 15 m to that for $h_e = 0$ m are about 0.88 and 0.66 for Figure 5.54 and 0.87 and 0.54 for Figure 5.55, whereas for the measured voltages the corresponding ratios are 0.78 and 0.50 for Figure 5.54 and 0.93 and 0.74 for Figure 5.55.

It follows from Figures 5.52 to 5.55 that the presence of nearby buildings causes reduction on lightning induced-voltages, as expected. The observed trend is in general agreement with that based on the measurement by Piantini *et al.* (2000, 2007). As mentioned in Section 5.6.1, buildings were represented by grounded aluminum boxes, which could lead to an overestimation of experimentally observed voltage reduction (Piantini *et al.* 2000, 2007). On the other hand, Baba and Rakov (2007, Table VI) showed that the electric field enhancement due to the presence of building is only slightly influenced by building conductivity ranging from 1 mS/m (dry concrete) to infinity.

Overall, the results presented in this section indicate that the effect of buildings on lightning-induced voltages on multiconductor systems with surge arresters and pole transformers can be studied using the FDTD method. In the simulations presented in this section, a personal computer with an OS of 64-bit Windows 7 and a CPU of 3.46-GHz Intel Core i7 was used. The time increment was set to $\triangle t = 0.5$ ns and the maximum observation time was set to 10 µs. The computation time needed for one run was about 24 h and the memory required was 2.4 GB.

5.6.4   Summary

We have presented lightning-induced voltages on multiconductor lines with surge arresters and pole transformers in the presence of nearby buildings computed using the 3D-FDTD method. This method uses a subgrid model in which spatial discretization is fine in the vicinity of the wires and coarse in the rest of the computational domain. The wires are simulated using the thin-wire representation. The magnitudes of FDTD-computed lightning-induced voltages are reduced in the presence of nearby buildings, as expected. The observed trend is in general agreement with measurements reported from their small-scale experiments by Piantini *et al.* (2000, 2007). This indicates that the effect of buildings on lightning-induced voltages in multiconductor networks with surge arresters and pole transformers can be studied using the FDTD method.

5.7   Lightning-induced currents in buried cables

5.7.1   Introduction

The use of shield wires to protect buried cables against lightning is a common and effective practice in the power and telecommunication industries. This practice consists of burying one or more bare conductors (known as shield or guard wires) above the cable, along the entire length where the protection is needed.

A theoretical model for the protective effect provided by a shield wire against direct lightning strikes was proposed by Sunde (1968), which predicts that the shield wire would carry a part of the lightning current that otherwise would flow through the cable. Sunde's model assumes that the effect of cable insulating outer cover is negligible, so that the shield wire is treated as if it was bonded to the metallic sheath of the cable through an electric arc. Douglass (1971) and Ungar (1980) have shown that the cable insulation is a key factor in determining the protective effect of the shield wire, and Chang (1980) has proposed a procedure for installing a shield wire on the basis of the electro-geometric model. More recently, Bejleri *et al.* (2004) and Barbosa *et al.* (2008) have carried out experiments with rocket-triggered lightning and shown that the shield wire can protect the cable from lightning whose current is delivered to the ground surface directly above or at some distance from the cable.

Besides protecting the cable against direct lightning strikes, the shield wire also provides some protection against lightning induced surges which may be dangerous to the cable insulation and to the equipment connected to the cable conductors. These surges are generated by the current flowing in the cable metallic sheath, due to the transfer impedance of the latter.

A comprehensive study of the lightning-induced currents in a buried cable has been carried out theoretically and experimentally by Petrache *et al.* (2005) and Paolone *et al.* (2005), respectively. The experimental results provided by Paolone *et al.* (2005) were used as reference for subsequent studies that evaluated the effect of ground stratification (Paulino *et al.* 2014, Paknahad *et al.* 2014b), propagation effects on the inducing fields (Paknahad *et al.* 2014a), and soil parameters (Paknahad *et al.* 2014c). However, these studies did not consider the effect of shield wire.

A theoretical study of the shield-wire effect on the lightning-induced currents has been carried out by Yang *et al.* (2012), using a two-step FDTD method (Yee 1966), where the inducing fields and the induced currents are computed separately. However, due to the lack of experimental data, Yang *et al.* validated their method only to a limited extend, using the results of Paolone *et al.* (2005) that did not include shield wires.

Presented in this section are experimental data that are used for validation of a simulation model based on the 3D-FDTD method (Yee 1966). It is worth mentioning that the 3D-FDTD method does not require the simplifying assumptions of the two-step FDTD computation. The 3D-FDTD model is subsequently used to analyze the effects of shield wire on lightning-induced currents in the cable, depending on the various model input parameters.

Section 5.7.2 presents the test setup and the experimental data, whereas Section 5.7.3 presents the model used for the FDTD simulations. Section 5.7.4 compares the results obtained using the 3D-FDTD model with the experimental ones. Section 5.7.5 presents a sensitivity analysis of some relevant parameters that may influence the shield-wire protective effect. Finally, Section 5.7.6 provides the conclusions.

5.7.2   Description of the experiment

The experimental data presented here were obtained at the rocket-triggered lightning test site operated from 2000 to 2007 in southwestern Brazil (Cachoeira

Paulista) (Barbosa *et al.* 2008, Barbosa and Paulino 2008, Paulino *et al.* 2009). The experimental data were collected when a lightning flash was triggered to the top of a telecommunication tower, so that the lightning current and its associated currents in the cable and in the shield wire could be recorded simultaneously. The test setup and the main experimental data collected are described in the following.

Figure 5.56 shows an overview of the test setup, where a 1-km-long buried cable and a 1-km-long shield wire were installed in front of a 30-m-high tele-communication tower, which was equipped with a rocket launcher and a measuring system for recording lightning currents. Another measuring instrumentation was placed in a box at a distance of 20 m from the closest point between the cable and the tower. The cable and the shield wire passed inside current probes that were connected to an oscilloscope.

Figure 5.57 shows a schematic side view of the shield wire and the cable. The cable is a shielded telecommunication cable that has 30 pairs of insulated copper wires with 0.64-mm diameter (53 Ω/km each), covered by a tubular metallic sheath and a 2-mm-thick plastic outer sheath (the plastic sheath diameter is $d_i = 20$ mm),

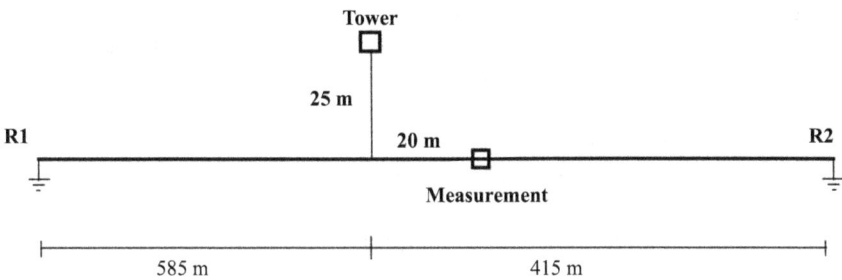

Figure 5.56   Plan view of the test setup showing a 30-m-high tower (lightning strike location), a 1-km-long buried cable, and a measurement box. 2016 ©IEEE. Reprinted, with permission, from Tanaka et al. (2016, Figure 1)

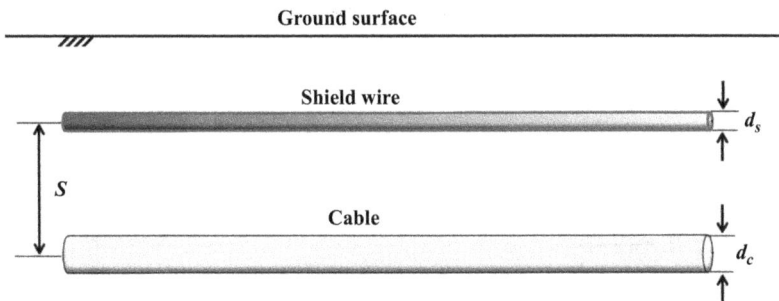

Figure 5.57   Schematic side view of buried shield wire and buried cable. 2016 ©IEEE. Reprinted, with permission, from Tanaka et al. (2016, Figure 2)

the latter having relative electric permittivity $\varepsilon_r = 2.9$. The outer diameter of the metallic sheath is $d_c = 16$ mm and its series resistance is 3.9 Ω/km. The shield wire is made of stainless steel and it is in contact with soil throughout its entire length. It has 2.5-mm diameter and 160-Ω/km series resistance. The shield wire and the cable were buried 0.3 and 0.6 m below the ground surface, respectively.

The ground resistivity was obtained from the measured resistance to ground of the shield wire at its extremities. The obtained value is 1850 $\Omega \cdot$m, which should represent the average ground resistivity along the shield wire route.

The lightning current was directed to the tower top by attaching the rocket wire to the launcher, which was connected to the tower structure by a single conductor. This conductor passed through a current probe (Pearson Current Monitor model 1330) that was connected to a nearby oscilloscope (Tektronix TDS 3014B). The rocket launcher and the oscilloscope were controlled through a fiber-optic cable connected to a computer in the control shelter. The triggering decision was made at the control shelter taking into account the intensity of lightning activity and the quasistatic electric field at ground surface measured by a nearby field mill.

In order to measure currents in the cable and in the shield wire, these conductors were run into a small section of insulating plastic tubes, which were placed inside a buried concrete box. The tubes passed inside current probes (Pearson Current Monitor, model 110), which were connected by shielded coaxial cables to a nearby battery-powered oscilloscope (Tektronix Model TDS 3014B) installed in a metallic box. The distance between the probes and the oscilloscope was less than 1 m.

Figure 5.58 shows the return-stroke current recorded at the top of the tower during one of the triggered lightning flashes. This is a negative flash, that is, the positive charge flows up from the ground to the cloud. As expected, the current

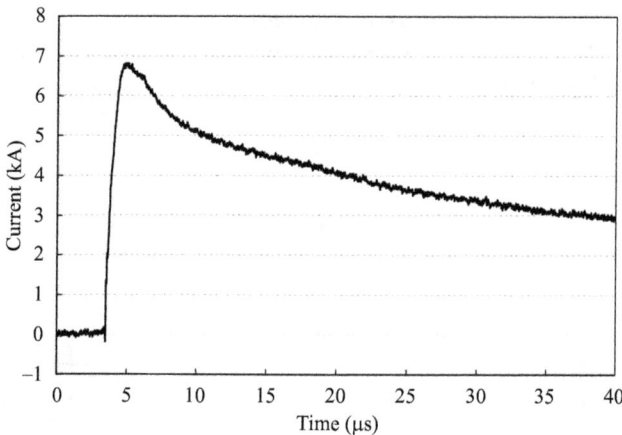

Figure 5.58   Lightning current recorded at the top of the 30-m-high tower. 2016 ©IEEE. Reprinted, with permission, from Tanaka et al. (2016, Figure 3)

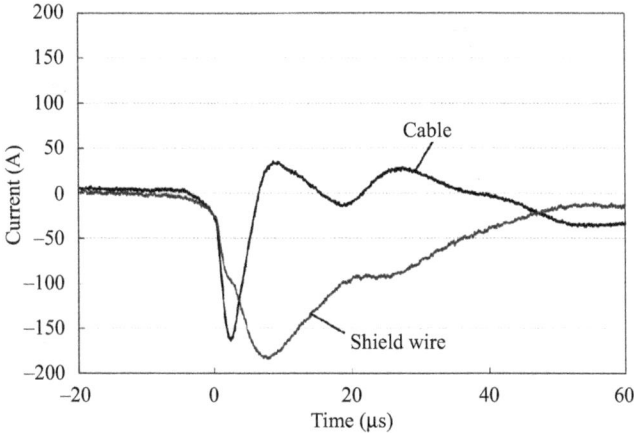

Figure 5.59 Measured currents in the buried cable and in the shield wire associated with the lightning strike whose current waveform is shown in Figure 5.58. 2016 ©IEEE. Reprinted, with permission, from Tanaka et al. (2016, Figure 4)

waveform is similar to the ones typically produced by subsequent strokes of natural flashes, showing a relatively short front time and a relatively low peak value.

Figure 5.59 shows currents in the buried cable and in the shield wire associated with the lightning strike whose current waveform is shown in Figure 5.58. It is seen that the width of cable current pulse is much shorter than that of the shield-wire current. Interestingly, the peak values of the currents are similar to each other, despite the great difference between the cable and shield wire characteristics (e.g., insulation, diameter, series resistance, and so on). Note that the shield-wire current includes both a lightning-induced component and a conducted component directly flowing from the lightning channel (tower) base through the ground.

5.7.3 Methodology

Figure 5.60 shows a 1-km-long cable with a 1-km-long shield wire, buried at depths 0.6 and 0.3 m, respectively, in a homogeneous soil, to be analyzed using the 3-D FDTD method. The working volume of 1050 m × 250 m × 1050 m (in the x-, y-, and z-directions) is divided nonuniformly, and surrounded by six planes of Liao's second-order absorbing boundary condition (Liao *et al.* 1984) to minimize unwanted reflections there. The cable and the shield wire are parallel to the x-axis. The minimum cell size is 20 mm × 3 mm × 3 mm, which is employed in the vicinity of the ends of the shield wire and the cable, and in the vicinity of the lightning channel. The cell size increases gradually as the distance from those regions increases: 0.07, 0.26, 1, 2.66, 3.9, 15.5, 36.2, and 54 m in the x-direction, 0.007, 0.02, 0.07, 0.26, 1, 1.9, 3.9, 9.7, and 15.5 m in the y-direction, and 0.007, 0.02, 0.07, 0.26, 1, 3, 3.9, and 15.5 m in the z-direction (the validity of this

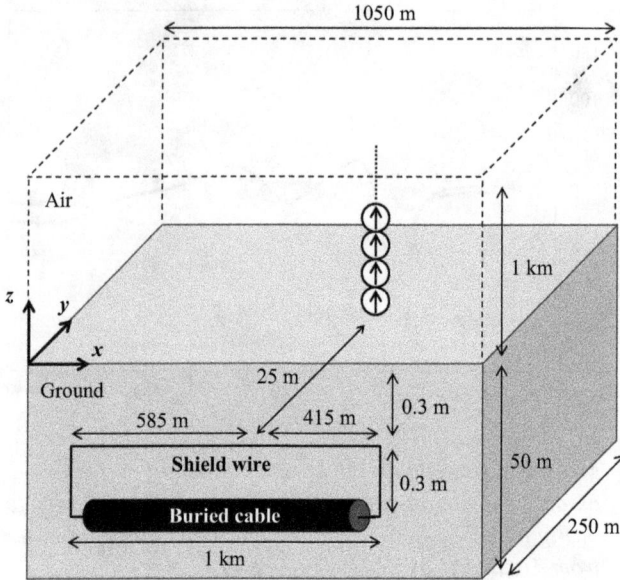

Figure 5.60   Model for representing the cable and the shield wire in the 3D-FDTD working volume. 2016 ©IEEE. Reprinted, with permission, from Tanaka et al. (2016, Figure 5)

nonuniformly discretized model is tested in Section 5.7.4). The total number of cells is 180 × 130 × 155. The time increment is set to 7.03 ps.

The thickness of the ground (distance between the ground surface and the bottom absorbing boundary) is set to 50 m. The ground resistivity value is 1850 Ω·m and its relative permittivity is 10.

The shield wire is modeled using the thin-wire representation proposed by Noda and Yokoyama (2002) and it is extended to lossy medium as proposed by Baba *et al.* (2005). Using this model, the equivalent radius of the shield wire is 0.69 mm. This value is obtained as $0.23\,\Delta s = 0.23 \times 3$ mm, where Δs is the radial-direction cell dimension at the shield wire location.

The cable (its tubular metallic sheath) is modeled as a single solid conductor covered by a dielectric sheath (individual insulated conductors inside the tubular metallic sheath (see Section 5.7.2) are not considered in this model). Figure 5.61 shows the cross-section of the cable model, where the cable (metallic sheath) is represented by a solid perfect conductor with a cross-section of 18 mm × 18 mm (6 cells × 6 cells), covered with a dielectric sheath whose thickness is set to 3 mm (one cell).

Since the dielectric sheath thickness (3 mm) in the FDTD model is different from that of the real cable (2 mm), the relative permittivity of the dielectric in the model is adjusted to $\varepsilon_{rm} = 3.7$ in order to provide the same capacitance to ground as

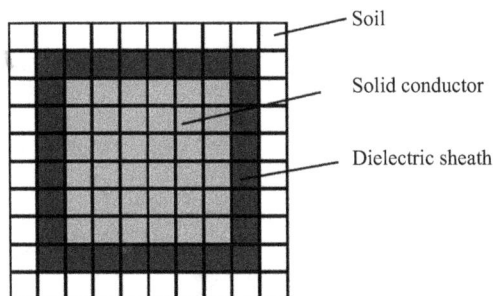

Soil

Solid conductor

Dielectric sheath

Figure 5.61   Cross-sectional view of the cable representation used in the FDTD model. The solid conductor represents the tubular metallic sheath (with insulated copper wires inside) and the dielectric sheath represents the plastic outer sheath of the cable. The single-cell layer surrounding the dielectric sheath represents soil. 2016 ©IEEE. Reprinted, with permission, from Tanaka et al. (2016, Figure 6)

Table 5.6   Parameters of the two Heidler functions used in the FDTD model of Tanaka et al. (2016)

I_1 (kA)	t_{11} (µs)	t_{12} (µs)	n_1	I_2 (kA)	t_{21} (µs)	t_{22} (µs)	n_2
6.2	0.6	2.9	2	4	3.8	50	2

that of the actual cable: $\varepsilon_{rm} = \varepsilon_r \ln (d_{im}/d_{cm})/ \ln (d_i /d_c)$, where $\varepsilon_r = 2.9$, $d_i = 20$ mm, and $d_c = 16$ mm are the relative permittivity of the dielectric sheath of the actual cable, its diameter, and the outer diameter of the tubular metallic sheath, respectively; $d_{im} = 30.5$ mm and $d_{cm} = 22.9$ mm are equivalent diameters of the dielectric sheath and the solid conductor in the FDTD model, respectively, which are obtained by dividing the perimeter of the square cross-section of each of them by π (see Figure 5.61).

A 1-km vertical lightning channel is located at a distance of 25 m from the cable and is represented by the modified TL model with linear current decay with height (MTLL) (Rakov and Dulzon 1987), with $H = 7000$ m. The return-stroke wavefront speed is set to 130 m/µs and the channel-base current is represented by the sum of two Heidler functions (Heidler 1985) having the parameters shown in Table 5.6. These parameters are adjusted to represent the recorded lightning current waveform shown in Figure 5.58. Figure 5.62 shows a comparison between the actual lightning current and the one used in the FDTD model; the two waveforms are essentially indistinguishable.

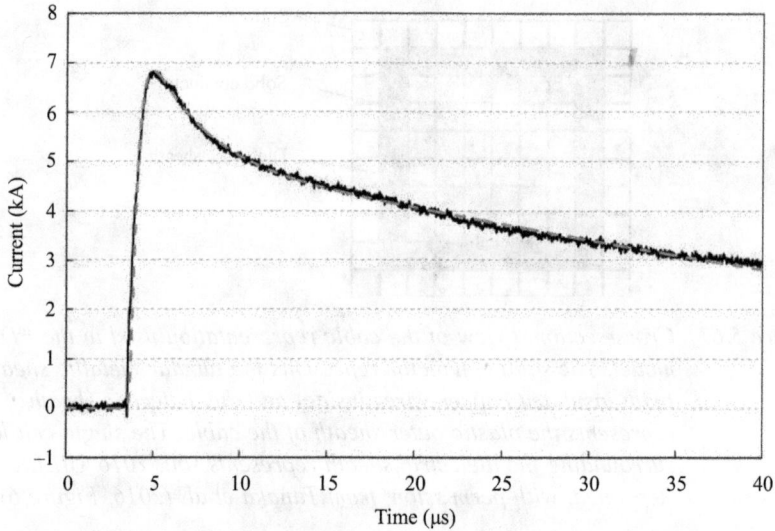

Figure 5.62   Lightning return-stroke current waveforms. Solid black line:
measurement; Dashed gray line: representation by Heidler functions
for the FDTD model. The two waveforms are essentially overlapping.
2016 ©IEEE. Reprinted, with permission, from Tanaka et al. (2016,
Figure 7)

5.7.4   Model validation

In this section, FDTD-computed results are compared with the experimental results
in order to validate the FDTD model. Figure 5.63 shows the FDTD-computed
waveform of lightning-associated current in the shield wire and the corresponding
measured one. The FDTD-computed waveform agrees reasonably well with the
measured one, although the FDTD-computed peak value is about 10% lower than
the measured one.

Figure 5.64 shows the FDTD-computed waveform of lightning-induced current in the cable and the corresponding measured one. The FDTD-computed
waveform agrees reasonably well with the measured one, although the FDTD-
computed peak value is about 20% lower than the measured one.

The relatively small discrepancies between FDTD-computed and measured
waveforms are probably due to some assumptions/approximations that did not
allow the capturing of all the actual experimental conditions by the FDTD-based
model. The overall reasonably good agreement between the FDTD-computed and
measured results is in support of the use of the FDTD model for simulating
lightning-associated currents in buried cables with shield wires.

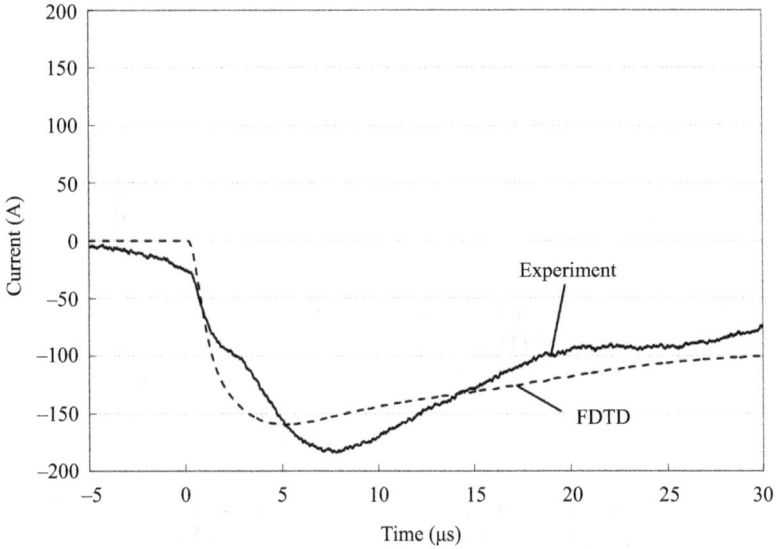

Figure 5.63 *FDTD-computed and measured waveforms of lightning-associated current in the shield wire. 2016 ©IEEE. Reprinted, with permission, from Tanaka et al. (2016, Figure 8)*

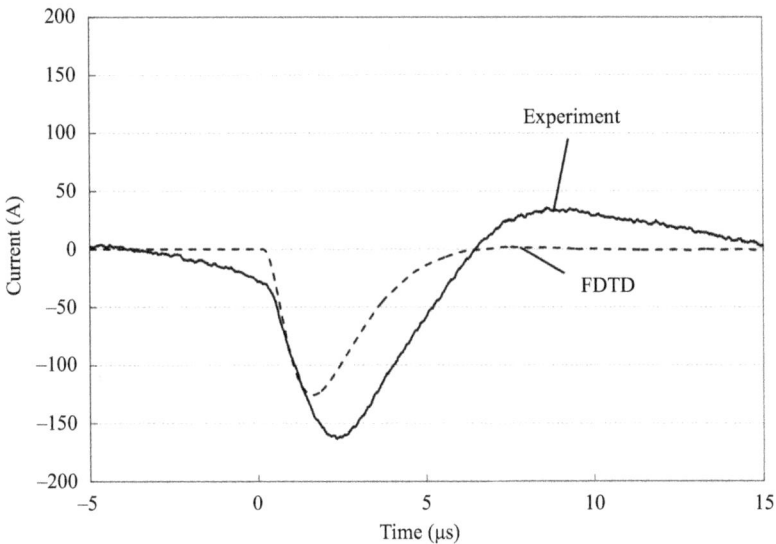

Figure 5.64 *FDTD-computed and measured waveforms of current induced in the cable. 2016 ©IEEE. Reprinted, with permission, from Tanaka et al. (2016, Figure 9)*

5.7.5   Sensitivity analysis

In this section, influences of different parameters on the effectiveness of shield wire are studied using the 3D-FDTD method.

5.7.5.1   Effect of shield wire

The presence of a shield wire is expected to reduce the current that would be induced in a cable below it. Figure 5.65 shows FDTD-computed waveforms of current induced in the cable in the presence of shield wire and in its absence. It follows from Figure 5.65 that the peak value of the cable current is reduced from 191 to 123 A by the presence of shield wire (36% reduction).

Figure 5.66 shows FDTD-computed waveforms of voltage generated across the 3-mm-thick cable insulation (dielectric sheath between the metallic sheath, represented by a solid conductor in the model, and the soil; see Figure 5.61) in the presence and absence of shield wire. In the presence of shield wire, the voltage peak is 11.8 kV (the corresponding electric field intensity is about 3.9 kV/mm); while in its absence, the voltage peak is 25.5 kV (the corresponding electric field intensity is about 8.5 kV/mm). Thus, the presence of shield wire caused a voltage reduction by more than a factor of 2.

The shield wire used in the experiment had a relatively high resistance (160 Ω/km), so that it is interesting to investigate the effect of this parameter on the cable and shield-wire currents. Figure 5.67 shows the FDTD-computed currents in the cable with a 160-Ω/km shield wire and with a perfectly conducting shield wire. It is

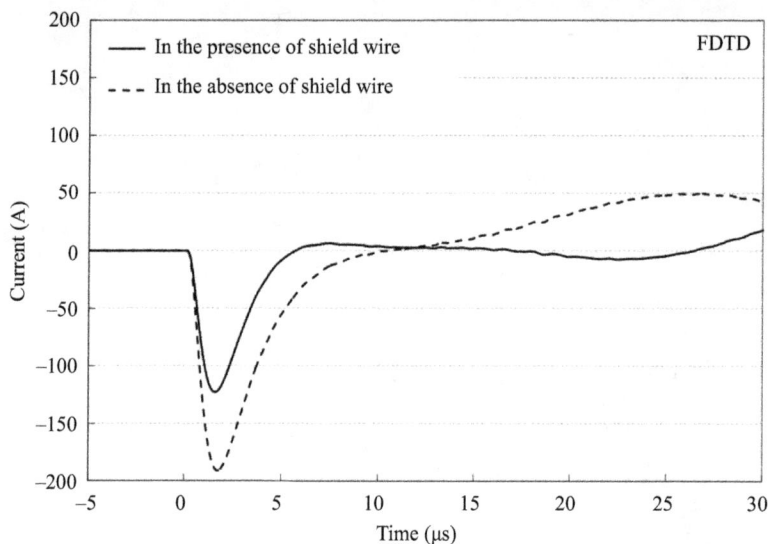

Figure 5.65   FDTD-computed waveforms of current induced in the cable in the presence and absence of shield wire. 2016 ©IEEE. Reprinted, with permission, from Tanaka et al. (2016, Figure 10)

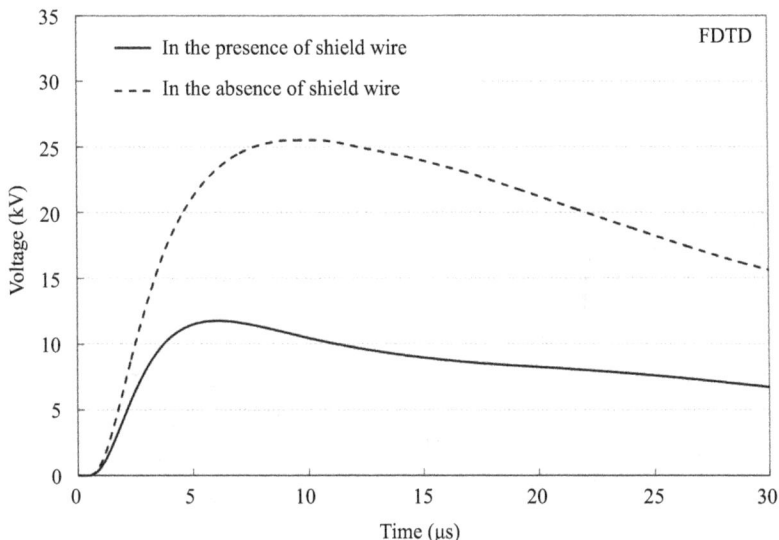

Figure 5.66 *FDTD-computed waveforms of voltage generated across the 3-mm-thick cable insulation (dielectric sheath between the solid conductor and the soil; see Figure 5.61) in the presence and absence of shield wire. 2016 ©IEEE. Reprinted, with permission, from Tanaka et al. (2016, Figure 11)*

seen from Figure 5.67 that making the shield wire perfectly conducting leads to an increase in the shield wire peak current by 9%, but has a negligible effect on the cable current. This result could be explained by the fact that the shielding effect is determined during the front of the cable currents, where the shield-wire series reactance is more important than its series resistance. This phenomenon can be better visualized by enlarging the initial part of the waveforms, as shown in Figure 5.68. Note that the cable current reaches its peak value at about 1.3 µs and, up to this time, current in the shield wire is practically not influenced by its series resistance. Therefore, as long as the induced current in the cable is concerned, it can be said that the shield-wire resistance can be neglected in the computations. As a consequence, the shielding factor can also be calculated by neglecting the intrinsic resistance of the shield wire.

5.7.5.2 Influence of the presence of 30-m-high strike object
In Sections 5.7.4 and 5.7.5.1, the presence of 30-m-high strike object (see Section 5.7.2) was not considered. To check if its effect is significant, we now include the strike object in FDTD simulations (Baba and Rakov 2005a). Lightning is represented by the MTLL model. It is assumed that the characteristic impedance of the lightning channel is 1000 Ω, the characteristic impedance of the strike object is 200 Ω, and the grounding impedance is zero. Therefore, the current reflection

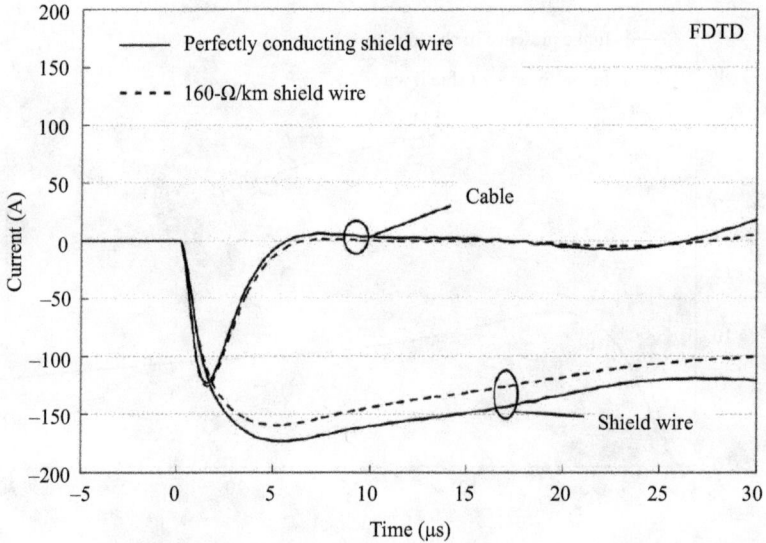

Figure 5.67   *FDTD-computed waveforms of currents in the cable and in the shield wire for 160-Ω/km shield wire and perfectly conducting shield wire. 2016 ©IEEE. Reprinted, with permission, from Tanaka et al. (2016, Figure 12)*

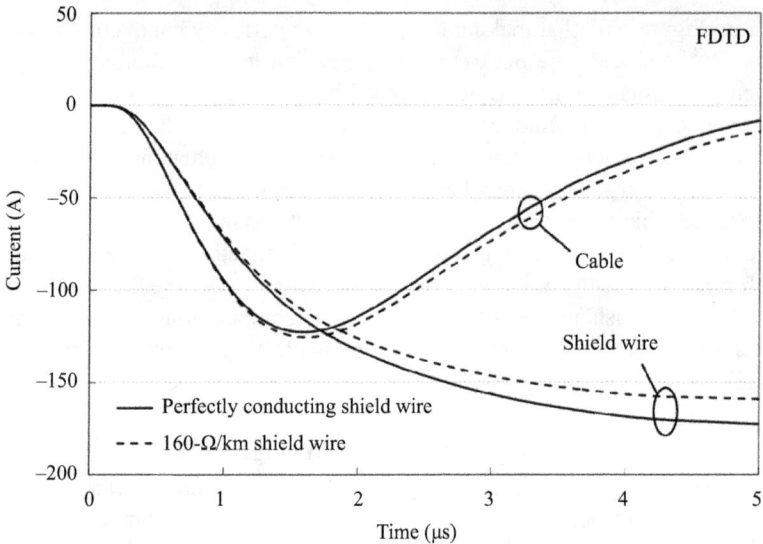

Figure 5.68   *Initial part (the first 5 μs) of the current waveforms shown in Figure 5.67. 2016 ©IEEE. Reprinted, with permission, from Tanaka et al. (2016, Figure 13)*

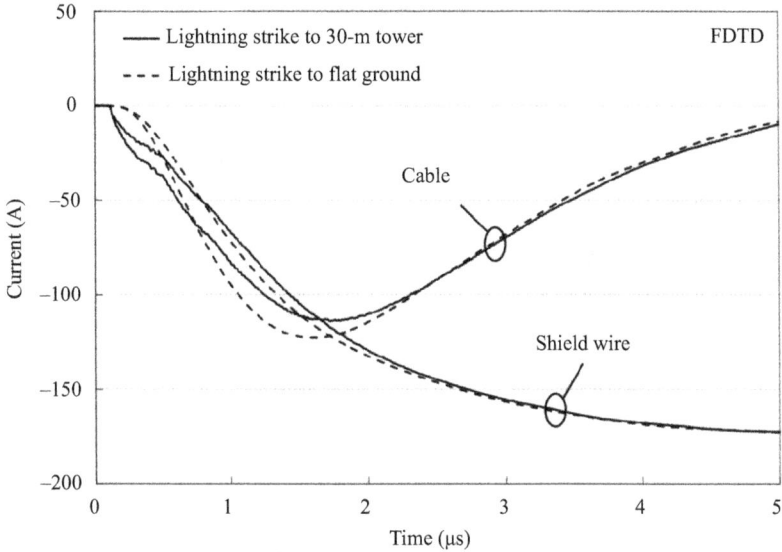

Figure 5.69 *FDTD-computed waveforms of currents in the cable and in the shield wire for the case of lightning strike to the 30-m-high object. Also shown are the corresponding waveforms for the case of lightning strike to flat ground. 2016 ©IEEE. Reprinted, with permission, from Tanaka et al. (2016, Figure 14)*

coefficient at the top of the strike object for upward-propagating current waves is $\rho_{top} = -0.67$, and the current reflection coefficient at the bottom of the object is $\rho_{bot} = 1$. The current-propagation speed along the strike object is set to the speed of light c, and the return-stroke wavefront speed along the lightning channel is set to $v = 130$ m/μs.

Figure 5.69 shows FDTD-computed waveforms of the currents in the buried cable and in the shield wire for the case of lightning strike to the 30-m-high object and for the case of lightning strike to flat ground. It appears from Figure 5.69 that the reduction of cable current due to the presence of 30-m-high tower is not significant (7%). This is because the risetime of lightning current (1.5 μs) is several times greater than the round-trip time of lightning current in the object (0.2 μs). The influence of the presence of strike object increases with increasing object height (e.g., Baba and Rakov 2005b, 2006).

5.7.5.3 Influence of lightning return-stroke speed

The return-stroke speed was set to $v = 130$ m/μs in the FDTD simulation in Section 5.7.4 (this parameter was not measured in the experiment described in Section 5.7.2). Here, we check if this parameter can significantly influence the cable and shield-wire currents. Figure 5.70 shows FDTD-computed current waveforms in the cable and in the shield wire for $v = 100$, 130, and 200 m/μs.

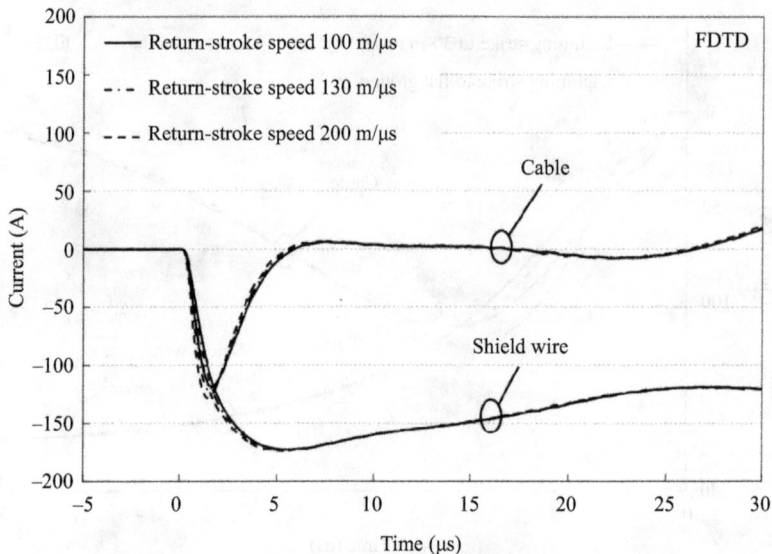

*Figure 5.70   FDTD-computed waveforms of currents in the cable and in the shield
wire for different lightning return-stroke speeds. 2016 ©IEEE.
Reprinted, with permission, from Tanaka et al. (2016, Figure 15)*

Magnitudes of currents in the cable and in the shield wire increase with increasing
v, but the difference is insignificant.

5.7.5.4   Influence of ground resistivity and relative permittivity

Figure 5.71 shows FDTD-computed waveforms of current in the cable and in the
shield wire for different values of ground resistivity (the inverse of ground conductivity): 100, 1000, and 1850 $\Omega \cdot$m. It is observed that both cable current and
shield-wire current increase with increasing the ground resistivity. The increase of
the cable current due to the increase of ground resistivity from 100 to 1850 $\Omega \cdot$m is
by a factor of about 2.5 (from 49 to 123 A), and that for the shield-wire current is by
a factor of 1.2. Both of these values are considerably smaller than the ratio of
ground resistivity values, 18.5.

Figure 5.72 shows FDTD-computed waveforms of current in the cable and in
the shield wire for different values of ground relative permittivity: 5, 10, and 15. It
appears from Figure 5.72 that the variation of ground permittivity from 5 to 15 has
a negligible effect on the cable and shield-wire currents.

5.7.5.5   Influence of cable-sheath and shield-wire bonding

Figure 5.73(a) shows FDTD-computed waveforms of lightning-associated currents in
the cable and in the shield wire for the case when the cable metallic sheath is bonded to
the shield wire both at the terminals and at the midpoint (at 500-m intervals) and those

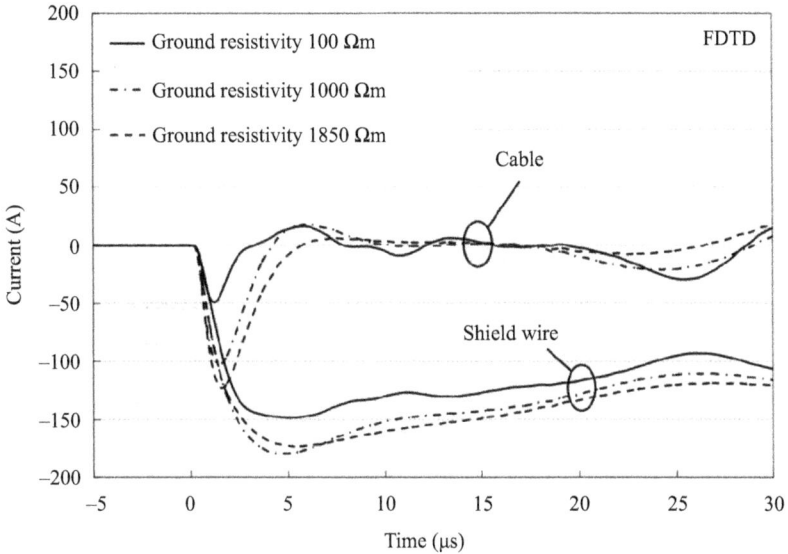

Figure 5.71 FDTD-computed waveforms of currents in the cable and in the shield wire for different values of ground resistivity. 2016 ©IEEE. Reprinted, with permission, from Tanaka et al. (2016, Figure 16)

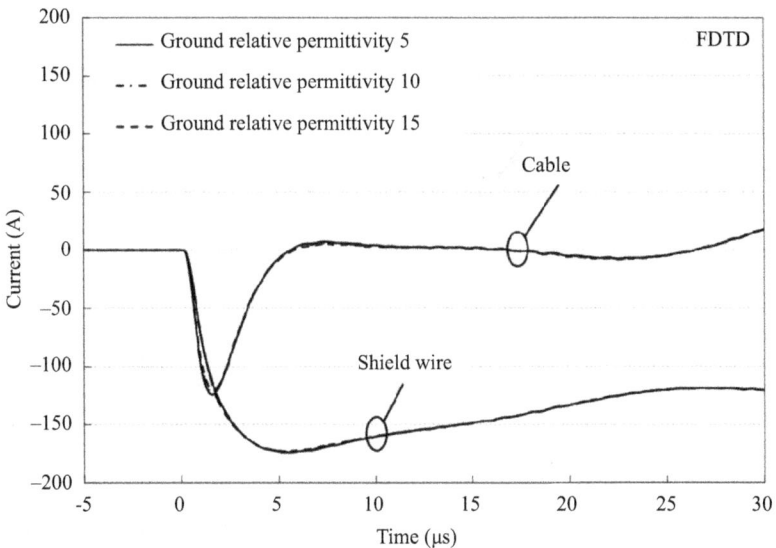

Figure 5.72 FDTD-computed waveforms of currents in the cable and in the shield wire for different values of ground relative permittivity. 2016 ©IEEE. Reprinted, with permission, from Tanaka et al. (2016, Figure 17)

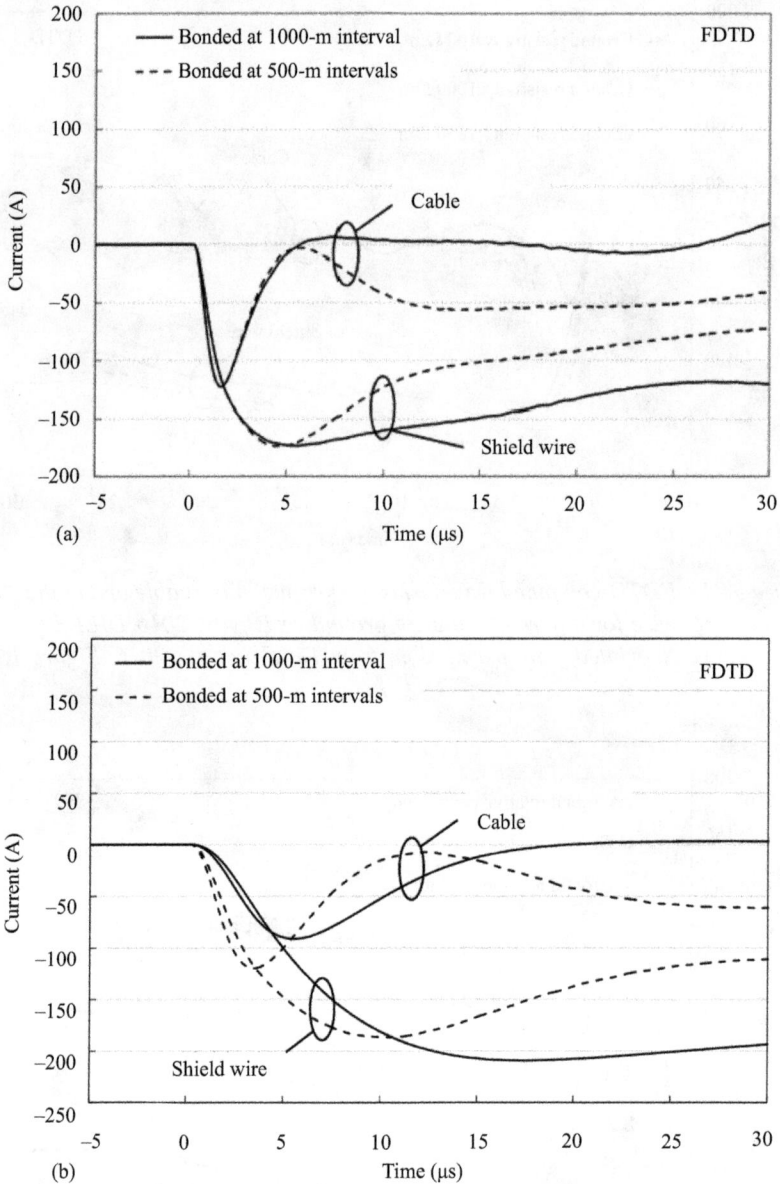

Figure 5.73   *FDTD-computed waveforms of currents in the cable and in the shield wire for two different bonding intervals. (a) 1.5-μs lightning current risetime; (b) 5-μs lightning current risetime. 2016 ©IEEE. Reprinted, with permission, from Tanaka et al. (2016, Figure 18)*

for the case when the cable metallic sheath is bonded to the shield wire only at the terminals (at 1000-m interval). Note that the current observation point is located 395 m from the right terminals of the 1-km-long cable and shield wire (and, therefore, 105 m from the midpoint (see Figure 5.56)). The cable and shield-wire current waveforms shown in Figure 5.73(a) are computed for the lightning current waveform presented in Figure 5.62, which has a risetime of 1.5 µs. Figure 5.73(b) is the same as Figure 5.73(a), but for the lightning current risetime of 5 µs. It appears from Figure 5.73(b) that some current flows in the cable metallic sheath from the shield wire via bonding points and the cable current increases.

Figure 5.74(a) shows FDTD-computed waveforms of the voltage generated across the 3-mm-thick cable insulation (dielectric sheath between the cable metallic sheath, represented by a solid conductor in the model, and the soil) for the above two bonding conditions and for the 1.5-µs-risetime lightning current. Figure 5.74(b) is the same as Figure 5.74(a), but for the lightning current risetime of 5 µs. It follows from comparison of Figure 5.74(a) and 5.74(b) that the peak voltage is reduced by decreasing bonding interval from 1000 to 500 m for the lightning current risetime of 5 µs, but not for 1.5 µs.

5.7.5.6   Influence of lightning strike location

Figure 5.75 shows the FDTD-computed waveforms of current in the cable and in the shield wire for two different lightning strike points: Strike Location A (same as in the simulations presented above) at a distance of 415 m from the right terminal (see Figures 5.56 and 5.60), and Strike Location B at a distance of 115 m from the right terminal (along the *x*-axis), both at a distance of 25 m from the cable (along the *y*-axis). The magnitude of the cable current for Strike Location B is similar to that for Strike Location A, but the polarity is opposite and the wavefront is longer. This trend is also seen for the shield-wire current. Clearly, the waveshape and polarity of the cable current and those of shield-wire current are dependent on the strike location relative to the cable/shield-wire terminals.

5.7.6   Summary

The presence of shield wire reduces the induced current in the cable metallic sheath and the voltage across its dielectric sheath. The series resistance of shield wire (160 Ω/km) has negligible influence on these effects. The presence of 30-m-high strike object slightly reduces the cable current. The higher the ground resistivity, the higher the cable current. The ground permittivity and the return-stroke speed have negligible effects on the cable current. The increase of the number of bonding points between the cable metallic sheath and the shield wire increases the current in the cable metallic sheath. It is also found to decrease the voltage across the cable dielectric sheath for a lightning current risetime of 5 µs, but not for 1.5 µs. The overall reasonably good agreement between the FDTD-computed and measured results supports the use of the FDTD model for simulating lightning-associated currents in buried cables with shield wires.

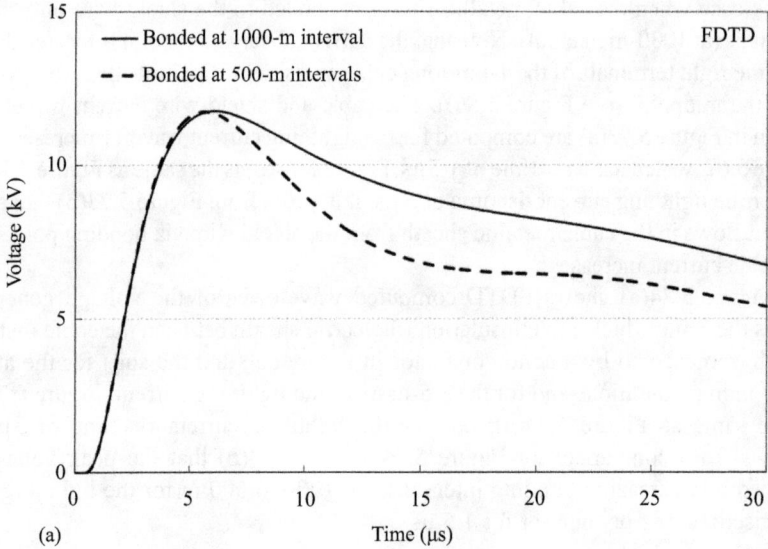

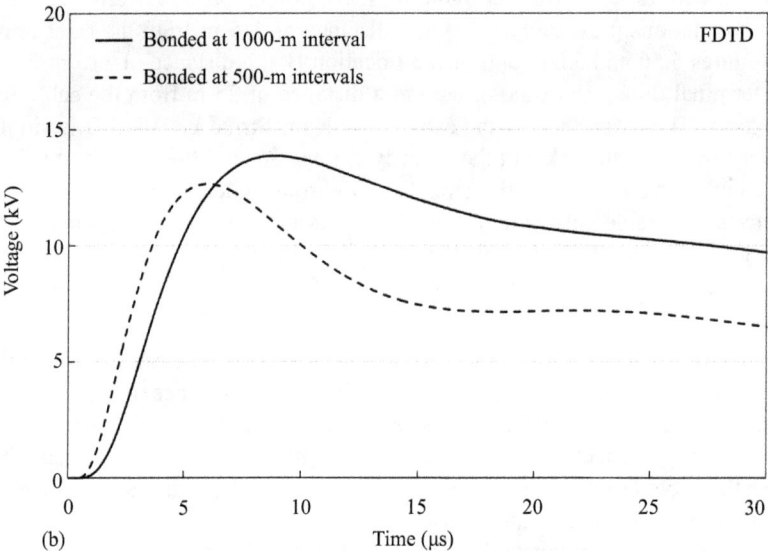

Figure 5.74   *FDTD-computed waveforms of voltage generated across the 3-mm-thick dielectric sheath of the cable for two different bonding intervals. (a) 1.5-µs lightning current risetime; (b) 5-µs lightning current risetime. 2016 ©IEEE. Reprinted, with permission, from Tanaka et al. (2016, Figure 19)*

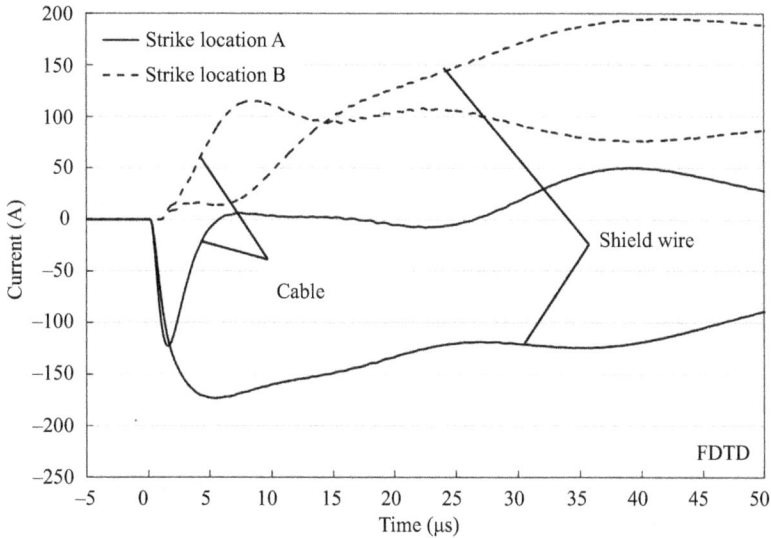

Figure 5.75   *FDTD-computed currents in the cable and in the shield wire for two strike points: A−415 m from the right terminal, B−115 m from the right terminal (along the x-axis, with the distance along the y-axis being the same, 25 m, in both cases; see Figures 5.56 and 5.60). 2016 ©IEEE. Reprinted, with permission, from Tanaka et al. (2016, Figure 20)*

5.8   Summary

In this chapter, about 30 journal papers published in the last 15 years, which use the FDTD method in simulations of lightning-induced surges, have been classified in terms of spatial dimension (2D or 3D), lightning channel representation, and application. About 30% of the simulations employed the 2D-FDTD method in the cylindrical coordinate system, and about 70% used the 3D-FDTD method. In the 2D-FDTD case, the method is employed to express distributed sources in the generalized telegrapher's equations in terms of incident electromagnetic fields illuminating overhead conductors. The source terms are different for different field-to-conductor coupling models, the most popular one being the model of Agrawal *et al.* (1980). In some 3D-FDTD simulations, the subgridding technique has been used to represent horizontal closely spaced thin conductors. The nonuniform gridding technique has been used for the same purpose. For representing the lightning return stroke, the TL model has been most frequently used, along with the MTLL and MTLE models. About 70% of the FDTD simulations have been concerned with induced surges associated with lightning strikes to flat ground and about 30% with induced surges associated with lightning strikes to the top of mountain, building, or tall object. Six representative works have been described in detail, which cover the following topics: (i) voltages induced on a single overhead conductor by lightning strikes to a nearby tall grounded object, (ii) lightning-induced voltages on an overhead two-conductor line, (iii) lightning-induced voltages on a single overhead conductor in the presence of corona, (iv) lightning-induced voltages on overhead multiconductor lines with surge arresters and pole transformers, (v) lightning-induced voltages on overhead multiconductor lines in the presence of nearby buildings, and (vi) lightning-induced currents in buried cables.

References

Agrawal, A. K., Price, H. J., and Gurbaxani, S. H. (1980), Transient response of multi-conductor transmission lines excited by a nonuniform electromagnetic field, *IEEE Transactions on Electromagnetic Compatibility*, vol. 22, no. 3, pp. 119–129.

Akbari, M., Sheshyekani, K., Pirayesh, A., *et al.* (2013), Evaluation of lightning electromagnetic fields and their induced voltages on overhead lines considering the frequency-dependence of soil electrical parameters, *IEEE Transactions on Electromagnetic Compatibility*, vol. 55, no. 6, pp. 1210–1219.

Baba, Y., and Rakov, V. A. (2003), On the transmission line model for lightning return stroke representation, *Geophysical Research Letters*, vol. 30, no. 24, 2294, doi:10.1029/2003GL018407.

Baba, Y., and Rakov, V. A. (2005a), On the use of lumped sources in lightning return stroke models, *Journal of Geophysical Research*, vol. 110, no. D03101, doi:10.1029/2004JD005202.

Baba, Y., and Rakov, V. A. (2005b), Lightning electromagnetic environment in the presence of a tall grounded strike object, *Journal of Geophysical Research*, vol.110, D09108, doi:10.1029/2004JD005505.

Baba, Y., and Rakov, V. A. (2006), Voltages induced on an overhead wire by lightning strikes to a nearby tall grounded object, *IEEE Transactions on Electromagnetic Compatibility*, vol. 48, no. 1, pp. 212–224.

Baba, Y., and Rakov, V. A. (2007), Electromagnetic fields at the top of a tall building associated with nearby lightning return strokes, *IEEE Transactions on Electromagnetic Compatibility*, vol. 49, no. 3, pp. 632–643.

Baba, Y., Nagaoka, N., and Ametani, A. (2005), Modeling of thin wires in a lossy medium for FDTD simulations, *IEEE Transactions on Electromagnetic Compatibility*, vol. 47, no. 1, pp. 54–60.

Barbosa, C. F., and Paulino, J. O. S. (2008), Measured and modeled horizontal electric field from rocket-triggered lightning, *IEEE Transactions on Electromagnetic Compatibility*, vol. 50, no. 4, pp. 913–920.

Barbosa, C. F., Zeddam, A., Day, P., and Burgeois, Y. (2008), Effect of guard wire in protecting a telecommunication buried cable struck by rocket-triggered lightning, Paper presented at the 29th International Conference on Lightning Protection, Uppsala, Sweden.

Barker, P. P., Short, T. A., Eybert-Berard, A. R., and Berlandis, J. P. (1996), Induced voltage measurements on an experimental distribution line during nearby rocket triggered lightning flashes, *IEEE Transactions on Power Delivery*, vol. 11, no. 2, pp. 980–995.

Bejleri, M., Rakov, V. A., Uman, M. A., Rambo, K. J., Mata, C. T., and Fernandez, M. I. (2004), Triggered lightning testing of an airport runaway lighting system, *IEEE Transactions on Electromagnetic Compatibility*, vol. 46, no. 1, pp. 96–101.

Bermudez, J. L., Rachidi, F., Rubinstein, M., *et al.* (2005), Far-field-current relationship based on the TL model for lightning return strokes to elevated strike objects, *IEEE Transactions on Electromagnetic Compatibility*, vol. 47, no. 1, pp. 146–159.

Burke, G. J., and Poggio, A. J. (1980), Numerical electromagnetic code (NEC) – method of moments, Technical Document 116, Naval Ocean Systems Center, San Diego.

Chang, H. T. (1980), Protection of buried cable from direct lightning strike, *IEEE Transactions on Electromagnetic Compatibility*, vol. 22, no. 3, pp. 157–160.

Chen, H., Du., Y, Yuan, M., and Liu, Q. H. (2018), Analysis of the grounding for the substation under very fast transient using improved lossy thin-wire model for FDTD, *IEEE Transactions on Electromagnetic Compatibility*, vol. 60, no. 6, pp. 1833–1841.

Chevalier, M. W., Luebbers, R. J., and Cable, V. P. (1997), FDTD local grid with material traverse, *IEEE Transactions on Antennas and Propagation*, vol. 45, no. 3, pp. 411–421.

Cooray, V. (1992), Horizontal fields generated by return strokes, *Radio Science*, vol. 27, no. 4, pp. 529–537.

Cooray, V. (1994), Calculating lightning-induced overvoltages in power lines: A comparison of two coupling models, *IEEE Transactions on Electromagnetic Compatibility*, vol. 36, no. 3, pp. 179–182.

Cooray, V. (2003), The Lightning Flash, p. 79, The Institution of Electrical Engineers, UK, 570 pages.

Darveniza, M. (2007), A practical extension of Rusck's formula for maximum lightning-induced voltages that accounts for ground resistivity, *IEEE Transactions on Power Delivery*, vol. 22, no. 1, pp. 605–612.

Diaz, L., Miry, C., Baraton, P., Guiffaut, C., and Reineix, A. (2017), Lightning transient voltages in cables of a large industrial site using a FDTD thin wire model, *Electric Power Systems Research*, vol. 153, pp. 94–103.

Diendorfer, G., and Schulz, W. (1998), Lightning incidence to elevated objects on mountains, Paper presented at the 24th International Conference on Lightning Protection, pp.173–175, Birmingham, UK.

Douglas, D. A. (1971), Lightning induced current surges, Paper presented at the Wire and Cable Symposium.

Dragan, G., Florea, G., Nucci, C. A., and Paolone, M. (2010), On the influence of corona on lightning-induced overvoltages, Paper presented at the 30th International Conference on Lightning Protection, Cagliali, Italy.

Du, Y., Li, B., and Chen, M. (2016), Surges induced in building electrical systems during a lightning strike, *Electric Power Systems Research*, vol. 139, pp. 68–74.

Fisher, R. J., and Schnetzer, G. H. (1994), 1993 triggered lightning test program: environments within 20 meters of the lightning channel and small area temporary protection concepts, Sandia Report, SAND94-0311/UC-706.

Fuchs, F. (1998), On the transient behaviour of the telecommunication tower at the mountain Hoher Peissenberg, Paper presented at the 24th International Conference on Lightning Protection, vol. 1, pp. 36–41, Birmingham, UK.

Harrington, R. F. (1968), *Field Computation by Moment Methods*, MacMillan Co., New York.

Hartmann, G. (1984), Theoretical evaluation of Peek's law, *IEEE Transactions on Industry Applications*, vol. 20, no. 6, pp. 1647–1651.

Heidler, F. (1985), Analitsche blitzstromfunktion zur LEMP-berechnung, Paper presented at the 18th International Conference on Lightning Protection, Munich, Germany.

Inoue, A. (1983), Study on propagation characteristics of high-voltage traveling waves with corona discharge, CRIEPI Report, no. 114 (in Japanese).

Ishii, M., Michishita, K., and Hongo, Y. (1999), Experimental study of lightning-induced voltage on an overhead wire over lossy ground, *IEEE Transactions on Electromagnetic Compatibility*, vol. 41, no. 1, pp. 39–45.

Ishii, M., Miyabe, K., and Tatematsu, A. (2012), Induced voltages and currents on electrical wirings in building directly hit by lightning, *Electric Power Systems Research*, vol. 85, pp. 2–6.

Janischewskyj, W., Shostak, V., Barratt, J., Hussein, A.M., Rusan, R., and Chang, J. -S. (1996), Collection and use of lightning return stroke parameters taking into account characteristics of the struck object, Paper presented at the 23rd International Conference on Lightning Protection, pp.16–23, Florence, Italy.

Kordi, B., Moini, R., Janischewskyj, W., Hussein, A. M., Shostak, V. O., and Rakov, V. A. (2003), Application of the antenna theory model to a tall tower struck by lightning, *Journal of Geophysical Research*, vol. 108, no. D17, doi:10.1029/2003JD003398.

Liao, Z. P., Wong, H. L., Yang, B. -P., and Yuan, Y. -F. (1984), A transmitting boundary for transient wave analysis, *Science in China*, vol. A27, no. 10, pp. 1063–1076.

Mata, C. T., Fernandez, M. I., Rakov, V. A., and Uman, M. A. (2000), EMTP modeling of a triggered-lightning strike to the phase conductor of an overhead distribution line, *IEEE Transactions on Power Delivery*, vol. 15, no. 4, pp. 1175–1181.

Michishita, K., and Ishii, M. (1997), Theoretical comparison of Agrawal's and Rusck's field-to-line coupling models for calculation of lightning-induced voltage on an overhead wire, *IEEJ Transactions on Power and Energy*, vol. 117, no. 9, pp. 1315–1316.

Michishita, K., Ishii, M., Asakawa, A., Yokoyama, S., and Kami, K. (2003), Voltage induced on a test distribution line by negative winter lightning strokes to a tall structure, *IEEE Transactions on Electromagnetic Compatibility*, vol. 45, no. 1, pp. 135–140.

Miyazaki, S., and Ishii, M. (2004), Influence of elevated stricken object on lightning return-stroke current and associated fields, Paper presented at the 27th International Conference on Lightning Protection, pp. 122–127, Avignon, France.

Namdari, N., Khosravi-Farsani, M., Moini, R., and Sadeghi, S. H. H. (2015), An efficient parallel 3-D FDTD method for calculating lightning-induced disturbances on overhead lines in the presence of surge arresters, *IEEE Transactions on Electromagnetic Compatibility*, vol. 57, no. 6, pp. 1593–1600.

Natsui, M. Ametani, A., Mahseredjian, J., Sekioka, S., and Yamamoto, K. (2018), FDTD analysis of distribution line voltages induced by inclined lightning channel, *Electric Power Systems Research*, vol. 160, pp. 450–456.

Natsui, M, Ametani, A., Mahseredjian, J., Sekioka, S., and Yamamoto, K. (2020), FDTD analysis of nearby lightning surges flowing into a distribution line via groundings, *IEEE Trans. Electromagnetic Compatibility*, vol. 62, no. 1, pp. 144–154.

Noda, T., and Yokoyama, S. (2002), Thin wire representation in finite difference time domain surge simulation, *IEEE Transactions on Power Delivery*, vol. 17, no. 3, pp. 840–847.

Noda, T., Ono, T., Matsubara, H., Motoyama, H., Sekioka, S., and Ametani, A. (2003), Charge-voltage curves of surge corona on transmission lines: Two measurement methods, *IEEE Transactions on Power Delivery*, vol. 18, no. 1, pp. 307–314.

Norton, K. A. (1937), The propagation of radio waves over the surface of the earth and in the upper atmosphere, *Proceedings of the Institute of Radio Engineers*, vol. 25, no. 9, pp. 1203–1236.

Nucci, C. A., Mazzetti, C., Rachidi, F., and Ianoz, M. (1988), On lightning return stroke models for LEMP calculations, Paper presented at the 19th International Conference on Lightning Protection, Graz, Austria.

Nucci, C. A., Diendorfer, G., Uman, M. A., Rachidi, F., Ianoz, M., and Mazzetti, C. (1990), Lightning return stroke current models with specified channel-base current: a review and comparison, *Journal of Geophysical Research*, vol. 95, no. D12, pp. 20395–20408.

Nucci, C.A., Guerrieri, S., Correia de Barros, M. T., and Rachidi, F. (2000), Influence of corona on the voltages induced by nearby lightning on overhead distribution lines, *IEEE Transactions on Power Delivery*, vol. 15, no. 4, pp. 1265–1273.

Paknahad, J., Behesthi, S., Sheshykani, K., and Rachidi, F. (2014a), Lightning electromagnetic fields and their induced currents on buried cables. Part I: The effect of an ocean–land mixed propagation path, *IEEE Transactions on Electromagnetic Compatibility*, vol. 56, no. 5, pp. 1137–1145.

Paknahad, J., Behesthi, S., Sheshykani, K., Rachidi, F., and Paolone, M. (2014b), Lightning electromagnetic fields and their induced currents on buried cables. Part II: The effect of a horizontally stratified ground, *IEEE Transactions on Electromagnetic Compatibility*, vol. 56, no. 5, pp. 1146–1154.

Paknahad, J., Behesthi, S., Sheshykani, K., Rachidi, F., and Paolone, M. (2014c), Evaluation of lightning-induced currents on cables buried in a lossy dispersive ground, *IEEE Transactions on Electromagnetic Compatibility*, vol. 56, no. 6, pp. 1522–1529.

Paolone, M., Petrache, E., Rachidi, F., *et al.* (2005), Lightning-induced voltages on buried cables. Part II: Experiment and model validation, *IEEE Transactions on Electromagnetic Compatibility*, vol. 47, no. 3, pp. 509–520.

Paolone, M., Rachidi, F., Borghetti, A., *et al.* (2009), Lightning electromagnetic field coupling to overhead lines: Theory, numerical simulations, and experimental validation, *IEEE Transactions on Electromagnetic Compatibility*, vol. 51, no. 3, pp. 532–547.

Paulino, J. O. S., Barbosa, C. F., Lopes, I. J. S., and Miranda, G. C. (2009), Time-domain analysis of rocket-triggered lightning-induced surges on an overhead line, *IEEE Transactions on Electromagnetic Compatibility*, vol. 51, no. 3, *Part II*.

Paulino, J. O. S., Barbosa, C. F., and Boaventura, W. C. (2014), Lightning-induced current in a cable buried in the first layer of a two-layer ground, *IEEE Transactions on Electromagnetic Compatibility*, vol. 56, no. 4, pp. 956–963.

Petrache, E., Rachidi, F., Paolone, M., Nucci, C. A., Rakov, V. A., and Uman, M. A. (2005), Lightning-induced voltages on buried cables. Part I: Theory, *IEEE Transactions on Electromagnetic Compatibility*, vol. 47, no. 3, pp. 498–508.

Piantini, A., and Janiszewski, J. M. (1998), Induced voltages on distribution lines due to lightning discharges on nearby metallic structures, *IEEE Transactions on Magnetics*, vol. 34, no. 5, pp. 2799–2802.

Piantini, A., and Janiszewski, J. M. (2000), Lightning induced voltages on distribution lines close to buildings, Paper presented at the 25th International Conference on Lightning Protection, pp. 558–563, Rhodes, Greece.

Piantini, A., and Janiszewski, J. M. (2003), The extended Rusck model for calculating lightning induced voltages on overhead lines, Paper presented at VII International Symposium on Lightning Protection, pp.151–155, Curitiba, Brazil.

Piantini, A., Janiszewski, J. M., Borghetti, A., Nucci, C. A., and Paolone, M. (2007), A scale model for the study of the LEMP response of complex power distribution networks, *IEEE Transactions on Power Delivery*, vol. 22, no. 1, pp. 710–720.

Pokharel, R. K., Ishii, M. and Baba, Y. (2003), Numerical electromagnetic analysis of lightning-induced voltage over ground of finite conductivity, *IEEE Transactions on Electromagnetic Compatibility*, vol. 45, no. 4, pp. 651–666.

Pokharel, R. K., Baba, Y., and Ishii, M. (2004), Numerical electromagnetic field analysis of transient induced voltages associated with lightning to a tall structure, *Journal of Electrostatics*, vol. 60, no. 2–4, pp. 141–147.

Rachidi, F., Janischewskyj, W., Hussein, A. M. (2001), Current and electromagnetic field associated with lightning-return strokes to tall towers, *IEEE Transactions on Electromagnetic Compatibility*, vol. 43, no. 3, pp. 356–367.

Rakov, V. A. (2001), Transient response of a tall object to lightning, *IEEE Transactions on Electromagnetic Compatibility*, vol. 43, no. 4, pp. 654–661.

Rakov, V. A. (2004), Lightning return stroke speed: a review of experimental data, Paper presented at the 27th International Conference on Lightning Protection, pp. 139–144, Avignon, France.

Rakov, V. A., and Dulzon, A. A. (1987), Calculated electromagnetic fields of lightning return stroke, *Tekh. Elektrodinam.*, vol. 1, pp. 87–89 (in Russian).

Rakov, V. A., and Uman, M. A. (1998), Review and evaluation of lightning return stroke models including some aspects of their application, *IEEE Transactions on Electromagnetic Compatibility*, vol. 40, no. 4, pp. 403–426.

Rakov, V. A., Uman, M. A., Rambo, K. J., *et al.* (1998), New insights into lightning processes gained from triggered-lightning experiments in Florida and Alabama, *Journal of Geophysical Research*, vol. 103, no. D12, pp. 14117–14130.

Ren, H.-M., Zhou, B.-H., Rakov, V. A., Shi, L.-H., Gao, C., and Yang, J.-H. (2008), Analysis of lightning-induced voltages on overhead lines using a 2-D FDTD method and Agrawal coupling model, *IEEE Transactions on Electromagnetic Compatibility*, vol. 50, no. 3, pp. 651–659.

Rizk, M. E. M., Mahmood, F., Lehtonen, M., Badran, E. A., and Abdel-Rahman, M. H. (2016a), Influence of highly resistive ground parameters on lightning-induced overvoltages using 3-D FDTD method, *IEEE Transactions on Electromagnetic Compatibility*, vol. 58, no. 3, pp. 792–800.

Rizk, M. E. M., Mahmood, F., Lehtonen, M., Badran, E. A., and Abdel-Rahman, M. H. (2016b), Induced voltages on overhead line by return strokes to grounded wind tower considering horizontally stratified ground, *IEEE Transactions on Electromagnetic Compatibility*, vol. 58, no. 6, pp. 1728–1738.

Rizk, M. E. M., Mahmood, F., Lehtonen, M., Badran, E. A., and Abdel-Rahman, M. H. (2017), Computation of peak lightning-induced voltages due to the typical first and subsequent strokes considering high ground resistivity, *IEEE Transactions on Power Delivery*, vol. 32, no. 4, pp. 1861–1871.

Rizk, M. E. M., Lehtonen, M., Baba, Y., and Ghanem, A. (2020), Protection against lightning-induced voltages: Transient model for points of discontinuity on multi-conductor overhead line, *IEEE Transactions on Electromagnetic Compatibility*, vol. 62, no. 4, pp. 1209–1218.

Rubinstein, M. (1996), An approximate formula for calculation of the horizontal electric field from lightning at close, intermediate, and long ranges, *IEEE Transactions on Electromagnetic Compatibility*, vol. 38, no. 3, pp. 531–535.

Ruehli, A. (1974), Equivalent circuit models for three-dimensional multiconductor systems, *IEEE Transactions on Microwave Theory and Techniques*, vol. 22, no. 3, pp. 216–221.

Rusck, S. (1957), Induced lightning over-voltages on power transmission lines with special reference to the over-voltage protection of low-voltage networks, Ph. D. dissertation, Royal Institute of Technology, Stockholm, Sweden.

Sadiku, M. N. O. (1989), A simple introduction to finite element analysis of electromagnetic problems, *IEEE Transactions on Education*, vol. 32, no. 2, pp. 85–93.

Sheshyekani, K., and Akbari, M. (2014), Evaluation of lightning-induced voltages on multi-conductor overhead lines located above a lossy dispersive ground, *IEEE Transactions on Power Delivery*, vol. 29, no. 2, pp. 683–690.

Sheshyekani, K., and Paknahad, J. (2015), Lightning electromagnetic fields and their induced voltages on overhead lines: the effect of a horizontally stratified ground, *IEEE Transactions on Power Delivery*, vol. 30, no. 1, pp. 290–298.

Silveira, F. H., and Visacro, S. F. (2002), Lightning induced overvoltages: The influence of lightning and line parameters, Paper presented at International Conference on Grounding and Earthing & 3rd Brazilian Workshop on Atmospheric Electricity, pp. 105–110, Rio de Janeiro, Brazil.

Silveira, F. H., Mesquita, C. R., and Visacro, S. (2002), Evaluation of the influence of lightning channel and return current characteristics on induced overvoltages, Paper presented at International Conference on Lightning Protection, Cracow, Poland.

Silveira, F. H., Visacro, S., Herrera, J., and Torres, H. (2009), Evaluation of lightning-induced voltages over a lossy ground by the hybrid electromagnetic model, *IEEE Transactions on Electromagnetic Compatibility*, vol. 51, no. 1, pp. 156–160.

Soto, E., Perez, E., and Younes, C. (2014), Influence of non-flat terrain on lightning induced voltages on distribution networks, *Electric Power Systems Research*, vol. 113, pp. 115–120.

Sumitani, H., Takeshima, T., Baba, Y., *et al.* (2012), 3-D FDTD computation of lightning-induced voltages on an overhead two-wire distribution line, *IEEE Transactions on Electromagnetic Compatibility*, vol. 54, no. 5, pp. 1161–1168.

Sunde, E. D. (1968), *Earth Conduction Effects in Transmission Systems*, 2nd Ed., Dover Publications, New York.

Tanaka, H., Baba, Y., and Barbosa, C. F. (2016), Effect of shield wires on the lightning-induced currents on buried cables, *IEEE Transactions on Electromagnetic Compatibility*, vol. 58, no. 3, pp. 738–746.

Tatematsu, A., and Noda, T. (2010), Development of a technique for representing lightning arresters in the surge simulations based on the FDTD method and its application to the calculation of lightning-induced voltages on a distribution line, *IEEJ Transactions on Power and Energy*, vol. 130, no. 3, pp. 373–382 (in Japanese).

Tatematsu, A., and Noda, T. (2014), Three-dimensional FDTD calculation of lightning-induced voltages on a multiphase distribution line with the lightning arresters and an overhead shielding wire, *IEEE Transactions on Electromagnetic Compatibility*, vol. 56, no. 1, pp. 159–167.

Thang, T. H., Baba, Y., Nagaoka, N., *et al.* (2012a), A simplified model of corona discharge on overhead wire for FDTD computations, *IEEE Transactions on Electromagnetic Compatibility*, vol. 54, no. 3, pp. 585–593.

Thang, T. H., Baba, Y., Nagaoka, N., *et al.* (2012b), FDTD simulation of lightning surges on overhead wires in the presence of corona discharge, *IEEE Transactions Electromagnetic Compatibility*, vol. 54, no. 6, pp. 1234–1243.

Thang, H. T., Baba, Y., Nagaoka, N., Ametani, A., Itamoto, N., and Rakov, V. A. (2014), FDTD simulations of corona effect on lightning-induced voltages, *IEEE Transactions on Electromagnetic Compatibility*, vol. 56, no. 1, pp. 168–176.

Thang, H. T., Baba, Y., Rakov, V. A., and Piantini, A. (2015a), FDTD computation of lightning-induced voltages on multiconductor lines with surge arresters and pole transformers, *IEEE Transactions on Electromagnetic Compatibility*, vol. 57, no. 3, pp. 442–447.

Thang, H. T., Baba, Y., Piantini, A., and Rakov, V. (2015b), Lightning-induced voltages in the presence of nearby buildings: FDTD simulation versus small-scale experiment, *IEEE Transactions on Electromagnetic Compatibility*, vol. 57, no. 6, pp. 1601–1607.

Uman, M. A., and McLain, D. K. (1969), The magnetic field of the lightning return stroke, *Journal of Geophysical Research*, vol. 74, pp. 6899–6910.

Ungar, S. G. (1980), Effects of lightning punctures on the core-shield voltage of buried cable, *The Bell System Technical Journal*, vol. 59, no. 3, pp. 333–366.

Visacro, S., Soares. J. A., and Schroeder, M. A. O. (2002), An interactive computational code for simulation of transient behavior of electric system components for lightning currents, Paper presented at the 26th International Conference on Lightning Protection, Cracow, Poland.

Visacro, S., and Soares, J. A. (2005), HEM: A model for simulation of lightning related engineering problems, *IEEE Transactions on Power Delivery*, vol. 20, no. 2, pp. 1206–1207.

Wagner, C. F., and McCann, G. D. (1942), Induced voltages on transmission lines, *AIEE Transactions*, vol. 61, pp. 916–930.

Wagner, C. F., Gross, I. W., and Lloyd, B. L. (1954), High-voltage impulse tests on transmission lines, *AIEE Transactions on Power Apparatus and Systems*, vol. 73, pp. 196–210.

Waters, R. T., German, D. M., Davies, A. E., Harid, N., and Eloyyan, H. S. B. (1987), Twin conductor surge corona, Paper presented at the 5th International Symposium on High Voltage Engineering, Braunschweig, Federal Republic of Germany.

Yang, B., Zhou, B.-H., Gao, C., Shi, L.-H., Chen, B., and Chen, H.-L. (2011), Using a two-step finite-difference time-domain method to analyze lightning-induced voltages on transmission lines, *IEEE Transactions on Electromagnetic Compatibility*, vol. 53, no. 1, pp. 256–260.

Yang, B., Zhou, B.-H., Chen, B., Wang, J.-B., and Meng, X. (2012), Numerical study of lightning-induced currents on buried cables and shield wire protection method, *IEEE Transactions on Electromagnetic Compatibility*, vol. 54, no. 2, pp. 323–331.

Yee, K. S. (1996), Numerical solution of initial boundary value problems involving Maxwell's equations in isotropic media, *IEEE Transactions on Antennas and Propagation*, vol. 14, no. 3, pp. 302–307.

Yokoyama, S., Miyake, K., Mitani, H., and Takanishi, A. (1983), Simultaneous measurement of lightning induced voltages with associated stroke currents, *IEEE Transactions on Power Apparatus and Systems*, vol. 102, no. 8, pp. 2420–2429.

Yokoyama, S., Miyake, K., Mitani, H., and Yamazaki, N. (1986), Advanced observation of lightning induced voltage on power distribution line, *IEEE Transactions on Power Delivery*, vol. 1, no. 2, pp. 129–139.

Yutthagowith, P., Ametani, A., Nagaoka, N., and Baba, Y. (2009), Lightning-induced voltage over lossy ground by a hybrid electromagnetic circuit model method with

Cooray-Rubinstein formula, *IEEE Transactions on Electromagnetic Compatibility*, vol. 51, no. 4, pp. 975–985.

Zhang, Q., Tang, X., Gao, J., Zhang, L., and Li, D. (2014a), The influence of the horizontally stratified conducting ground on the lightning-induced voltages, *IEEE Transactions on Electromagnetic Compatibility*, vol. 56, no. 2, pp. 435–443.

Zhang, Q., Zhang, L., Tang, X., and Gao, J. (2014b), An approximate formula for estimating the peak value of lightning-induced overvoltage considering the stratified conducting ground, *IEEE Transactions on Power Delivery*, vol. 29, no. 2, pp. 884–889.

Zhang, Q., Chen, Y., and Hou, W. (2015a), Lightning-induced voltages caused by lightning strike to tall objects considering the effect of frequency dependent soil, *Journal of Atmospheric and Solar-Terrestrial Physics*, vol. 133, pp. 145–156.

Zhang, Q., Tang, X., Hou, W., and Zhang, L. (2015b), 3-D FDTD simulation of the lightning-induced waves on overhead lines considering the vertically stratified ground, *IEEE Transactions on Electromagnetic Compatibility*, vol. 57, no. 5, pp. 1112–1122.

Zhang, L., Wang, L. Yang, J. Jin, X. and Zhang, J. (2019), Effect of overhead shielding wires on the lightning-induced voltages of multiconductor lines above the lossy ground, *IEEE Transactions on Electromagnetic Compatibility*, vol. 61, no. 2, pp. 458–466.

Zhang, J., Zhang, Q., Hou, W., *et al.* (2019a), Evaluation of the lightning-induced voltages of multiconductor lines for striking cone-shaped mountain, *IEEE Transactions on Electromagnetic Compatibility*, vol. 61, no. 5, pp. 1534–1542.

Zhang, J., Zhang, Q., Zhou, F., Ma, Y., Pan, H., and Hou, W. (2019b), Computation of lightning-induced voltages for striking oblique cone-shaped mountain by 3-D FDTD method, *IEEE Transactions on Electromagnetic Compatibility*, vol. 61, no. 5, pp. 1543–1551.

Index

www.ingramcontent.com/pod-product-compliance
Lightning Source LLC
Chambersburg PA
CBHW060248230326
41458CB00094B/1537